Success in Economics

Success in
ECONOMICS

Derek Lobley BA
Lecturer in Economics
Thurrock Technical College
Essex

John Murray

Printed in Great Britain by
Cox & Wyman Ltd
London, Reading and Fakenham

Cased 0 7195 2902 6
Paperback 0 7195 2901 8

Foreword

This book is for anyone who wants a basic, fully comprehensive course in economics related to the needs and activities of the world we live in. The course is also designed for students working for their A level examinations or for the ONC/OND in Business Studies. It is particularly appropriate to the new two-year Economics syllabus of the ONC/OND courses, and for the Economics syllabuses of many of the professional bodies such as the Institute of Chartered Accountants and the Association of Certified Accountants, the Institute of Bankers, the Institute of Chartered Secretaries and Administrators, the Institute of Cost and Management Accountants, the Institute of Freight Forwarders, the Institute of Transport, the Institute of Marketing and the Institute of Export.

The book is divided into Units of study, each defined in its own area, with self-testing questions at the end of the Unit. The Units follow a logical order, but this can be varied by students working on courses which follow somewhat different patterns.

In general the book progresses from micro-economic principles to macro-economics, with Units on money and the financial institutions forming a link between them. The exception is Unit Two (macro-economics: The National Income) which comes early in the book as it is the key to much of the material which follows.

Whatever the order in which you read the book, you will find the system of frequent cross-referencing helpful. These references are either to remind you of topics which have been discussed earlier, or to indicate where a topic will be dealt with more fully later.

An additional aid to study is the 'Suggested Further Reading' given on page 355. It is divided into Units to correspond with the text, and chapter references within books are also given where appropriate. The list of books recommended has been kept within workable limits, and you will see that a number of the titles contain material which is relevant to more than one Unit.

The self-testing questions are mainly of the essay type, so no one answer can be provided. Where a question does have a short answer, these are given on page 358.

Early in the book we have introduced diagrams to illustrate economic concepts. If you are working for an examination it is especially important for you to understand these and to copy some of them. It is a great asset to be able to include diagrams when answering examination questions.

D.L.

Acknowledgments

In writing this book I have been grateful for the assistance of many people. Figures from Government publications, especially those from the 'Blue Book' on *National Income and Expenditure*, are reproduced by permission of Her Majesty's Stationery Office; figures relating to the Bank of England and the National Debt are reproduced by permission of the Bank of England; and tables from *The Times 1000* are reproduced by permission of Times Newspapers Limited.

I would like to thank Jock Oliver and my colleague Geoffrey Whitehead, both of whom read and criticized the book at various stages and made many valuable suggestions. Finally I must thank my wife, Brenice, who typed the book and without whose patience the whole project would have been impossible.

D.L.

Contents

Unit One

The Economic Problem

1.1 Introduction

Most of us have some idea of what economics is about even if we have never studied the subject. Newspapers, radio and television bombard us with information about economic problems. There are very many of these: industrial disputes over wages, crises on the Stock Exchange when share prices fall, parliamentary debates about taxation, official directives about bank lending, international discussions about the balance of payments and trading relationships, housewives' grievances over rising prices, businessmen's complaints over falling profits, and government expenditure on defence and welfare. All these and many more economic problems are covered by the media and clearly represent part of the subject matter of economics.

Although this seems straightforward and we appear to be discussing things we know something about, we must be extremely cautious in this assumption. The problems we have discussed are aspects of *applied economics*; what we have to consider at the beginning of a study of economics is *basic economic theory*. Economics is a science and has to be studied in a scientific manner, although it cannot be said to be anywhere near an exact science. Problems are considered, analysed, simplified and perhaps subjected to mathematical treatment as they are solved. There are many specialized words used in economics, as in any science, and as they have precise meanings they have to be learnt as they occur and used in their correct manner. The difficulty here is that many technical economic words are used in everyday speech in a similar but usually much broader sense. It is essential that the precise economic meaning is learnt and used for all such words.

What do all the economic problems we mentioned have in common? The most obvious link between them is their concern with the use of *resources*, or perhaps more accurately in this case *monetary resources*, which are limited in supply. It is the use of resources, or their allocation and organization, that constitutes the subject matter of economics. If there were an infinite supply of everything that mankind required there would be no need for the study of economics for it is essentially the study of things in short supply. Let us consider a factory worker who has £300 to spend and must decide whether to spend it on a newer car or on taking his family to the Mediterranean for a holiday. His resources (£300) are limited so he must make a choice. The government is also faced with such choices, perhaps between extra expenditure on national defence or the building of more schools. Even its resources are limited.

All individuals and institutions are faced with the same basic problem: there are many desirable things that they would like to do, but their resources will not allow them to achieve them all at the same time, and so they are forced to consider alternatives and make a choice.

1.2 Economic Resources

Economic resources include all those things which are available and can be used to produce goods and services. For convenience we may divide them into three categories:

(a) Land

This is 'land' in its broadest sense and includes virtually all non-human natural resources: agricultural land and building land, mines and quarries, rivers, oceans and the atmosphere and everything in them.

(b) Labour

This is work performed by the world's population, and includes not only labour performed in exchange for a wage or salary payment but also work undertaken within the family or household. All kinds of human occupation and effort are included here, both physical work and skilled mental effort. It was once usual to distinguish one particular type of labour, *entrepreneurism*, from the others. The entrepreneur was the person who owned and controlled a business in his own right, for example a shopkeeper. Most economists are now agreed that since the entrepreneurial or business functions are divided between many different people in the modern economy it is appropriate that they should be included in the general category of labour (see Section 7.7).

(c) Capital

Most people regard capital as wealth, but this is not its meaning in the economic sense. It is a word with many meanings in economics, but can be most easily described as the stock of physical assets accumulated by society which facilitate the production of goods and services. Plant and machinery are the most obvious kind of capital, but roads, schools and hospitals also represent capital and make an important contribution to the production of goods and services.

Obviously we cannot regard the boundaries between these three categories as arbitrary. We might, for instance, consider a newly-discovered region purely as land. But if we use it for agriculture the land may well contain a considerable amount of capital in the form of artificial fertilizers. Or suppose the land contains rich mineral deposits? We might then consider these as part of the capital. Labour too contains an element of capital: the skills that the apprentice learns in training may be regarded as stored-up knowledge that will be used in future production.

1.3 Costs

The resources of land, labour and capital may be combined in a number of ways to produce the goods and services we need. Problems arise because there are insufficient resources to meet all requirements, and economics is the study of how such problems are resolved. If a given set of factors of production is employed to produce motor cars desired by one group of consumers, it cannot also be producing the television sets required by another group. As resources are limited we must choose what is to be produced, and the true cost of the one opportunity chosen is the loss of the alternative opportunity which is not chosen. This is one of the fundamental ideas of economics, the doctrine of *opportunity* or *alternative costs*.

Opportunity costs derive from the fact that we have to choose what to do with our resources, and this reflects the view of many commentators that economics is basically the science of choice—choice that becomes necessary because of the scarcity of resources. The existence of opportunity costs frequently leads the economist to regard costs in a different way from that used by the accountant. An accountant is largely concerned with the money spent on the production of a particular item, while an economist would probably want to know what alternatives had been foregone in order to produce the item. You might, for instance, paint the outside of your house, and this would be done by foregoing the opportunity of painting the inside.

It is worth noting here too that a distinction may be made between *private costs* and *social costs*. If a road haulage company sends a vehicle from London to Birmingham its costs include depreciation, tax and insurance on the vehicle, the cost of fuel, loading and unloading and wages. These are the private costs borne by the operator. There are in addition social costs which include traffic congestion to which the vehicle contributes, and cost to the general public in the form of environmental pollution caused by carbon monoxide and noise. These are the social costs of the journey, borne not by the operator but by society as a whole. To a great extent economics is a study of the ways by which these various costs can be kept to a minimum while resources are organized in the most efficient way.

1.4 Production and Consumption

Costs of all kinds are incurred in the *production* of goods and services and much of economics is concerned with choice in relation to production. We have seen that the production of one commodity by certain factors precludes the production of another commodity by those same factors. The choice of what is produced depends very largely upon the consumers' willingness to buy. Manufacturers will be able to sell goods which give *satisfaction* to the consumers. Another way of looking at this is to say that consumers will buy goods only if they will be able to derive some *utility* from them. Utility is used here in a technical economic sense and means the ability to satisfy consumers' wants or desires. A commodity which has utility is not necessarily useful. Women wear decorative jewellery not because it is useful, but because it gives them great satisfaction. Although many goods, technically known as *producer goods*, are not bought by consumers, but by manufacturers, the same argument can be applied. A firm will buy a machine only if it can yield satisfaction by producing the goods that the firm wishes to sell.

We may summarize and say that the object of production is to convert factors of production into goods and services which yield utility to consumers or producers. If goods cease to give utility or satisfaction they will soon cease to be produced as they would not be bought. For example, in the high-fashion industry tastes change rapidly and last season's clothes no longer give utility so are no longer produced as they would not be sold.

You must, of course, realize that there are many obstacles to the smooth working of this mechanism. The whip hand lies with the producer of goods and services, and we shall see later (Section 5.6) that the power of the consumer and his choice are limited. For example, the facts of our economic system are such that there are only two major producers of household detergents in this country and consumers are forced to buy one or another of their products.

1.5 Income and Wealth

Another way of looking at economic activity is to say that it is concerned with the production of *wealth*, which may be regarded as accumulated income. If a man spends all his annual income on food, drink, clothes and rented accommodation, his wealth will not increase over the year, but if he saves 20 per cent of his income then he has increased his wealth. Wealthy people do not necessarily have a high income but they, or their forebears, have been able to save out of their previous income. The same is true of society as a whole: wealthy societies are those which have in the past saved some of their current income, and converted it into productive assets such as machinery or transport facilities. They can then increase their current income in subsequent periods. Once again choice is involved both for the individual and for society as a whole; maximum consumption at the present time prevents higher consumption in the future. A careful person will save some income to establish a store of wealth, either as an insurance against future difficulties or for the express purpose of bettering his situation. Perhaps he will move out of his rented rooms into a house of his own. A thrifty community must also be sure not to live beyond its means; it must not devote all its resources to producing food and drink and none to improving its stock of capital assets.

Consider, for example, the mechanization of agriculture. Using primitive, labour-intensive methods, perhaps aided by oxen and a wooden plough, many communities in South-East Asia barely produce enough to feed themselves. If they were able, by some miracle, to produce a surplus which could be exchanged for a tractor, some balanced fertilizers and high-yield seeds, their productivity would increase greatly. In fact many such communities have to rely on other people's savings, in the form of international aid, to achieve such a breakthrough; but the point is fundamental. *If we wish to increase output we must be prepared to set aside some income for the purchase of productive assets.* We shall need to look carefully at the question of increasing output especially in relation to time. For instance, should we aim to increase output as much as possible this year or in five or ten years time?

1.6 The Nature of Economic Choices

We have seen so far that in one way economics is the science of choice: the householder may choose between saving and spending; the businessman between the manufacture of different goods; the government between greater expenditure on defence and heavier investment in education. It is very unlikely that any of these choices will be of the all-or-nothing variety; it is not a question of whether a consumer spends all his income on food or all on clothes. The decision he has to make is normally whether to spend an extra pound on food or on clothes. He will already be spending a considerable proportion of his income on both commodities. He now has to decide whether an extra amount of food or of clothes (or some other commodity) will give him the greater increase in satisfaction. It is a decision to be made at the margin, that is a decision about a small increase in one or another commodity. Most of the consumer's economic choices are of this nature and are called *marginal choices*. The same type of thinking has to be used by businessmen and governments. It is not a question of whether to build hospitals or schools (we obviously need both); the point at issue is whether to build one more hospital or one more school. *The margin is thus the point of decision in economics.* Economics is very largely concerned with such marginal choices, for example we are not concerned with whether people eat food or not; we know they do. What we want to find out is why they eat an extra unit of food in preference to smoking an extra unit of cigarettes. Similarly we know that a factory employs workers, the important issue is: where does the employer draw the line? Does he employ 99 or 100 men? Much of our time will be spent examining such questions.

1.7 Production and Exchange

We have seen that economic activity is concerned with the production of those goods and services desired by both private consumers and business enterprises. The production of goods by one group of people for use by another group implies the existence of some sort of trade or exchange. This is another important aspect of economics. While it is true that there is no point in producing goods for which there is no demand, it is equally true that the level of production depends partly upon the number of customers that can be reached. In primitive societies with no contact with the outside world, there is clearly no scope for the production of output which would be surplus to the requirements of the society itself. As communications improve trade and exchange of goods takes place and so the production of surpluses becomes worth while. It may even reach the stage where *international trade* becomes a possibility. Thus the development of roads, canals and railways in England during the seventeenth, eighteenth and nineteenth centuries encouraged trade to expand and surpluses to be exchanged between areas. The more recent development of refrigerated shipping has permitted the export of meat from New Zealand and Australia to the United Kingdom. In addition to an analysis of the allocation of resources and of the supply and demand for goods we must also investigate the functioning of markets and the principles of exchange including the importance of money (the oil of the economy), and the great range of financial institutions associated with it.

1.8 The Economist and Society

The economist as a social scientist faces very special difficulties—difficulties that are quite different from those of the physical scientist. When investigating the operation of the economy, he is dealing with people and they are not only different, they are unpredictable. He cannot gather them together in a laboratory, eliminate friction, gravity or idiosyncrasy, observe their behaviour and, on the strength of those observations, make firm assertions about the effect of various economic policies. What he must do is to examine society as it is, examine its reaction to past experiments in economic policy, and on the basis of this try to forecast its response to new initiatives of policy. He may be assisted by the construction of economic models, designed to highlight certain aspects of economic life. An economic model is an authentic or mathematical formulation of the essential features of an economy, but with many features held constant so that the relationship between other features can be closely examined. Sometimes such models (more usually the elementary examples) may seem very far removed from the real world, but they really do convey a great deal of information on the way in which the economy works. You should look very carefully at the models used later in this book and make the greatest efforts to understand them.

Once the economist has constructed his model and drawn his conclusions about its economic implications, he must consult other social scientists before putting his conclusions into perspective. According to the branch of the subject he is studying he will need to consider the views of geographers, historians, lawyers, sociologists, accountants, planners, engineers and many others. It is part of the excitement of economics that its frontiers are shared with all other subjects. As you proceed with your study of economies you will develop your awareness of the place it has in the study of society as a whole.

1.9 Questions and Exercises

1. Describe the *economic* problems likely to face the sole survivor of a nuclear explosion.
2. Take your daily newspaper. Make a list of all the economic problems reported and discussed in it.
3. Distinguish between opportunity costs, accounting costs and social costs.
4. 'Consumers buy goods because they yield satisfaction.' Explain this statement.
5. Distinguish between utility and usefulness.
6. What are the opportunity costs of being a full-time student?
7. How do you think the economic problems facing a £17·00 per week farm labourer (a) correspond to (b) differ from those facing a successful pop-singer?
8. Write an essay describing the principles that govern the allocation of your personal expenditure.
9. The object of economic activity is to transform raw materials into goods and services. Make a list of the products derived from crude oil.
10. Why are economic choices necessary? Give examples of economic choices made recently by (a) the Government (b) a nationalized industry (c) yourself.

The National Income

2.1 Introduction

We saw in Unit One that the purpose of economic activity is the production of goods and services, and the output of a modern economy is an endless flow of such utilities. As the creation of both goods and services is counted as production by the economist then no distinction is made between the work of a farm labourer, a skilled physician, a shop girl or a lorry driver. Thus a manufactured commodity is 'produced' in the economic sense at every stage of its journey from its basic ingredients to its sale across a counter. Potatoes, wheat, fish, coal, natural gas, pig-iron, sheet steel, motor cars, tractors, roads, medical services and a million other items as well as a host of services are produced in the United Kingdom in any year. It would be possible, with a great deal of effort, to draw up an inventory of the goods and services produced in one actual year, but such a list would be of little economic significance as it would be so complex that comparison with earlier years or other economies would be impossible. If all the goods and services produced in a given year were reduced to their monetary value then they could be added to give the value of total output for that year. This result is called the *gross national product* (GNP). This concept is one of the most important economic indicators, and is frequently mentioned in parliament and the press. Its correct economic definition is: *the aggregate value of the goods and services produced during the year by the factors of production within the economy plus the net income from abroad.* The GNP is more popularly known as the national income and it is occasionally called the national expenditure. The terms are derived by seeing the same end result from a different viewpoint, but whichever term we use, the importance of the concept is absolutely clear. Disputes between trade unions and employers are frequently arguments about the respective shares of labour and shareholders in the national product, and when the government talks of economic growth it is really talking about the size of the national product.

As either the size or the distribution of the gross national product is at the heart of most economic problems it is appropriate that we establish quite clearly what is meant by this phrase. In this Unit we shall therefore examine its meaning and consider the difficulties of measuring it and make an assessment of its significance.

2.2 Production as a Flow

The economy can be seen as a vast production line with raw materials and labour being fed in at one end and finished products flowing out at the other. The process is continuous and with no regard to time, so it is only for convenience that we cut into the flow to measure its value every twelve months. Unless it is specifically stated to the contrary, GNP figures refer to a calendar year, although there is no reason why daily, weekly, monthly or even five-yearly measurements should not be taken. Whatever the time interval selected there are bound to be difficulties. For example at the end of any period there will be some goods in the course of production, some finished but awaiting delivery. Some goods in stock at the beginning of the year may still be stock; the value of stocks may have changed. Those responsible for computing the national income must devise ways for overcoming all these, and many other problems.

If we consider any particular product, the final flow probably depends upon the co-operation and co-ordination of thousands of individuals and institutions. However complicated the process and however numerous the intermediate stages, the cycle of production and indeed the economy as a whole may be represented as a continuous flow, as in Fig. 2.1. These matters are considered in more detail in Unit Five.

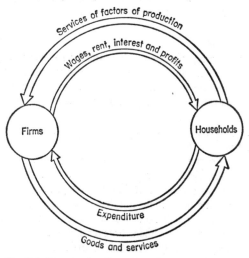

Fig. 2.1 The circular flow of production and income

This is a simplified outline of the economy which should be remembered as it will take on greater sophistication in later chapters. It contains the essential ingredients—the household (or personal) sector which owns factors of production and supplies them to the firms (corporate sector) in exchange for money; and the firms themselves which combine the factors of production to produce goods. As well as helping to produce goods and services by supplying factors of production, the household has another important function: that of spending the income it has earned on the goods and services it has helped to produce. Firms thus acquire the money necessary to finance the next cycle in the process. In any industry or economy the basic cycle may be repeated any number of times in one year and there will always be three ways of measuring the value of the goods and services produced in the economy in that year.

2.3 The Measurement of Gross National Product (GNP) in a Simple Economy

Let us imagine an economy in which the following transactions occur in the course of a year:

1. Farmer Adam sells wheat grown by his own and his employees' efforts to miller Brown for £100.
2. Brown grinds the wheat and sells the flour to baker Clarke for £150.
3. Clarke sells the flour when made into bread, to retailer Davis for £200.
4. Davis sells the bread to the actual consumers for £250.

In this economy, as in our own very complex economy, the national product may be measured in three ways:

(a) The output method

As we are endeavouring to measure the value of the national product, the most obvious approach to this is to establish the output of every firm and thus, by addition, obtain the GNP. The gross value of the output of the four firms in the above example economy amounts to £700, but it would be quite wrong to conclude that this is the value of the national product, for the community has only £250 of bread to show for its efforts. The mistake which has been made is that the wheat has been measured several times. If we add together both Adam's £100 of output and Brown's £150 we are counting the value of the wheat twice. In order to avoid this we must add on only the value added at each stage; that is the value of output minus the cost of raw materials. If we assume that Adam obtained his seed free then the value of the national product becomes £100 + £50 + £50 + £50 or £250, which is the amount spent by the consumers on bread, the finished product. This answer leads us fairly obviously to the second method of measuring GNP.

(b) The expenditure method

Instead of standing, as it were, at the farm or factory gate measuring the value of output, it is possible to ask everyone how much they have spent on goods and services during the past year. Once again care must be taken to avoid double-counting and only final expenditure must be counted; all intermediate expenditure must be omitted. The expenditure of Brown, Clarke and Davis must not be added as well as the final expenditure of the consumers, or we shall again arrive at a GNP of £700. By counting only final expenditure we obtain a GNP of £250—the value of the bread produced.

(c) The income method

The third method of measuring the GNP is to establish the incomes of the owners of factors of production. Adam and his employees have £100 to share among themselves; the miller and his employers have £50 to share, the baker will also have £50 to share between himself and his employees, as will the retailer and his staff. Thus total incomes amount to £250, which is the value of the gross national product. We see from these measurements that GNP comes to the same amount when determined by the three different aspects of the economy. In its simplest form this relationship can be expressed as the following identity:

national output ≡ national expenditure ≡ national income.

2.4 Real World Complications

In a real economy the measurement of the value of goods and services is a good deal more complicated than in our simple model, for five basic reasons.

(*a*) The government undertakes a wide range of economic activities, imposing taxes and also providing goods and services. It even runs some industries. Thus there are three sectors to consider in the economy, the personal, the corporate and the public sectors.

(*b*) Many people receive an income although they do not produce anything. As we are trying to measure the value of goods produced, the calculated gross national product would be exaggerated if these 'non-productive' incomes were included. The most obvious such income is that of old age pensioners who provide no economic service in exchange for their pensions during the period in which they receive them. The same is also true of students receiving grants, and men taking unemployment benefit. (We are not concerned here with the rightness or wrongness of these benefits, only with the fact that they do not represent production.) All these incomes are *transfer payments,* the transfer of money from one source to another without the donor receiving any productive service in return. Further examples are the pocket money given by a father to his son, and the money won on a football pool. Such wins involve basically a transfer of money from all the losers to all the winners.

(*c*) Many goods and services do not reach the market-place, and are not therefore automatically included in calculations of gross national product. They therefore become the subject of arbitrary decisions, which may vary from time to time or between countries, and make comparisons difficult. For example the many hours of service undertaken by housewives are not included in the national product, and neither are any other services provided within the family unit. Part of the increase in the gross national product of the United Kingdom during the twentieth century is attributable to the tendency to buy outside the home, services which were once provided within the family. Such goods and services are likely to be a much more serious source of confusion in poorer countries where so many more of them do not enter the market.

In one important case, however, the *imputed* value of a non-marketed service is included in the national product and this is the calculated rent of owner-occupied houses. If such a house were rented from a landlord rather than owned by the occupier, then the occupier would have to pay rent and this would appear as part of the gross national product. When compiling the final accounts it is normal to include the imputed rent of such dwellings. If this were not done, then the size of the national product would depend partly on the number of owner-occupiers, which would give a wrong impression.

(*d*) Our economy, like all others, is really an open one, depending heavily upon importing, exporting and international investment. The national income statisticians must therefore take this factor into consideration. For example, much of the expenditure by United Kingdom residents is on imported goods produced in other countries; while many goods produced in this country are sold abroad. Adjustments have to be made for both these factors. In the same way, a part of many incomes is interest or rent obtained from shares or property held overseas. Allowance must be made for this when computing by the expenditure or output method if the results are to be compatible.

(*e*) Timing also creates minor problems. It is obviously not possible to

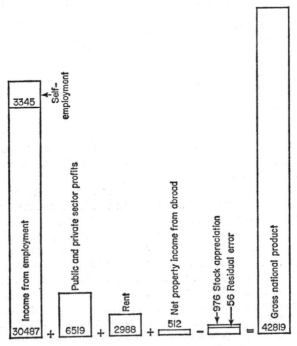

Fig. 2.2 Income in the gross national product

stop the economy at a given time, such as midnight on 31 December each year, and make the necessary calculations for measuring the gross national product. At any one time many goods will have been produced but not sold (it is convenient to regard these as having been purchased by the producer); some incomes earned by producing goods this year will not be received until the following year. When the necessary adjustments have been made for all these factors, the three versions of GNP may be represented as in Figs. 2.2, 2.3 and 2.4.

2.5 The Income Method

(a) Income obtained from employment includes both wages and salaries which are measured at their gross value, as this gives the cost of employment of labour to the individual firm. Taxation on incomes is thus ignored, for if this were not the case the level of national product would vary with changes in the level of income tax.

(b) Income from self-employment includes an element of wages and profits which have accrued to the self-employed person.

(c) The gross profits obtained from trading must be allocated in various ways. A portion of the profits becomes the personal income of the trader and his associates. This is often in the form of dividends paid on shares held in the company. A part of the profits must be handed to the Government in the form of company taxation and yet another part is kept by the company to build its reserve fund. In order to clarify our understanding of this treatment of taxation let us imagine an economy operating without money. In this economy a firm produces 1000 tractors and is taxed at the rate of 10 per cent on its output and keeps 5 per cent of its output in reserve. It will therefore have to hand 100 tractors to the Government as a tax and it will keep 50 tractors in reserve. When assessing the firm's contribution to the national product these 150 tractors must be included or else the product will be underestimated. In exactly the same way the money value of output paid in tax or kept in reserve must be included.

(d) Rent includes not only rent received by private land or house-owners, but also rent received by the Government from various sources and the *imputed rent* of owner-occupied houses.

(e) Property income from abroad is included as it represents money earned by British investment in industry and property held overseas.

(f) Another modification which must be made is deduction for stock appreciation. For example goods produced earlier in the year and valued at £5 at the time of production, may have increased their value to £5·50 by the end of the year. To include such goods at £5·50 would exaggerate their value so stock appreciation of £0·50 must be deducted.

(g) This method of measuring the national income does not give the same results as the other two methods so an adjustment has to be made for errors and omissions. This is called the *residual error*, and it is very small, about one-third of 1 per cent,

This gives us the gross national product at factor cost which represent the actual cost of production of the goods, and it must be distinguished from gross national product at market prices for this includes indirect taxes and subsidies (see Section 2.6).

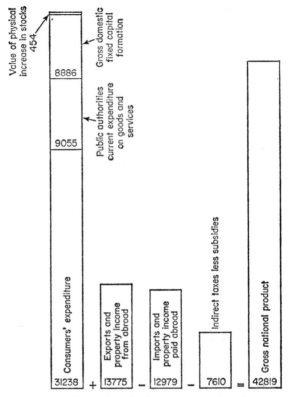

Fig. 2.3 Expenditure in the gross national product

2.6 The Expenditure Method

The second method of calculating national income is by counting up total expenditure.

(*a*) Consumers' expenditure accounts for the bulk of total expenditure. One difficulty here is that many items have indirect taxes added to their price while others are subsidized, that is, sold at less than their cost price. Adjustments have to be made for these amounts. To make the results compatible with those obtained by other methods an allowance must be made for imputed rents.

(*b*) Central and local government bodies frequently buy goods and services and, of course, this expenditure must also be included.

(c) Much of the expenditure taking place during a year is not for the acquisition of goods and services needed for immediate use, but for the purchase of plant, machinery, roads and schools and other fixed assets needed to facilitate the subsequent production of goods and services. These items are included here as what is known officially as *gross fixed capital formation,* and represent, in the main, expenditure by companies and public authorities. Fixed capital formation simply means the purchase of fixed productive assets, for which money has been committed and cannot easily be retrieved. This is more generally called investment: fixed capital formation is really investment in plant and machinery of all kinds. The gross (or total) value is given rather than the net value for this latter would allow for the depreciation or wearing out of assets during the year. For example, gross fixed capital formation in 1971 was £9923 million, but during the year existing capital assets were estimated to have depreciated or worn out to the extent of £5012 million, leaving a net investment or capital formation of £4911 million. An exceptional kind of capital formation is individual consumers' expenditure on the purchase of a new house, which is regarded as investment in the national accounts.

(d) As we said earlier (Section 2.4) at the end of the year some goods will

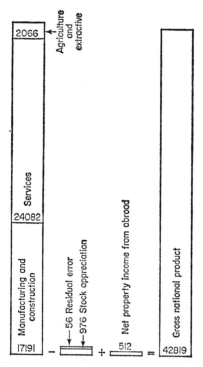

Fig. 2.4 Output in the gross national product

have been produced but will be held in stock prior to selling. The value of the physical increase in stocks must be added in our calculation. People will have earned incomes by producing these goods during the year and the value of these incomes will have been included in calculations by the first method.

These four items give gross domestic expenditure at market prices, but in order to reconcile this with gross national product at factor cost the following adjustments are necessary.

(i) Many of the incomes included in the first calculation will have been earned in the production of exports which are naturally not bought by consumers in the United Kingdom and so are not subject to United Kingdom expenditure. Therefore the value of exports must be added; so must rents and share dividends received from property held abroad. A corresponding deduction must be made for UK imports and rents and share dividends paid to non-UK residents.

(ii) Finally, when goods are sold, their price may include an element of indirect taxation or of subsidy. In the case of indirect taxation this must be deducted from the price to establish the factor cost, and in the case of a subsidy it must be added on to the price. For example, cigarettes retailing at 30p may bear 20p tax which must be deducted from the price to obtain the factor cost; while milk retailing at 5p per pint might be subsidized by 1p a pint, so this 1p must be added on to establish the factor cost. Once this is all done we will confirm the calculations made in finding the GNP at factor cost, for the GNP found by the incomes method is exactly the same, allowing again for the small residual error.

2.7 The Output Method

The total output of all industries gives the gross national product, but this is not, however, equal to the national output. Modifications have to be made to make sure that only the net output of each enterprise is included (see Sections 2.1, 2.2 and 2.3), and to ensure that there is no double-counting. Allowances must be made for the factors mentioned in Sections 2.5 (d), (e) and (f) above. Before we look at the significance of the gross national product and the way it has changed in recent years, it is useful to consider some other ways of measuring the national product which are in fairly common use.

2.8 Other Measurements of the National Product

We have been largely concerned with the size of gross national product at factor cost. We have noticed that there is a difference between factor cost and market prices and this stems from the existence of indirect taxes and subsidies:

Gross National Product at Factor Cost = Gross National Product at Market Prices − Indirect Taxes + Subsidies.

For some purposes, however, both these concepts are misleading, for during the course of producing the current year's national product, some capital assets will have been partly used up, that is, they will have depreciated in value. If we then subtract an allowance for depreciation (which will be fairly arbitrary) we obtain the *net national product* (NNP) *at factor cost or at market prices.*

If the net property income from abroad is excluded from the gross or net national product as measured above we arrive at either *the gross or the net domestic products* which measure the value of goods and services produced by the factors of production located in this country. The gross or net domestic products may be estimated at either factor cost or market prices.

It is worth noting here that the term *national income* is often used very loosely in connexion with the national product. It should be reserved for the measurement of *net national income at factor cost.* Throughout this book it will be used in that sense, and when we speak of the *national product* we will be referring to *gross national product at factor cost;* that is to say, we shall not be making any allowance for depreciation.

2.9 Why the National Product is Measured

It is obvious that the measurement of the national product is a very complicated business. In the United Kingdom a year's figures do not finally become available until the following autumn when they are published under the title *National Income and Expenditure,* popularly referred to as the *Blue Book.* The main justifications for the effort involved may be summarized as follows:

(*a*) The Government needs the information obtained so that it can plan the economic policy. The national income accounts are not the only indicators of the state of the economy, but they do reveal trends—the level of investment or consumption, for example, or the rate at which the economy is growing (or declining). It is these *trends* that are likely to be of importance in planning rather than the absolute figures for any item for any particular year.

(*b*) The figures act as a basis enabling assessments and comparisons to be made in one country over a number of years, or between several countries in a given year or over a number of years. Such comparisons may throw some light on differences in economic performance. It may be possible to see that the standard of living in country A is higher than that in country B as the former devotes more of its resources to investment, or because the proportion of national income absorbed by profits is greater, or because less assistance is given to overseas countries, or because government expenditure is lower. While national income statistics do not provide absolute answers they will give many clues that enable governments to formulate appropriate policies.

2.10 The Limitations of National Income Statistics

The object of economic activity is the production of goods and services and the size of the gross national product gives an indication of the success of the economy in this direction. We must, however, be very wary of interpreting every rise in the GNP as an improvement in the standard of living (itself an ambiguous term open to various interpretations). There is a tendency to regard a 3 per cent rise in the GNP as tantamount to a 3 per cent rise in welfare or living standards, but this is a dangerous assumption to make as there are several reasons why it may be invalid.

(*a*) If prices rise by 3 per cent while the same goods and services are produced as before, then the GNP will appear to have grown by 3 per cent. Obviously it will not have done so and it is necessary to eliminate such price changes when comparing two different years. The greater the gap between the years of comparison the more important such an adjustment becomes.

(*b*) Over a long period the population may grow more rapidly than the national income. In this case national income per head will have fallen and living standards may be lower. Thus when comparing changes in the standards of living the average income per head may be a more valuable indicator than total national income.

(*c*) The structure of the national product is also very important. If the gross national product rises from £30,000 million in year *x* to £40,000 million in year *y* but the proportion available for personal consumption falls from 80 per cent to 50 per cent, then living standards will probably not have risen. Such a state of affairs may be caused by a war effort, an export drive or when a developing country is trying to accumulate capital.

(*d*) We have already noticed that many goods and services are not exchanged for money and are not therefore noted in official statistics. For example, if in year one everyone paints his own house, but in year five everyone employs a painter to do the job, then there will be an apparent rise in the gross national product. We must be careful to make allowance for this sort of thing when using the statistics, though once again it is only likely to be of great significance over a fairly long period.

(*e*) The above difficulties apply to comparisons made within a country over a number of years. When we try to use national income statistics to compare living standards or growth rates of different countries these difficulties still apply, but there are also additional problems. The basis of the statistics may vary from country to country; the proportion of goods and services not reaching the market is likely to be much greater in a poor country than in a rich one; expenditure on clothes and fuel in hot countries is lower than in cold ones, but this does not imply lower living standards.

(*f*) Finally, whether we are making domestic or international comparisons, we should remember that the standard of living is not measured merely by the amount of material goods available to the economy or the individual. Non-economic factors such as political or religious freedom, the incidence of disease or the expectation of life may all contribute to the standard of living, and material measurements may need modification in the light of such matters. In addition we should make some allowance for the disutilities of growth such as noise and pollution.

2.11 The Demand for Goods

The national income statisticians can only begin their work after the economy has run its course for the year. We must now proceed to examine the forces and institutions which determine production in various parts of the economy. This is known as the study of *micro-economics*. Then we shall return to consider the relationship between the national product and the level of employment, the rate of inflation and the balance of payments. These broader matters form the study of *macro-economics*.

2.12 Questions and Exercises

1. Distinguish between the following: gross national product at factor cost, net national product at market prices, gross domestic product at market prices and net national income.
2. What is the gross national product? What are the main difficulties associated with its measurement?
3. Discuss the relationship between the standard of living and the national income.
4. Which of the following is not a transfer payment?
   (*a*) Unemployment pay.
   (*b*) Housekeeping allowance to a wife.
   (*c*) Housekeeper's wages.
   (*d*) Gambling win.
   (*e*) Student's grant.
5. Which of the following items should be included in the GNP?
   (*a*) Wages paid to civil servants.
   (*b*) Government purchase of stationery.
   (*c*) Government expenditure on old age pensions.
   (*d*) The Prime Minister's salary.
   (*e*) A loan of £500 from a private citizen to the Government.
6. True/False:
   (*a*) Stock appreciation indicates the amount by which stocks have risen during the year.
   (*b*) Transfer payments are excluded from national income because they do not represent goods or services.
   (*c*) Exports are not included in the gross domestic product.
   (*d*) The incomes earned by producing goods will always equal the factor cost of the goods.
   (*e*) GNP at factor cost = gross domestic product at market prices − depreciation.
7. The following transactions occur in an economy:
   (i) *A* sells raw materials to *B* for £100; (ii) *C* sells raw materials to *B* for £50; (iii) *B* sells finished products to *D* for £250; (iv) the Government imposes a

B

purchase tax of £30 on D's goods when he sells them to final consumers for £330 (including tax). There are no other transactions. Calculate

(a) GNP at factor cost.

(b) GNP at market prices.

8. Show the relationship between GNP at market prices and net national product at factor cost.

9.

Country	GNP in 1950	GNP in 1970
A	£10,000 million	£30,000 million
B	£5,000 million	£50,000 million

'Country B now has a higher standard of living than country A.' What additional information would you require to verify this statement?

10. How do you account for the item 'net property income from abroad' in the national income statistics?

11. Obtain a copy of the current edition of *National Income and Expenditure* (all reference libraries should have one). Use it to answer the following questions:

(a) By how much has consumers' expenditure increased in the last 10 years?

(b) What proportion of the Gross National Product was generated by consumers' expenditure in the latest year for which figures are available?

(c) By what percentage has the Gross Domestic Product risen since 1963?

Unit Three

Demand

3.1 Introduction

In Unit Two we saw that a great number of goods and services are produced each year to satisfy the requirements of consumers. The organization of production will be examined later in Units Five to Ten for this is not the side of the market with which most people are familiar. In this Unit we shall examine the principles which determine the level of demand. The sum of all individual demands is the composite demand of the population which producers must attempt to satisfy. When economists define 'demand' they cannot equate it with 'want', for man's wants are unlimited, but his demand for a particular item is often limited by the level of the price he must pay for it. The demand for goods or services can therefore be defined as that quantity of the goods or services which purchasers will be prepared to buy at a given price, in a given period of time. We must now try to understand how the level of demand for a particular product is determined.

3.2 The Main Influences on Demand

The level of demand for a particular product may be considered from the point of view of an individual (consumer demand) or of all consumers together (market demand). In either case there will be a number of factors affecting the level of demand at any particular time.

(a) The price of the product

Most consumers would probably recognize that price was the main influence determining their demand for a product. Some goods would be very expensive and so would be outside the price range of most consumers and would be dismissed from their consideration. The remaining goods do compete for the consumer's attention, and their price greatly affects his demand for them. Experience and market research indicate that in general, the higher the price the lower the quantity of goods demanded. When price rises occur some consumers will think that the good is no longer worth buying at the price asked, and so will reduce their purchases. If the price falls then more people will consider the utility of the goods and will buy them if they think them worth the money. The significance of price changes varies from product to product, as will be shown in Section 3.7.

(b) The level of income

An individual's income is often decisive in determining his demand for a product. Many goods are very expensive and beyond the reach of most consumers; however much they may desire a Rolls-Royce, or a country mansion, they have no demand for it in an economic sense. Economic demand is always effective, or supported by money. At the other end of the scale a person's income may be so high that he has no demand for cheap bottled coffee essence, as he prefers and can afford to buy more expensive selected coffee beans. As incomes change patterns of demand change.

(c) Tastes
All people are different and even members of the same family have different tastes and preferences. Non-smokers have no demand for cigarettes; the owners of motor cars have relatively little demand for public transport. A consumer's tastes are important in determining his demand for a product. If two products are exactly the same price he will always buy *A* rather than *B* just because he likes it better. If his tastes change then his pattern of demand will probably change. Indeed the object of commercial advertising is to influence tastes in favour of the advertiser's product; conversely adverse publicity (for example, an unfavourable report in *Which*) will tend to reduce the demand for a product.

(d) Prices of other goods
In many cases the demand for one commodity will depend partly upon the prices of other commodities, especially if they are close substitutes. If the price of colour television sets was to fall from £300 to £120, there would be a significant reduction in the demand for monochrome sets. By paying a little more, a potential purchaser could buy the better value colour set. If the price of motor cars was to fall by 50 per cent very many more would be sold and there would be a heavy increase in the demand for petrol.

In addition to these four influences on demand there are dozens of others which might be mentioned. Here are three of them.

(i) Population.
Demand is also influenced by the size and structure of the population, and it will be strong if the population has ample purchasing power. Changes in demand are not likely to be significant over a short period but over a number of years the pattern may change considerably. In the United Kingdom it is possible to trace the effects of the post-war bulge in the birth rate on the demand for various goods and services. Thus in the late 1940s there was a heavy demand for nursery equipment and toys and this has been followed by increased demand for housing and consumer durables and will probably eventually lead to an increased demand for geriatric facilities.

(ii) Government policy.
The demand for some goods is affected by Government policy towards them. Motor cars and motoring are taxed by the Government; this increases the price of cars and, to a certain extent, reduces demand. Some foods are subsidized by the Government so their market prices are reduced and the demand for them increases.

(iii) **Seasonal factors.** The demand for many goods and services varies according to the time of year. Thus the demand for ice cream and paddling pools rises in summer and falls considerably in the winter.

All these influences can work on demand at the same time so it becomes almost impossible to establish their relative importance. For example, if we try to determine the effect of a fall in price on the demand for gramophone records, we should have to isolate this one change. If at the same time consumer income rose, the price of tape recorders doubled and the record companies undertook a wide publicity campaign then it would be quite impossible to estimate the effect of cheaper records. In our examination of markets and demand we are mostly interested in the effects of a change in price on the quantity demanded. We therefore simplify matters by assuming that all other factors are constant. These other factors are known as *the conditions of demand,* and by not allowing them to change we can, for example, eliminate the effect of changes in the prices of other goods and so are able to measure the effect of price changes on the demand for one particular product. This will seem an unrealistic approach, but we must simplify things at the beginning so that relationships will be clearly shown. It is possible to re-introduce changes in the conditions of demand at a later stage when our understanding is greater. We always have to move from the simple to the more sophisticated study of a situation.

3.3 Individual Demand and Market Demand

When an individual visits his local shopping centre he is confronted by a vast array of goods and services which he may buy if he has some money. There are two sets of decisions for him to make: whether or not he will buy certain goods and, if he does buy, in what quantity. Some goods have no appeal to the consumer as he would derive no satisfaction from using or consuming them. If a man is not interested in shooting he will scarcely visit a gunsmith; but if he plays golf he may well call at a sports outfitters and buy some equipment, for he will hope to derive satisfaction from its use. This ability of a good to yield satisfaction to the consumer is known as its utility and is the underlying determinant of individual demand. The consumer, of course, may not realize this, but it is true to say that unless a consumer expects to derive utility from a commodity he will not buy it. Such commodities are not necessarily useful; if a good is useful to us then it also has utility for us, but the fact that a commodity yeilds us utility does not imply that it is useful. For example when one has a flat battery a starting handle is useful and yeilds utility, while a print of a vintage car is scarcely useful but may nevertheless give utility (or satisfaction) to the person hanging it on his wall.

The utility obtained from a particular item will vary with time and place. A block of ice cream is likely to be more appealing to a child in London on a sweltering July day than to a child in Snowdon in January. This shows to some extent how demand for some goods varies during the year.

Let us now return to our golfer. Suppose his wife has bought him a set of golf clubs and a season ticket for the local course but forgotten that he will need some golf balls. What will determine the number of golf balls he will buy, if the conditions of demand are taken as fixed? Clearly the rest of his golfing equipment will be useless without at least one ball and he would be prepared to pay a high price for one ball as he would expect a large amount of utility from it. As he is a beginner at the sport his aim is likely to be erratic, and he would probably lose his ball so he will be prepared to buy perhaps two or three balls at the prevailing price, as he would expect to get considerable utility from them. If the retailer, however, offered him one hundred balls at the same price he would be unlikely to accept the offer as he would not use many of the balls for a long while so they would yield him no satisfaction in the foreseeable future. If the golfer had to buy the balls one at a time at an auction or by bargaining for them, the price he would be prepared to offer would be related to the utility he expected to get from each successive ball. Thus he would be ready to pay heavily for the first ball but probably not quite so highly for the next and when he had sufficient to meet his requirements he would buy no more, as he would prefer to spend his money on other things. His approach to the problem is summarized in Table 3.1 and illustrated in Fig. 3.1.

Table 3.1 Diminishing marginal utility

No. of golf balls	Total utility (pennyworths)	Marginal utility (pennyworths)	Maximum price offered
1	40	40	40
2	71	31	31
3	97	26	26
4	119	22	22
5	139	20	20
6	157	18	18
7	173	16	16
8	188	15	15
9	202	14	14
10	$214\frac{1}{2}$	$12\frac{1}{2}$	$12\frac{1}{2}$
11	226	$11\frac{1}{2}$	$11\frac{1}{2}$

The term *marginal utility* can be defined as the satisfaction received from possessing one extra unit of a commodity, or the satisfaction lost by giving up a unit. As we said in Unit One most economic decisions are made at the margin and are concerned with the satisfaction gained from one unit more of commodity A or one more of commodity B. The golfer buying his balls is an example of the operation of the *Law of Diminishing Marginal Utility*. This law states that the more of a commodity we have, the smaller the

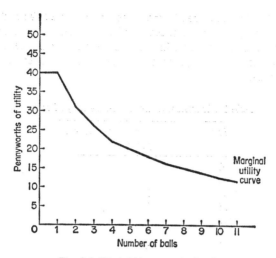

Fig. 3.1 Diminishing marginal utility

utility we are likely to derive from an extra unit of it. Many examples spring to mind: as winter approaches the possession of a warm anorak is highly desirable and a consumer may be prepared to pay £8·00 to £9·00 for such a garment. One anorak will be sufficient to keep out the cold and the consumer would be unlikely to derive as much utility from a second anorak as from the first. He would be unlikely to pay so much for a second one. Similar considerations apply to almost all items of everyday consumption.

Another point emerges from Table 3.1. There are no special units in which to measure utility. We can either talk about units of utility or about pennyworths of utility, and here we follow the latter practice. As we see if the golfer has three balls it would be irrational for him to offer more than 22p for a fourth ball, since he expects to obtain only 22 pennyworths of utility from it. If the price were 23p he would obtain more utility by spending his 23p in other ways. In order to maximize his satisfaction a consumer must always behave in this way, ensuring that marginal utility is proportional to price.

In practice, of course, the consumer will not have to pay more for the first ball than the second. He will have to pay the prevailing market price for each one, and it is up to him to decide how many he wants and can afford. If the price is 20p he will buy five, as the price will correspond with the marginal utility he gets from the fifth ball. A fall in price to 15p would induce him to buy eight balls. This does not imply that his requirements are fully met, but that at prevailing prices a ninth ball would yield too little utility or, to use the common phrase, will not be value for money. If some enterprising child on a golf-course offers to sell him a ball for 14p or less he will certainly be prepared to take it.

We may now take the important step of converting our table of *diminish-*

ing marginal utility into what is known as an *individual demand schedule*. We are preparing a demand schedule of our golfer for golf balls. This schedule assumes that changes in demand are entirely due to changes in the price of the commodity, since all other conditions are assumed to remain the same.

Table 3.2 Individual demand schedule for golf balls

Price (pence)	Quantity demanded (per month)
50	0
40	1
31	2
26	3
22	4
20	5
18	6
16	7
15	8
14	9
$12\frac{1}{2}$	10
$11\frac{1}{2}$	11

As we have established that he is prepared to pay a price equivalent to the marginal utility he derives we can say that when price is, for example, 18p per ball he will buy six balls. If the price rises to, say 50p, he will refuse to buy any balls, and will either use old ones or temporarily abandon the game. The same information is shown graphically in Fig. 3.2 which is

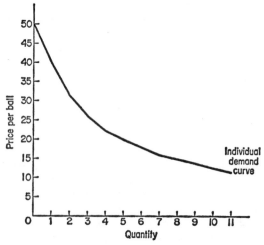

Fig. 3.2 *An individual demand curve for golf balls*

known as an individual demand curve. This measures price on the vertical
axis and the quantity demanded on the horizontal axis.

From the diagram we can find the number of balls that on average our
golfer will buy at any particular price. For example, if the price were 19p
he would buy on average five and a half balls per week. Both the table and
the diagram confirm the fact that as price falls the quantity demanded
increases. We can now see the reason for this. As consumption rises the
marginal utility to be derived from an extra unit falls and consumers can
only be persuaded to buy more units if the price is reduced. There is no
point in spending, say 8p, to get 5 pennyworths of utility, and no sense in
buying goods when a further supply would be a nuisance and lead to less
satisfaction.

There are, of course, thousands of other purchasers of golf balls and
each player has his own demand curve for balls. Some examples are given
in Fig. 3.3.

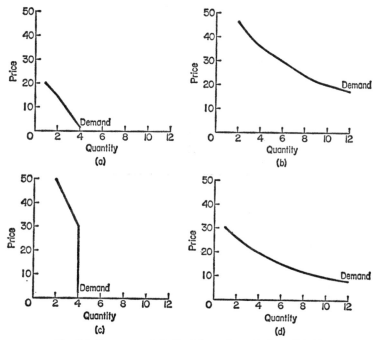

Fig. 3.3 *Some more individual demand curves for golf balls*

Each curve is representative of a golfer of distinctly different habits from
the others. In 3.3 (*a*) the player doesn't come into the market until price is
as low as 20p, while in 3.3 (*b*) the consumer is prepared to buy two balls at
49p and would no doubt still be in the market when price rose above 50p.
So would the golfer in 3.3 (*c*) although he cannot be tempted to buy more

than four balls per month, whereas the man in 3.3 (*d*) would buy twelve if price fell to 8p even though he won't enter the market till price falls to 30p.

The economist is not, however, normally much concerned with the demand schedule of a single individual for a single commodity. He is much more interested in the overall level of demand for a product. This can be established by adding together the demands of all individuals at each price. We thus obtain the *market demand schedule* which can then be converted into the *market demand curve* for the product, as shown in Table 3.3 and Fig. 3.4. The schedule does not include prices above 60p, or those below 5p, but there is no reason to suspect that its trend would alter at either end.

It is obviously important that manufacturers of golf balls should know the actual level of demand for their product at various prices. As we are

Table 3.3 **The market demand schedule for golf balls**

Price	Quantity (000's) per week
60	2
55	3
50	5
45	6
40	8
35	10
30	12
25	15
20	20
15	30
10	50
5	90

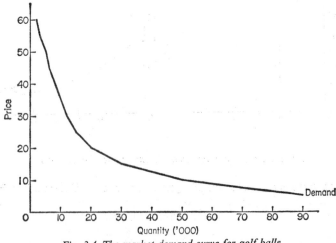

Fig. 3.4 The market demand curve for golf balls

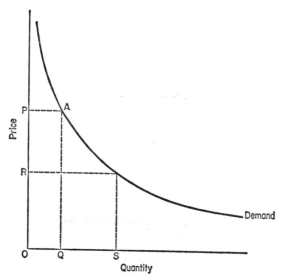

Fig. 3.5 A generalized demand curve

discussing the general principles of demand such numerical precision is unnecessary, and we will normally use a generalized demand curve like that shown in Fig. 3.5, using letters to indicate the various prices and quantities.

In this way it is possible to discuss the general principles of demand without being concerned with the details for particular products. This curve (Fig. 3.5) characterizes all normal demand curves, although they do not all have the same slope or run so smoothly. For example, in Fig. 3.5 when the price is OP a quantity of OQ will be demanded. Price OR will be sufficiently low to persuade consumers to buy OS units. Total consumer expenditure at price OP is shown by the rectangle OPAQ, which is price multiplied by quantity.

3.4 A Digression on Consumer's Surplus

The observant reader may not have been entirely happy with our discussion of Table 3.1, part of which is included here as Table 3.4.

Table 3.4 Marginal and Total Utility obtained from use of golf balls

No. of balls	Total utility (pennyworths)	Marginal utility (pennyworths)
1	40	40
2	71	31
3	97	26
4	119	22
5	139	20
6	157	18

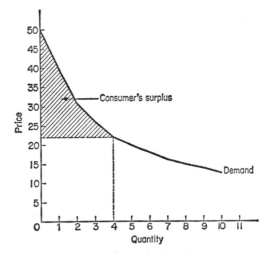

Fig. 3.6 Consumer's surplus

We have already seen that if the price of balls is 22p our consumer will buy four balls (equating price and marginal utility), thus spending 88p. Four balls, however, provide him with 119 pennyworths of utility (the sum of the individual marginal utilities) and if the balls had been sold at an auction he would have been prepared to pay 119p for them. This extra 31 pennyworths of utility is known as the *consumer's surplus* and it is shown as the shaded area in Fig. 3.6. The consumer's surplus is the utility or satisfaction which the consumer receives but does not pay for. It is frequently the target of tax-collectors, and in this context we shall consider it again in Section 12.9 (*b*).

3.5 Changes in the Conditions of Demand

We noted earlier that in order to isolate the effect of price changes on demand it was necessary to exclude other influences altering demand. These influences may be called the *conditions of demand* and are as follows: incomes, the price of other goods, and changes in taste or fashion. If these conditions of demand change then the demand curve changes so that at every price a different quantity will be demanded. The following examples demonstrate this point.

(*a*) A rise in incomes
At one level of income the market demand curve for the product is shown by DD in Fig. 3.7, and at a price OP consumers are prepared to buy OQ units, thus spending OPAQ (price × quantity) on the product. If incomes in general rise then it is likely that existing purchasers will buy more of the commodity, and that potential purchasers who had previously considered

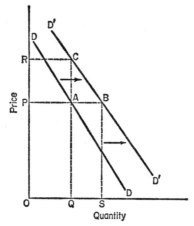

Fig. 3.7 A change in the conditions of demand

the product outside their price range will now come into the market so that
at OP, OS units will be bought, with consumers prepared to spend OPBS.
We must not infer that they will be able to acquire OS units; that depends
upon the supplies being available.

An alternative view of this change in demand is that the original quantity
OQ can now be sold at a price OR rather than OP so that expenditure will
be ORCQ. Later we will show that the eventual price paid is likely to be
between OP and OR and the eventual quantity bought between OQ and OS.
There are some commodities to which this analysis does not apply. It is
highly unlikely, for example, that an increase in incomes will move the
demand curve for salt very much to the right.

A fall in incomes, other things remaining equal, will be expected to move
the demand curve to the left.

(b) A change in the prices of other goods

The effect of a change in the prices of other goods depends upon the relationship between those goods and the commodity under consideration. There
are three possibilities. The first case is that in which goods have a neutral
relationship to each other. Except in the sense that all goods are competing
for the consumer's attention the demand for some goods is completely
independent of the demand for some other goods, thus it is very unlikely
that a change in the price of salt would have any influence on the demand for
petrol. The other possibilities are that the goods are *substitutes* for each
other, such as two brands of petrol, or that they are *complements* to each
other and jointly demanded such as fountain pens and ink.

(i) **Substitute goods.** If the price of one brand of petrol rises because of an
increase in costs, the demand curve for other brands of petrol may be
expected to move to the right as people change to the relatively better

bargain. (Diagrammatically the move is the same as in Fig. 3.7.) In a similar way the curve will move to the left if the prices of other brands fall.

(ii) **Complementary goods.** If the price of fountain pens rises, the demand for ink will tend to fall as consumers change to using ball-point pens. In this case the demand curve for ink moves to the left indicating a lower quantity being demanded at each price.

The extent to which demand curves shift in response to changes in the price of other goods depends upon the degree of complementarity or substitutability and the relative costs of the goods. For example a heavy fall in the price of motor cars will move the demand curve for petrol to the right, but it does not follow that a fall in the price of petrol has the same effect on the demand curve for cars. This is because the cost of a gallon of petrol is very small compared to the cost of a motor car. The importance of the distinction between substitutes and complements will be made clear when we begin our analysis of markets and price determination in Unit Twelve.

(c) A change in tastes

The tastes of consumers are extensively moulded by advertising and other forms of publicity. The object of commercial advertising is to move the demand curve for a product to the right so that people buy more of it at each price. Thus a successful advertising campaign will move the demand curve for a product to the right, while success for the Government's anti-smoking campaign will move the demand curve for cigarettes to the left, with the effects illustrated in Fig. 3.7.

3.6 Extensions and Increases in Demand

There is no law-making body giving definitive names to economic concepts, but it prevents confusion if the same terms are always applied to the same concepts. Changes in demand provide a clear example of this. In Fig. 3.8 demand can move from OQ to OS in two different ways.

The change may be the result of a fall in price from OP to OR in which case the consumer moves along the curve DD from point A to point B. Such a move is conventionally called an *extension of demand* and occurs without the conditions of demand changing. The reverse move when price rises from OR to OP is referred to as a *contraction of demand*. The move from Q to S could also have been accomplished, however, by shifting to the new demand curve D' D' following a change in the conditions of demand. In this case consumers move from point A to point C, thus acquiring a greater quantity at the existing price OP. Such a move is called an *increase* in demand; the move from S to Q (when the conditions of demand change) being a *decrease* in demand.

An alternative, and equally acceptable, approach is to refer to the exten-

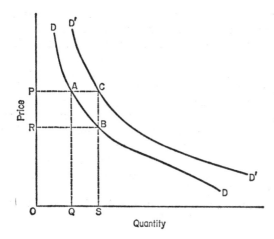

Fig. 3.8 An extension and an increase in demand

sion or contraction of demand simply as a movement along the demand curve, and the increase or decrease of demand as a movement of the demand curve.

3.7 Price Elasticity of Demand

We have already become familiar with the basic principle that a fall in price leads to an extension of demand, but we have said nothing about the rate at which demand rises when price falls. In some cases the response is quite marked: for example a small reduction in the price of one brand of petrol will induce many consumers to buy it in preference to other brands. A similar change in the price of all brands of petrol, however, would have little effect on the demand for petrol as a whole. The concept of *price elasticity of demand* has been developed to allow us to measure the degree to which demand responds to a change in price. We are thus able to compare the sensitivity of the demand for various goods to a change in price.

Price elasticity of demand is important to suppliers of goods as well as to the consumers. Let us consider the actions of a housewife doing the weekly shopping. She has a shrewd idea of the prevailing prices of the goods that interest her and will be keen to find bargains. She knows the usual price of sugar is 5p per pound and she normally buys 2 lb per week. She is unlikely to refuse to buy if the price rises to 5½p or 6p but might try to economize if it reaches 7p or 8p. Similarly she is unlikely to buy in bulk if the price falls to 4½p or 4p. Most consumers behave in this way and so suppliers will expect to sell about the same amount of sugar each week even though prices may fluctuate a little. We can, however, consider the demand for sugar from a slightly different aspect. If our housewife notices another grocer selling identical sugar on 'special offer' for one week only at 2½p per pound, she is likely to buy her sugar at the new grocers rather than at her normal store. Many others will behave in a similar way and because of the lower price the second retailer will have a considerable rise in demand, at the expense of other retailers.

Thus when we are talking of elasticity of demand we must define our terms carefully. Taking the above example, we must distinguish between the demand for sugar in general and the demand for sugar at a particular outlet as the latter is likely to be more responsive to a change in price. We must also distinguish, for example, between the demand for soap-powder in general and the demand for a particular brand of soap-powder. In this case the latter is likely to be more responsive than the former to a change in price. It is true to say that the more widely we define a product the less sensitive its demand is likely to be to a change in price.

To revert to our example, let us suppose that the experience of the second retailer is summarized in Table 3.5.

Table 3.5 Hypothetical change in demand for sugar at a single retailers

Price (p)	Quantity per week (lbs)	Total outlay by consumers
5	400	£20·00
2½	1000	£25·00

To measure the price elasticity of demand for sugar we use the formula

$$E^D = \frac{\% \text{ Change in quantity demanded}}{\% \text{ Change in price}}$$

and in the case of our retailer we obtain

$$E^D = \frac{150}{50} = 3$$

It is most important to remember that it is the percentage changes that are relevant, not the absolute changes.

The grocer's total revenue from sugar rises from £20·00 to £25·00 when he reduces his price. An increase in revenue always occurs when elasticity is greater than one. The same thing is shown graphically in Fig. 3.9, where the reduction in price results in a *loss* of revenue of £10·00 on the 400 lb that could have been sold at 5p (this is represented by rectangle *A*) and a gain in revenue of £15·00 on the extra 600 sold as a result of the change in price in price (rectangle *B*).

When we measure elasticity it may be anywhere between zero and infinity. There are six possiblities which are explained and illustrated in Fig. 3.10.

(*a*) Elasticity of demand = zero (Fig. 3.10 (*a*)). In this case the quantity demanded neither expands nor contracts when price changes from OP

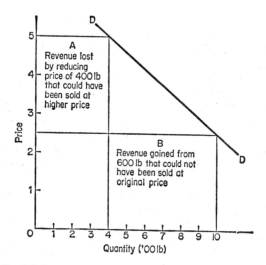

Fig. 3.9 *Diagrammatic illustration of elasticity of demand*

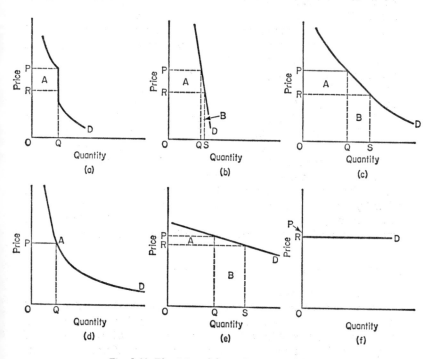

Fig. 3.10 *Elasticity of demand: six possibilities*

to OR. A producer will be free to increase his price in such circumstances, without losing customers. The demand for petrol or addictive drugs may be inelastic over moderate price ranges. As we see in Fig. 3.10 (a), the demand curve is vertical over the relevant range. A fall in price would lead to a fall in revenue since rectangle B which we saw in Fig. 3.9 does not exist. *Demand in this case is said to be perfectly inelastic.*

(b) Elasticity of demand is greater than 0 but less than 1 (Fig. 3.10 (b)). In this case a fall in price does result in an extension of demand but a less than proportionate one. The net result is that total revenue (for the producer) or expenditure (for the consumer) falls when the commodity price falls. Rectangle B is smaller than rectangle A. *Demand is said to be inelastic* for this commodity.

(c) and (d) Elasticity of demand = 1 (Fig. 3.10 (c)). In this instance we find that the percentage fall in price is exactly matched by the percentage extension in demand and so total revenue remains constant. Rectangles A and B are equal in area. *Demand is said to have unit elasticity.* There is no particular reason why elasticity should equal 1 any more than it should equal 0·1 or 1·1, but we should appreciate the shape of the curve in Fig. 3.10 (d) which has elasticity of one over its whole length for it often becomes important when the Government or some other body wishes to stabilize producers' receipts. Any rectangle drawn from this curve will be equal in area to OPAQ and consumers' expenditure remains constant.

(e) Elasticity of demand is greater than 1 but less than infinity (Fig. 3.10 (e)). The quantity of goods demanded changes by a greater proportion than the change in price and accordingly total revenue rises when price falls (rectangle B is greater than A). *Demand is elastic.*

(f) Elasticity of demand equals infinity (Fig. 3.10 (f)). This is where there is more use in the construction of economic models than in practice. The demand curve is horizontal indicating that at one price OP the demand is zero but if the price falls fractionally the producer can sell as many units as he can produce. Again a fall in price leads to a rise in revenue. Rectangle A does not exist; the size of B depends upon the amount sold by the producer. *Demand is said to be perfectly elastic.*

Most demand curves are likely to have elasticities corresponding to cases (b) or (e) above. The technique of comparing total revenue at two different prices can only indicate whether elasticity is greater or less than one. If greater precision is needed the method described above has to be used.

3.8 Some Pitfalls Surrounding Elasticity

(a) Arithmetic measurement

When we measure elasticity we are trying to measure the slope of the demand curve at a particular point. Since price changes in the United Kingdom must of necessity be of at least ½p, which may represent quite a high percentage of the price, we have to measure elasticity over a *range* of points. No great difficulty is involved so long as we are consistent in measuring for a fall or for a rise in price. Confusion can arise if measurements are not consistent. When using the data in Table 3.5 we found that the elasticity was 3 when the price fell. In this case a 50 per cent fall in price led to a rise of 150 per cent in quantity demanded. But suppose we regard the price as having risen from $2\frac{1}{2}$p to 5p. Now the quantity demanded falls by 60 per cent owing to a rise in price of 100 per cent and the elasticity is measured by

$$\text{Elasticity of demand} = \frac{60}{100} = 0\cdot6$$

This is tantamount to warning a motorist that he is about to descend a 1 in 7 hill, and then on the return journey telling him that it is 1 in 4! It is the same hill and the same slope. The demand curve seems to behave likewise. In our example, the difference we obtained was large because of the large percentage price change. But suppose we examine a section of a different demand schedule in Table 3.6.

Table 3.6 Hypothetical demand schedule for refrigerators

P (£)	Q (per week)
100	500
99	502

If the price falls from £100 to £99 the price elasticity of demand may be calculated as follows:

$$E^D = \frac{\%\ \text{Change in Q}}{\%\ \text{Change in P}} = \frac{+0\cdot4}{-1} = -0\cdot4$$

On this occasion we have included plus and minus signs to show the direction of the change. Since, with very few exceptions, there will always be a plus sign and a minus sign (price and quantity change in opposite directions) price elasticity is normally negative. By convention, however, the minus sign is omitted as it is elsewhere in this chapter.

Let us now suppose that the price of refrigerators rises from £99 to £100. Elasticity is measured as follows:

$$E^D = \frac{\%\ \text{Change in Q}}{\%\ \text{Change in P}} = \frac{\frac{2}{502}}{\frac{1}{99}} = \frac{0\cdot398}{1\cdot01} = 0\cdot394$$

This is slightly lower than the previous figure but the difference is much more acceptable than that which we had in connexion with Table 3.5. The difference occurs because we are dealing with a much smaller percentage change in price. If the reader were to substitute £99·50 for £99·00 in Table 3.6 he would find that the difference in measured elasticity would be even smaller, when a price fall and a price rise are considered.

(b) Optical illusions
Consider the two demand curves shown in Fig. 3.11. A quick glance suggests that 3.11 (a) is elastic while 3.11 (b) is inelastic. A more careful examination shows that they convey exactly the same information and thus have the same

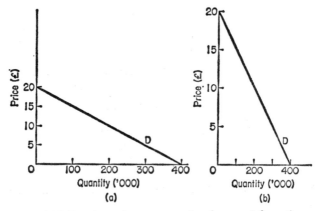

Fig. 3.11 Two demand curves conveying the same information

elasticity! We must be very careful of labelling curves elastic or inelastic without considering the scales involved. But we shall often find it useful to refer to one of two curves as the more elastic or the less elastic.

(c) Changing elasticity

Only in rare circumstances as in Fig. 3.10 (d) will a demand curve have constant elasticity over its entire length. The elasticity of the curves in Fig. 3.11 varies from infinity to zero as we move along them from top to bottom. A mathematical proof of this is possible but too complex for this book. An indication of the reason is given by the following argument. When the price of a good is £20·00 a reduction of £1·00 would increase sales by 20 units, and when the price is £5·00 a reduction of £1·00 would have the same absolute effect. But at the lower price £1·00 is a greater percentage change in price, and 20 units is a smaller percentage change in quantity than at the higher price. Accordingly, elasticity of demand is lower and it continues to fall as price falls.

3.9 The Determinants of Elasticity of Demand

Having established that demand curves slope downwards at different rates we must now consider the main factors that determine the elasticity of demand for any particular commodity.

(a) The availability of substitutes

This is the most important consideration. If we have a commodity labelled food, which includes everything we eat, demand for it is likely to be inelastic; but if we have a commodity called brussels sprouts the demand for it is likely to be fairly elastic as there are other green vegetables which can be substituted for it if price rises.

The promotion campaigns that characterize the marketing of petrol and household detergents are based on the belief that the elasticity of demand for an individual brand is high and that by reducing price (or offering free gifts which the consumer believes equivalent to a price reduction) one supplier can attract custom from another even though the demand for the product as a whole may be inelastic.

(b) The consumer's budget

In general if a consumer spends only a small proportion of income on a commodity his elasticity of demand for it will be low. Even a doubling of the price of a box of matches would be unlikely to reduce demand significantly. Similarly if a person buys one newspaper a day out of his income of £30·00 per week a modest rise in price will not cut his demand, whereas he might reduce his purchases if he has been in the habit of buying five or six papers per day.

3.10 Two other Elasticities of Demand

While price elasticity of demand is the most important indicator of the sensitivity of demand and has many important applications which are discussed in Unit Twelve, we must not overlook two other types of elasticity, which are associated with changes in the conditions of demand.

(a) Cross elasticity of demand

It may often be necessary to have an estimate of the effect of a change in the price of one good on the demand for another good. This is particularly true where goods are either substitutes for one another or else complementary goods. The response of the demand for one good to changes in the price of another is called the *cross elasticity of demand*. It is measured by the formula:

$$\text{Cross elasticity of demand} = \frac{\% \text{ Change in quantity of } A \text{ demanded}}{\% \text{ Change in price of } B}$$

In the case of substitutes cross elasticity will be positive, a fall in the price of B will be followed by a fall in the quantity of A demanded. If the goods are complementary in demand, cross elasticity will be negative. This concept may be useful to producers when they try to estimate the effects of changes in the prices of competitors' goods on the demand for their own. A high positive cross elasticity will indicate the desirability of maintaining stable prices when competitors increase theirs, as consumers readily switch from one product to another.

(b) Income elasticity of demand

A change in consumer income will normally lead to a change in the level of demand for some goods, and the income elasticity of demand indicates the extent of this change. It is measured by the formula:

$$\text{Income elasticity of demand} = \frac{\% \text{ Change in quantity}}{\% \text{ Change in income}}$$

Income elasticity, which will normally be positive, is hardly of great importance as far as individual consumers are concerned, but it does have significance for particular industries. One would expect the income elasticity of demand for food to be low (hence explaining in part the difficulties of developing countries in expanding the markets for their agricultural products) and the income elasticity of demand for leisure activities to be high (explaining in part the growth of private transport and continental holidays). Unfortunately demand analysis does not allow us to illustrate cross elasticity and income elasticity precisely, but we will return to these concepts in Unit Four.

3.11 Regressive Demand Curves

While the majority of demand curves behave as those described above and slope down from left to right, occasionally we find examples like those shown in Fig. 3.12 which do not fit into the pattern. These curves form a very tiny minority, but they are interesting special cases.

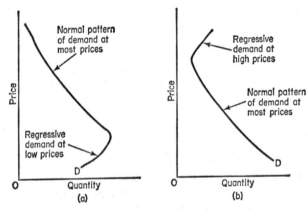

Fig. 3.12 Regressive demand curves

(a) Giffen goods (Fig. 3.12 (a))

This is the term applied usually to those basic foodstuffs consumed by people on very low incomes. The demand for these goods rises when the price rises so that the demand curve runs from south-west to north-east for some of its length. The reason for this phenomenon is that when the price of the good rises, consumers on very low incomes maintain their former level of purchases, and finding that they can no longer afford other goods actually increase their purchases of the Giffen good. The term 'Giffen goods' is derived from the economist Sir Robert Giffen who noticed this behaviour among the Irish peasants during the potato famine in the 1840s.

(b) **Goods of ostentation** (Fig. 3.12 (b))
These may show the same tendency in the upper price ranges. There are people who believe that you have to pay a high price to get anything worthwhile, and in some circumstances it may be possible for a producer to exploit this characteristic by raising the price of goods and actually increasing his sales by attracting consumers who would otherwise reject the goods as too cheap. Such goods are purchased not for their intrinsic value, but to emphasize status.

3.12 Questions and Exercises

1. Outline the main factors influencing the demand for colour television sets.
2. Show with the aid of diagrams how changes in the conditions of demand will influence the position of the demand curve.
3. Distinguish between utility and marginal utility.
4. Why is marginal utility more important in the theory of demand than total utility?
5. Define price elasticity of demand. Draw diagrams to show curves with elasticity of demand of (a) unity (b) zero (c) infinity.
6. Calculate price elasticity of demand for the following changes in price and quantity (when price falls):

P (pence)	Q (quantity)
10	1000
9	1100
8	1500
7	1500

7. Explain the connexion between total expenditure on a good and elasticity of demand for the good.
8. What are the main determinants of elasticity of demand? Illustrate your answer.
9. Some demand curves do *not* slope down from left to right. Why not?
10. What do you understand by demand? What are the main influences on the demand for bread?
11. The demand for various products varied in the following way between the years 1971 and 1972.

	A	B	C	D	E
1971	10,000	15,000	150,000	70,000	10,000
1972	11,000	12,000	150,000	71,000	100,000

During this period the size of the Gross National Product rose by 5 per cent. Calculate the income elasticity of demand for each of the products. Comment on the proposal of a manufacturer to cater for a similar increase in demand in each case for 1973, when the GNP is again expected to rise 5 per cent.
12. The price of good A is fixed at 15 pence. The demand for this product varies in the following way when the price of good B changes:

Price of B	20	15	12	10
Demand for A	700	200	100	75

Calculate the cross elasticity of demand at each price.

13. Given that the price elasticity of demand for a product is 1·5, and that total revenue at today's price of 12p is £20·40, calculate the total sales and total revenue (a) when price falls to 10p; (b) when price rises to 14p.

14. What factors are likely to influence (a) the price elasticity of demand; (b) the income elasticity of demand for each of the following goods and services over the next 5 years?
   (a) launderettes; (b) public transport; (c) gramophone records; (d) meat; (e) margarine.

15. A company can vary its output to any level between 100 units and 1000 units per week, without altering the cost per unit of production which is 9p. The demand schedule for the product is:

Price	18	16	14	12	10	8
Quantity	50	120	250	600	900	1500

What price should be charged (a) to maximize revenue; (b) to maximize the difference between total costs and revenue?

Unit Four

Consumer Choice: Indifference Curves

4.1 Introduction

In Unit Three we examined the demand for a particular commodity, but said little about the pattern of demand, or the forces which influence a consumer when comparing commodities with one another. The demand curve is really more useful to the firm or industry analysing the probable effects of a change in the price of its product, than to an economist trying to explain consumer behaviour, or to consumers themselves. This is particularly true as the consumer is not able to measure utility quantitatively and will not make adjustments to his pattern of consumption every time there is a fractional change in the price of one of the commodities he consumes. In our previous discussion of demand we measured utility in 'pennyworths', but such measurement is not really practical. A consumer is more likely to say that one commodity or group of commodities gives him more satisfaction or utility than does another. You can test this for yourself: you will not be able to say precisely how much utility you derive from a unit of cheese or a unit of petrol, but you will almost certainly be able to say that one of them yields more utility than the other. This is called an ordinal approach to utility (putting commodities or groups of commodities in order of preference), while assigning a numerical value to the utility derived from each unit consumed is called the cardinal approach.

Another weakness of demand curve analysis is that it rests heavily on the rather artificial concept of constant conditions of demand. It does not conveniently allow, for example, prices and incomes to change at the same time. Nor when the price of a commodity falls does the demand curve analyse the causes of the extension of demand that follows. It cannot show whether consumers buy more because they feel they are now getting a better bargain, or because they now have a higher real income as they can buy their former quantities and still have money remaining for further purchases. It is more likely to be a combination of these factors, which are known respectively as the *substitution effect* and the *income effect*.

4.2 Indifference Curves

To overcome some of the difficulties in connexion with demand curves the technique of indifference-curve analysis has been developed. For this technique a pattern of indifference curves is built up, each of which records a number of choices made between alternatives which yield equal satisfaction. Let us assume that a consumer confines his expenditure to two commodities, food and clothes, and that he is at present buying 25 units of food and 20 units of clothes, a combination which gives him a certain amount of utility or satisfaction. If one unit of food is taken away his total utility is reduced but can be restored to its former level by the addition of one or more units of clothes; perhaps he would require $1\frac{1}{2}$ units of clothes to compensate for the loss of one unit of food. In such a situation he is said to be indifferent between the two combinations, 25 units of food and 20 units of clothes, and 24 units of food and $21\frac{1}{2}$ of clothes. Table 4.1 shows a number of other combinations of food and clothes giving the consumer the same amount of utility as the original combination. Such a list is known as the consumer's *indifference schedule*.

Table 4.1 A consumer's indifference schedule

Units of food	Units of clothes
25	20
24	$21\frac{1}{2}$
23	$23\frac{1}{2}$
22	26
21	29
20	33
19	39
18	50

The schedule agrees with our earlier discussion of diminishing marginal utility. As the consumer is allowed less and less food he demands an increasing number of units of clothes to compensate for the loss of one unit of food. Thus when he has 25 units of food $1\frac{1}{2}$ units of clothes compensates for the loss of one unit of food, but when he had only 20 units of food he needs 6 units of clothes to compensate him for the loss of one more unit of food.

This is the principle of *the diminishing marginal rate of substitution, which states that the greater the amount of commodity a consumer has the more easily he may be compensated for the loss of a marginal unit.* This explains the shape of indifference curves, convex to the origin as shown in Fig. 4.1, and as fully explained in Section 4.3.

Both the schedule and the curve (Fig. 4.1) apply to only one consumer. Another consumer with the same initial combination of food and clothes might well have had a different scale of preference. He might, for example, be a much heavier eater and require more than $1\frac{1}{2}$ units of clothes to compensate for the loss of one unit of food. In this case the indifference schedule and curve are different but display the same principle of a diminishing marginal rate of substitution.

Fig. 4.2 presents two indifference curves relating to different consumers. The slopes of the two curves are different because of the difference in tastes of the two consumers. At point *a* they are each consuming OM of food and ON of clothes. But when each of them is deprived of the same amount of

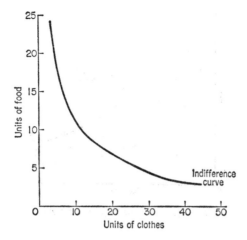

Fig. 4.1 A consumer's indifference curve

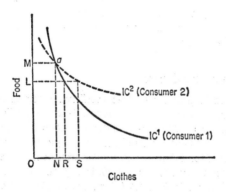

Fig. 4.2 Indifference curves relating to different consumers

food (ML), they require different amounts of clothes in compensation. Consumer 1 is not a big eater and requires only NR of clothes to maintain his satisfaction, while consumer 2 is less willing to sacrifice food and it requires NS of clothes to induce him to do so.

If our original consumer had started from a situation in which he had one more unit of either food or clothes or of both he would have been gaining more utility than when he had 25 units of food and 20 of clothes. The indifference curve would then have been upwards and to the right of the first curve. Each of these curves, representing various combinations of goods between which the consumer is indifferent, is part of a family of such

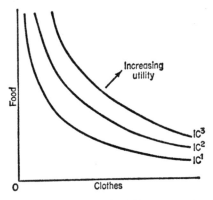

Fig. 4.3 A family of indifference curves

curves unique to the consumer to whom they apply. A selection of curves from such a family is shown in Fig. 4.3.

Total utility is constant along the length of any one curve, but as the curves move away from the origin the total utility increases. These curves are just a selection from a whole multitude of curves that could be drawn for any one consumer. In this case each begins with a slightly different combination of food and clothes.

4.3 The Shape of the Curves

A process of elimination shows that indifference curves must be drawn convex to the origin. Fig. 4.4 (*a*) is impossible as an indifference curve, as the rational consumer cannot be indifferent between points *a* and *b*. He must derive more utility at *a* where he has the same amount of clothing but a greater amount of food than at *b*. A similar argument applies to Fig. 4.4 (*b*) and to the U-shaped curve in Fig. 4.4 (*c*). Indifference curves cannot double back upon themselves. In Fig. 4.4 (*d*) *a* must be preferable to *b* since the consumer has more of both commodities. None of these curves consists entirely of points of indifference. In Fig. 4.4 (*e*) the law of diminishing marginal utility and therefore of the diminishing marginal rate of substitution is not operating. At point *a* where the consumer has a large amount of food and relatively few clothes he demands the same amount of compensation for the loss of a given amount of food as he does at *b* where he has relatively little food and should therefore value it more highly and demand more clothing in exchange.

In Fig. 4.4 (*f*) the position is even more absurd for the consumer is more reluctant to sacrifice food at *a* where he has an abundance than he is at *b* where he is running short. He demands more compensation when he has a good stock of food than when his stock is much lower.

It is only in Fig. 4.4 (*g*) that we find a rational illustration of the

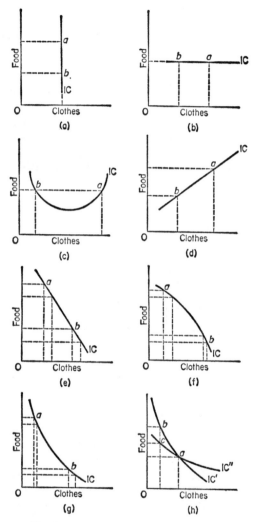

Fig. 4.4 The shape of indifference curves

consumer's behaviour. At *a* a small amount of clothing compensates for the loss of a unit of food of which, at this point, the consumer has a considerable amount. As we move down the indifference curve towards the point *b*. we find that increasing amounts of compensation must be offered to replace successive units of food, if he is to retain the same amount of satisfaction.

A further characteristic of indifference curves is illustrated in Fig. 4.4 (*h*); they cannot cross one another. The figure is supposed to show that the consumer is indifferent between points *a* and *b* and between points *a* and *c*. If this were truly so he would be indifferent between *b* and *c* but he cannot be as *b* offers him more food and therefore greater satisfaction.

It is unrealistic to assume that a consumer is confined to the purchase of two goods, but it is impossible to display any more on a two dimensional diagram. A convenient way of overcoming this difficulty is to plot the amount of a single commodity along one axis and to regard all the others together as money income and plot this on the other axis, as in Fig. 4.5.

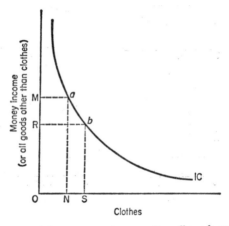

Fig. 4.5 *An indifference curve incorporating all goods consumed*

In this case if the consumer finds himself at point *a* we know that he is buying ON units of clothing and once he has bought this he still has OM units of income to spend on all the other goods. We must not make the mistake of thinking that he pays OM units of income to obtain ON units of clothes. We are not dealing with demand curves! If the consumer is at *b*, buying OS of clothing and leaving OR to spend on other things, then we know that he has sacrificed MR of money to obtain NS of clothes.

4.4 The Budget Line

The rational consumer seeks to maximize the satisfaction he gets from spending his income. It is interesting to see how he goes about this. We assume that he spends all of his income (if he wants to save we must talk of disposable income) and we know that indifference curves yield successively higher utility the further they are from the origin, so the consumer will want to be on as high an indifference curve as possible.

Let us consider first, Fig. 4.6. Here the consumer is allowed an income of £20, and we can plot his position on the y axis if he spends nothing.

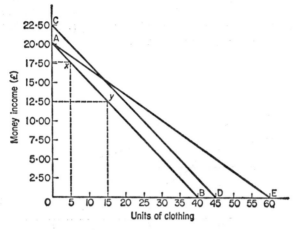

Fig. 4.6 The budget line

If, however, he decides to spend all of his income on clothing which costs 50p per unit he can in the first instance buy 40 units. The line AB which joins these two points is known as the consumer's *budget line*. This shows all the possible combinations of clothing and money income (which is spent on other goods) that the consumer can enjoy, assuming that he spends all his income.

The consumer could, of course, be placed anywhere within the triangle OAB, but he would then have some unspent income and would not be maximizing his satisfaction. Possible positions for him are at x where he has 5 units of clothes and £17·50 worth of other goods; and at y where he has 15 units of clothes and £12·50 of money income.

(a) The effect of a change in income

If the consumer's income rises by £2·50 his budget line will move away from its original position to a parallel line at CD. This shows that he can now obtain more satisfaction by spending more income on other goods, or by buying more clothes, or by a combination of both. A reduction in income has the opposite effect and causes the budget line to move towards the origin.

(b) The effect of a change in price

If the price of clothes falls the consumer can buy more of the now cheaper commodity. This is shown in Fig. 4.6 by the budget line AE which is not so steep as AB. In this case the price has fallen to 33⅓p which enables the consumer to buy a maximum of 60 units of clothing.

4.5 Consumer Equilibrium

We are now in a position to describe the way in which a consumer reaches an equilibrium position where he maximizes his utility or satisfaction. In Fig. 4.7 we have combined a budget line with three curves taken from a large family of similar indifference curves.

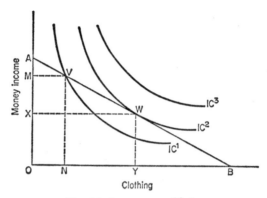

Fig. 4.7 Consumer equilibrium

The consumer is able to take up a position anywhere along his budget line AB. Let us see how his satisfaction changes as he goes from A to B. At A he buys no clothing and is on an indifference curve to the left of IC¹ and so gains less satisfaction than he obtains if he could reach IC¹. As he moves down AB he eventually reaches IC¹ at V, so by giving up an amount AM of other goods he is able to buy some clothing. A glance at the figure shows that he could be in a better position if he moved beyond V. Indeed as soon as he moves to the right of V he is on a higher curve than IC¹; the spaces between the curves shown in the figure could be filled with hundreds of others. As he continues to buy clothes at the expense of other goods he eventually arrives at W where the budget line just forms a tangent to IC² and at this point he buys OY units of clothes for XA units of income, and retains OX units of income to spend on other things. W is the best position he can reach with an income of OA for as soon as he moves to the right of W he is on a lower indifference curve which will give him less satisfaction. He cannot possibly reach a higher curve unless his income rises or the price of clothing falls. These two possibilities will be examined in Sections 4.6 and 4.7.

We may take another view of consumer equilibrium. Suppose food is priced at 50p per unit and clothes at £1·50. If the aim of the consumer is to maximize satisfaction should he consume each commodity in such proportions that the marginal utility derived from each is equal? Suppose he tries to do this? We can easily show that he is not maximizing utility and is not therefore in equilibrium. By giving up one unit of clothes he can buy three units of food, and, since the marginal utility from a unit of food and a unit of clothes is the same, his total utility must increase when he exchanges one unit of clothes for three of food. But by how much should he reduce his consumption of clothes, and increase his consumption of food if he is to maximize his satisfaction? Suppose he exchanges clothes for food to the point where the marginal utility of clothes is four times as high as that of food? (Remember that the marginal utility of clothes rises as consumption falls and that of food falls as consumption increases.) At this point he would be beyond the point of maximum satisfaction for, by spending £1·50 on clothes instead of food, he can increase his utility. He should continue until the marginal utility of clothes is only three times that of food. At this point he would be in equilibrium since he could not increase his total utility by altering his pattern of consumption. At this point too, we can express his equilibrium by saying that

$$\frac{\text{Marginal utility of food}}{\text{price of food}} = \frac{\text{Marginal utility of clothes}}{\text{price of clothes}}$$

and indeed this is the general condition for consumer equilibrium. This is the case at point W in Fig. 4.7. For if the consumer moves towards B the marginal utility of clothes declines by more than the rise in the marginal utility of other goods.

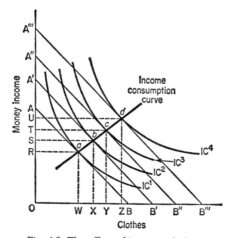

Fig. 4.8 The effect of increases in income

4.6 A Change in Income

As income rises the budget line moves progressively further from the origin thus enabling the consumer to increase his purchases of clothing from OW units to OZ units and at the same time he is able to increase his purchases of other goods from OR to OU. The line joining the points of

c

tangency, *a, b, c* and *d* is known as the *income consumption line* and it shows how the consumer changes his purchases as his income rises. In this case the curve slopes upwards from the origin in a straight line, showing that the consumption of a commodity rises roughly in proportion to income, and also that consumption of other goods rises as well. There are other possibilities which are illustrated in Fig. 4.9.

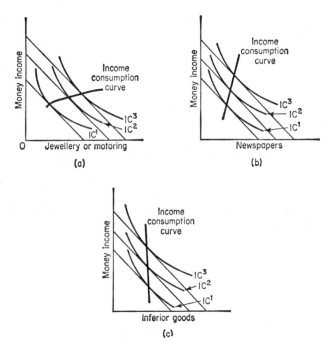

Fig. 4.9 Three other income consumption curves

In Fig. 4.9 (*a*) the income consumption curve is rather flat and tends to level off. This is the curve of a luxury good like jewellery or a prestige motor car for which demand rises more than proportionately with income (see below). In Fig. 4.9 (*b*) the steepness of the income consumption curve shows that consumption of the commodity measured along the *x* axis does not increase in proportion to income. In this case the axis is labelled 'newspapers' but it might as easily have been 'salt' or 'food'. Fig. 4.9 (*c*) is a rather special case; for the income consumption line actually slopes backwards showing that consumption of this commodity declines as income increases. Such commodities are known as *inferior goods*. A special type of these—*Giffen Goods*—was discussed in Section 3.11. Inferior goods are those we buy when our income is relatively low, but discard as our income rises. Margarine is a frequently quoted example, since people commonly prefer butter and buy more as they become richer, but similar changes occur between simulated fur and real fur, *vin ordinaire* and vintage wines, 'cork' cricket balls and leather cricket balls. It should be clearly understood that Giffen goods are those for which demand rises when their price rises, while inferior goods are considered in relation to the consumer's income.

The income consumption curve has a further use as it shows how consumption changes with income. It enables us to measure income elasticity of demand:

$$\text{Income elasticity of demand} = \frac{\%\ \text{Change in quantity}}{\%\ \text{Change in income}}$$

This is something that demand curves do not permit. In almost every case the income elasticity will be positive as the quantity demanded rises when income rises. The main exception is in the instance of inferior goods illustrated in Fig. 4.9 (c) where the quantity demanded falls as income rises, and the income elasticity of demand is negative.

While an individual's income elasticity of demand may not be of great significance the combined income elasticities of all consumers of a particular good or service may be very important to producers and planners. For example, if it is known that the income elasticity of demand for colour televisions is high and a rapid rise in incomes is anticipated then there is every reason for producers to install extra plant to meet the anticipated higher demand for their product.

4.7 A Change in Price

Fig. 4.10 shows what happens when the price of a commodity falls.

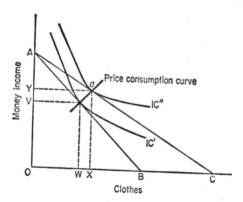

Fig. 4.10 The effect of a fall in price on consumer equilibrium

In this case the budget line becomes less steep reflecting the lower price of clothes, and once again the consumer can increase his satisfaction by moving to IC" this time at *a*. In this particular case, he not only increases his consumption of clothing from OW to OX but also spends a greater amount, OY rather than OV on other goods. As his money income is constant we know that he must be spending less on clothing than he was at the higher price, so we know that his demand for clothing is inelastic. (See Fig. 3.10 (*b*): when price falls, total expenditure is lower.)

The curve joining the points of maximum satisfaction is referred to as the *price consumption curve* and other possible curves are shown in Fig. 4.11.

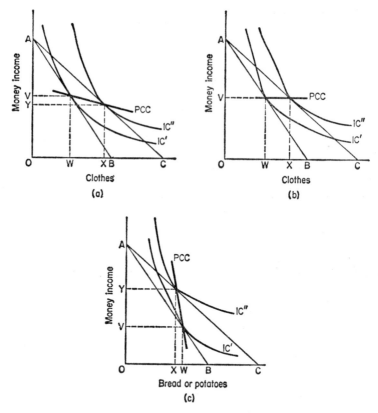

Fig. 4.11 Some other price consumption curves

In Fig. 4.11 (*a*) the price consumption curve slopes down. This is because this particular consumer finds clothes such a good buy relative to other goods when price falls, that he actually retains *less* money income (OY rather than OV) to spend on other goods. Thus, more money is spent on clothes when the price falls and the price elasticity of demand is greater than one.

In Fig. 4.11 (*b*) a fall in price leads to increased consumption of clothing again (OX units instead of OW), but in this case the money income retained to spend on other goods remains constant at OV. As income is fixed at OA we know that total expenditure on clothing must remain constant and we see that in this case the price elasticity of demand is equal to unity.

Fig. 4.11 (c) shows what happens in the case of Giffen goods. A fall in price reduces consumption from OW to OX units, as consumers prefer to spend a greater amount of income on other goods even though the bread or potatoes have fallen in price. The reason for this is that the consumer's real income has thus been increased and the increase enables him to enjoy a more varied diet than he had before. This is one instance where the price elasticity of demand is positive, since a fall in price (minus sign) is associated with a fall in demand (minus sign).

4.8 Income Effects and Substitution Effects of a Fall in Price

The changes in demand observed in Fig. 4.8 were the result of changes in money income and also (since prices were assumed to be constant) real income. Such a change in demand is referred to as *the income effect,* and except in the case of inferior goods we may expect it to lead to an increase in demand.

The changes noted in Fig. 4.10 and 4.11 were the result of a fall in price. A fall in price (with money incomes constant) leads to an increase in real incomes as the consumer is able to purchase as many goods as he did at the previous price and still have money left to spend on extra goods. Part of the increase in demand that occurs when prices fall may thus be ascribed to the income effect. The remainder of the increase is attributable to the *substitution effect.* This is the tendency of the consumer to increase purchases when the commodity price falls as it then represents a better bargain when compared with other goods. Demand curve analysis does not allow us to distinguish between the income and substitution effects of a fall in price, but with the aid of indifference curves we can do so.

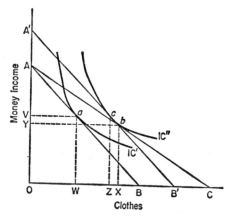

Fig. 4.12 The income and substitution effects of a fall in price

With the price of clothing at $\dfrac{OA}{OB}$ in Fig. 4.12 the consumer maximizes his satisfaction when at a by buying OW units of clothing and OV units of other goods. A fall in price to $\dfrac{OA}{OC}$ enables him to reach b on IC″ and buy OX units of clothing and OY units of other goods. He now obtains more utility or satisfaction than was possible at the higher price, but he could have reached IC″ by another route, and, by definition, have been deriving as much satis-faction as he does at b. If the price of clothing had remained constant but his income had increased by an amount AA′ his budget line would have moved to A′B′ enabling him to settle at c on IC″. The move from a to c would allow consumption of clothes to increase from OW to OZ and this would be the income effect of the fall in price—attributable to a higher real income. The move along IC″ from c to b allowing consumpton to rise from OZ to OX is the substitution effect of the fall in price.

The substitution effect will always be positive, moving the consumer down the indifference curve, and will normally be reinforced by a positive income effect. The explanation of the fall in demand for Giffen goods when price falls is attributable to the positive substitution effect being outweighed by a negative income effect.

4.9 Cross Elasticity of Demand

If we return to our original practice of examining the consumer's demand for two goods (as in Fig. 4.1) we can obtain a better idea of the concept of cross elasticity of demand than is possible with other techniques. In Fig. 4.13 two possibilities are examined.

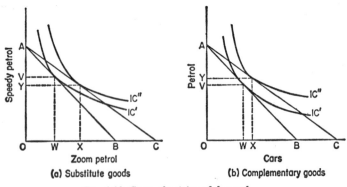

Fig. 4.13 Cross elasticity of demand

We have seen that the cross elasticity of demand is measured by the formula:

$$\text{Cross elasticity of demand} = \frac{\% \text{ Change in quantity of A}}{\% \text{ Change in price of B}}$$

In Fig. 4.13 the indifference curves are taken to be the community indifference curves (the sum of individual curves). In Fig. 4.13 (*a*) the price consumption curve slopes down and shows us that if the price of Zoom petrol in terms of Speedy falls enabling consumers to buy OC rather than OB if they so wish, they actually increase their purchases from OW to OX, thus reducing their demand for Speedy from OV to OY. Thus a fall in the price of Zoom leads to a fall in the demand for Speedy, and by attaching numerical values to the axes it would clearly be possible to establish the precise cross elasticity of demand for Speedy. (It will be positive as the price of Zoom and the quantity of Speedy move in the same direction.)

In Fig. 4.13 (*b*) the price consumption curve slopes upwards. A fall in the price of cars leads to a greater demand for cars but lower expenditure on them (AY instead of AV). As people are buying more cars they will also buy more petrol and the cross elasticity of demand is negative (an *increase* in demand associated with a *fall* in price).

4.10 Some Applications of Indifference Curves

Indifference curve techniques were not developed just to confuse students of economics! They do offer a more penetrating analysis of consumer demand than simple demand curves, and they are of considerable importance in the study of advanced economic theory. So it is worth while to make the effort and really try to understand them. It is convenient at this point to examine two uses, other than the analysis of effects of changing prices and incomes, that may be made of the curves.

(*a*) Inflation

Indifference curves demonstrate the effects of inflation or the situation in which prices and incomes are rising. When prices rise consumers must secure a rise in money incomes in order to maintain their real income and their standard of living. A 10 per cent rise in prices has to be accompanied by a 10 per cent rise in money income if consumers are not to suffer a fall in real income. Fig. 4.14 reveals a more subtle change.

With his original income OA the consumer had a budget line AB and chose to buy OW units of clothes and spend OV on other goods. If his income rises by 10 per cent to compensate for a 10 per cent rise in prices he can still buy a maximum of OB units of clothes but his budget line moves to CB, enabling him to buy OX units of clothes and retain OY units of money. He therefore moves to a higher indifference curve, even though his real income is constant. The higher money income gives greater satisfaction. Although real incomes are not higher, as the money incomes will buy only the same quantity of real goods, the consumer is deluded into buying more as the extra money has less utility, and he thinks the residue larger than it actually is. This is one way in which inflation distorts the pattern of expenditure. Other effects of inflation are considered in Unit Twenty-three.

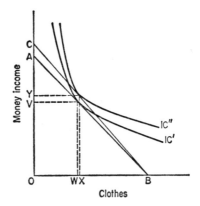

Fig. 4.14 Inflation and constant real income

If the price of clothes rose and there was no compensating rise in income the budget line would have become steeper and forced the consumer to a lower indifference curve. This would reduce his living standards, and this is the normal effect of inflation.

(b) Taxation

In Fig. 4.15 a comparison is made between the relative effects of income taxes and expenditure taxes.

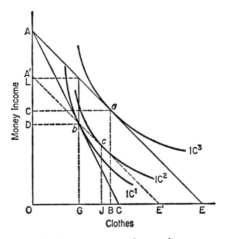

Fig. 4.15 Income taxes and expenditure taxes

In the absence of taxation we can assume that the consumer is at *a* buying OB units of clothes and OC units of other goods. If a tax is imposed on clothes, their price is raised and the budget line moves to AC and the consumer to point *b* on IC^1. At this point he buys OG units of clothes and spends OD units on other things. Before the tax was imposed the consumer could have combined OG units of clothes with OL units of money income or other goods as we can see from the budget line AE. The tax has therefore reduced his real income by LD. This could equally well have been achieved by the imposition of an income tax equivalent to LD, when the consumer was at *a*. If this tax had been imposed, the original budget line would have moved downwards to A'E' by an amount LD but remaining parallel to AE and passing through *b*. This shift would have enabled the consumer to move to *c* on IC^2 which is preferable to the position *b* that the expenditure tax leaves him in. He is able to enjoy GJ more of clothing than he could when clothes were taxed.

While this is true for the individual whose indifference map we have drawn it is not necessarily true for all consumers, and so we cannot on the basis of this analysis argue that income taxes are preferable to expenditure taxes.

4.11 Questions and Exercises

1. The theory of indifference curves assumes that consumers try to maximize satisfaction. How do *you* try to maximize the satisfaction you derive from your expenditure?
2. Explain what is meant by the law of diminishing marginal utility. Show how it determines the shape of indifference curves.
3. Give examples to show the operation of the principle of the diminishing marginal rate of substitution.
4. With the aid of a diagram show how a consumer will derive maximum satisfaction from his expenditure on two goods.
5. 'A rise in the price of one of the goods a consumer buys causes his satisfaction to fall.' Explain and illustrate.
6. Distinguish between the income effect and the substitution effect of a fall in price. Illustrate your answer.
7. Inferior goods are those for which demand falls as income rises. Giffen goods are those for which demand rises when price rises. Illustrate each of these by means of indifference curves.
8. The following sets of combinations of meat and drink represent two indifference schedules for a consumer.

Units of meat	16	12	9	7	5	4	3
Units of drink	1	2	4	6	9	13	20
Units of meat	16	13	10	7	5	3	1
Units of drink	3	4	6	10	16	23	31

Plot the indifference curves on graph paper. If the consumer spends all his income on meat he can buy 19 units, and if he spends it all on drink he can buy $12\frac{1}{2}$ units. Draw his budget line. At what combination of the two goods does he maximize satisfaction?

If the price of meat rises so that it is only two-thirds that of drink, at what quantities will the consumer now be in equilibrium?

9. A consumer has £10 to spend on books and records. The price of books is 30p and records 50p. He allocates his money in such a way that the marginal utility of books is equal to 15 and that of records 30. How could he increase total utility?

10. Plot the following indifference curves on graph paper.

Units of bread	Units of milk	Units of bread	Units of milk
20	$3\frac{1}{2}$	20	9
16	5	18	10
14	7	11	16
10	12	10	21
6	20	6	44

The price of bread is twice as high as that of milk, and if all income is spent on milk, 40 units can be bought. Draw the budget line. (*a*) What quantities are consumed in equilibrium? (*b*) If the price of milk remains fixed, to what extent must the price of bread rise to put the consumer in equilibrium on indifference curve 1? (*c*) What quantities are now consumed?

Unit Five

Specialization and Production

5.1 Introduction

We have considered the main determinants of the level of individual and market demand in some detail and must give our attention to the factors governing the production and supply of goods. In this section of the book we shall examine the organization of production and the costs incurred in the creation of utilities.

5.2 The Meaning of Production

In Unit One we likened the economy to a giant production line with raw materials and factors of production being fed in at one end, and goods and services of all kinds flowing off at the other end. This is precisely the meaning of the term production: *it is the process of converting raw or basic materials originally extracted from the earth, sea or atmosphere (perhaps one day the moon) into forms which yield utility or satisfaction to consumers, at a time and place convenient to the consumers.* Land, labour and capital have to be combined in various proportions to bring about this transformation. In some circumstances it is appropriate to produce goods by employing a high proportion of labour to other factors of production. Thus in relatively poor, but highly populated, areas textiles are produced by *labour-intensive* methods. In wealthier countries each unit of labour may enjoy the assistance of a large amount of capital when producing textiles. Such industries are said to be *capital intensive*. We need not consider only finished goods; the provision of services often reflects the availability of factors of production. Housing estates built in the inter-war and immediate post-war period reflect a lack of concern with the problem of land shortage for a considerable amount of land was allocated to each dwelling. More recently the realization of the shortage of land has led to its more intensive use, partly by the erection of individual houses with smaller gardens but more especially by accommodating people in multi-storey blocks of flats. Similarly the towering skyline of New York derives from, and is dictated by, the shortage of land. As a final example, a comparison can be made between British and African agriculture. This shows that the British farmer is assisted by a far greater amount of capital in the form of machinery, scientifically developed seeds and artificial fertilizers than is his African counterpart. One represents capital-intensive agriculture and the other, labour-intensive agriculture. Methods of production are dictated largely by factor endowments.

5.3 Types of Production

The distinction between these two forms of production corresponds to the division that is normally made between *indirect production* and *direct production*.

(a) Direct production
Under this system the economic unit is normally very small and may well be self-sufficient. It consists of a single family or tribe producing all it needs for itself. A great deal, though a declining proportion, of African agriculture for example, is still of the *subsistence* variety—that is that the unit of production, usually the family, succeeds in producing enough food to survive on but not enough to bring to market. Using antiquated techniques and primitive tools, ignorant of the need for soil conservation, the peasant farmer is trapped in a seemingly endless cycle of poverty, for he is unable to produce a surplus for the market that might give him the resources to improve his methods. The outstanding feature of his difficulties is that his *productivity* is very low. By productivity we normally mean output per head though in this case it would not matter if we took output per head or output per acre: we should still find that productivity is much lower in Asia and Africa than in the United States or Europe. Nor is this fact confined to agriculture; indeed differences are even more marked in manufacturing industries, and the differences in terms of gross national product per head are very clear. The reason for such low productivity is the very fact that the economies concerned rely upon or have not moved far away from a system of direct production towards the system of indirect production that characterizes all the advanced economies.

(b) Indirect production
No one, not even the wealthiest man, can regard himself as self-sufficient in the modern industrial economy. Everything that he enjoys, from the basic necessities to the greatest luxuries, come to him through the endeavours of hundreds perhaps millions of his fellow men both in this country and abroad. Each producer acquires the goods he needs indirectly, by exchanging his surplus output for that of other producers. So vast is the range of economic activity that it is convenient to break it into three broad categories each emphasizing different aspects of production.

(i) **Primary production.** This is the *extraction* of basic materials from the land, sea or air. Those engaged in agriculture, mining or quarrying, and fishing are the clearest examples. Without them and others engaged for example in drilling for oil, none of the subsequent stages of production could take place. The transformation of their product is the role of other workers.

(ii) Secondary production. The outstanding feature of the modern economy is the vast range of manufactured goods that it produces. Processed food, literature, televisions, motor cars, aeroplanes, even moon rockets are all the result of workers of varying skills and abilities engaged in the secondary sector and changing the basic materials extracted by the primary producers into forms acceptable to final consumers. While you may live in the United Kingdom without knowing anyone engaged in primary production it is very unlikely that you do not know anyone engaged in secondary activities since vast numbers of people are employed to produce the material goods that are generally taken for granted in twentieth-century Europe.

(iii) Tertiary production. Even when goods reach the end of the production line in a factory they are often of little use to final consumers. They have to be stored, transported, insured in transit, perhaps stored again by whole-salers and retailers before they reach their final destination, and all of these operations have to be financed. Those engaged in the tertiary (third) sector of the economy provide all these commercial services and many others. Without such services primary and secondary production would be point-less, for what use is a steadily rising mountain of coal at the pithead, or the motor industry's output of over a million vehicles if they merely surround the factories in which they are produced? *Production does not cease until*

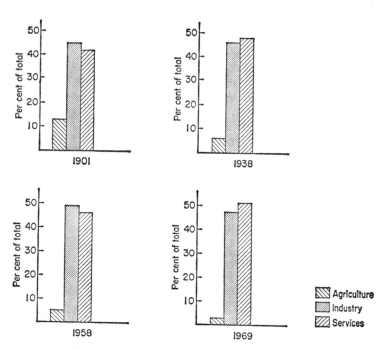

Fig. 5.1 Structure of employment in the United Kingdom, 1901–69

the goods reach their final consumer, and it is not necessary for there to be a tangible change in materials for production to occur—transformation through time and space are just as productive as transformation of materials. In addition to these important commercial services there are many *direct* services such as those of the hairdresser or the pop-singer which fall within the tertiary sector and are, like other services, productive.

As a general rule the balance between the three sectors gives an indication of the stage of development of the economy. Fig. 5.1 shows the structure of civilian employment by sector in the United Kingdom for various years.

We can see how an increasing proportion has found its way into the tertiary sector. This is one of the recognized signs of an advanced economy. By way of comparison, consider the employment profiles in Fig. 5.2. The less developed economies have a greater proportion of the working population in agriculture, while the United States has the largest proportion of all in the 'service occupations'. We see that the way in which the working force *specializes* in various activities alters as the economy passes through different stages of economic development. The concept of specialization must be studied in some detail.

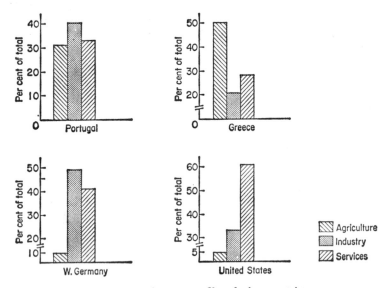

Fig. 5.2 Employment profiles of other countries

5.4 Specialization and Division of Labour

Adam Smith drew attention to the importance of division of labour or specialization in his book *The Wealth of Nations*. He was primarily concerned with the increase in productivity of a particular industry when the manufacture of its products was broken down into many specialized activities. There are other forms of specialization and we shall consider these before examining specialization or the division of labour in greater detail.

We shall notice that specialization may occur at any level—between countries or companies, for instance. Such specialization could occur without division of labour at plant level. The latter term is normally reserved for the practice of dividing a particular job into several smaller jobs.

(a) International specialization

Geographical, geological and climatic factors give some areas of the world advantages over others in the production of certain goods. Tropical fruits cannot be produced economically in Europe, and copper cannot be mined where nature has not deposited it. Thus some countries are capable of manufacturing much the same range of products, but it will often be very much to a country's advantage to concentrate its resources on a specific area of production. This will be explained further in Unit Twenty-four.

(b) Regional specialization

Similar factors often lead to specialization within a country. Factor endowments and economic history have often combined to concentrate industries in particular areas with the result that it is generally very difficult for competitive plants to be established outside those areas. For example, such are the natural and acquired advantages of Sheffield as a centre for the production of cutlery that it would be an optimistic man indeed who built his cutlery factory in Cornwall or Norfolk. Regional specialization is more fully discussed in Unit Nine.

(c) Specialization between industries

An economy includes many industries each producing different commodities. While it is very common to speak of distinct industries, for example, the motor manufacturing industry, it is in practice very difficult to draw boundaries between industries. When talking of the motor industry should we include the manufacturers of electrical components or the firms supplying upholstery for car seats? If such firms are included as part of the motor manufacturing industry then where should that part of their output absorbed by the ship-building or domestic appliance industries be included?

An attempt has been made to overcome such difficulties and the authorities have established a *Standard Industrial Classification* (SIC) to be used in the collation of all official industrial statistics. The whole economy is divided into 24 broad groups as shown in Table 5.1. The problem mentioned in the previous paragraph of one firm being in two or three different industries is generally overcome under SIC rules by allocating each firm to a group on the basis of its main product.

Table 5.1 The standard industrial classification

I	Agriculture, Forestry, Fishing
II	Mining and Quarrying
III	Food, Drink and Tobacco
IV	Coal and Petroleum Products
V	Chemicals and Allied Industries
VI	Metal Manufacture
VII	Mechanical Engineering
VIII	Instrument Engineering
IX	Electrical Engineering
X	Shipbuilding and Marine Engineering
XI	Vehicles
XII	Metal Goods not specified elsewhere
XIII	Textiles
XIV	Leather, Leather Goods and Fur
XV	Clothing and Footwear
XVI	Bricks, Pottery, Glass and Cement
XVII	Timber, Furniture
XVIII	Paper, Printing and Publishing
XIX	Other Manufacturing Industries
XX	Construction
XXI	Gas, Electricity and Water
XXII	Transport and Communication
XXIII	Distributive Trades
XXIV	Insurance, Banking, Finance and Business Services
XXV	Professional and Scientific Services
XXVI	Miscellaneous Services

Source: *The Standard Industrial Classification*, HMSO

Each classification or *order* is divided into various trades and each trade may itself be subdivided. Reference to the *Annual Abstract of Statistics* will show that the SIC is very widely used in the presentation of official statistics.

(d) Specialization between firms
Industries are composed of a number of firms which can be regarded as the *units of control*. The SIC makes no provision for firms as a single firm will often have interests in more than one industry. But the firm is an entity which can take decisions, through its board of directors, that can have a considerable influence on the economy. Firms are discussed more fully in Unit Seven, but here it should be noted that within an industry there are many highly specialized firms.

(e) Specialization between factories
One firm will often control many factories, and these are usually referred to as *plants* and are the units of production. What they produce will be determined by decisions taken within the firm and it is not at all unusual for a firm to arrange for its plants to specialize in the production of particular

parts of the final product. A motor manufacturer might find it economic to build engines in one plant, axles in another, car bodies in a third and so on, and then transport all the parts some considerable distance to yet another plant for final assembly.

(f) Specialization within plants

Within plants there may be two kinds of specialization. In the first case the plant may produce more than one item and in this instance it is helpful to regard the plant as two or more working side by side. The second and more important fact is that within every plant there is considerable specialization of labour. In a manufacturing plant some employees will be receiving and storing raw materials and components, the majority will be engaged in the manufacturing processes, others will be checking and packing the finished product while others will deal with its dispatch to the customers. In the largest plants thousands of workers co-operate to produce the finished product, and in all probability not one of them would have the knowledge or skill to have produced it by himself. This is the advantage of the division of labour.

5.5 The Advantages of Division of Labour

The most important advantage of specialization is production through division of labour, but there are other considerations as well. Let us examine them.

(a) With a system of indirect production a given set of factors of production can produce a far greater volume of goods and services than is possible with methods of direct production. This fact is of overwhelming importance. Adam Smith told the story of a small factory in which ten men could produce perhaps 48,000 pins per day by dividing the work between them, whereas each might have had difficulty in producing 20 pins a day if he had to do the entire task. The British shoe industry produced 96 million pairs of shoes in 1970 and employed approximately 98,000 people to do the jobs. To each person was allocated a small fraction of the work involved in making each pair of shoes. Doubtless not a tenth of this output could have been achieved had each worker been left to make an entire pair alone. It is no exaggeration to say that without industrial division of labour economic growth would be minimal.

The remaining advantages of the specialization of labour are really all factors contributing to the vast increase in output associated with modern productive methods.

(b) Each person is endowed with different aptitudes and skills, and an economic system that breaks work down into a variety of different jobs gives individual members of society the opportunity to exploit their particular talents to the full. Those with considerable manual dexterity may be best suited to the assembly of electronic components, or to operations in the hosiery trade, while those who are especially numerate may be suited to careers in the expanding computer industry. Not all people will find their optimal positions, but they are more likely to do so when there is specialization.

(c) Once a person establishes himself in a job he usually discovers that his skill at that job increases with practice. Have you ever marvelled at the smoothness of a wall left by a skilled plasterer; or at the precision of a joint made by a carpenter? Their first attempts at such tasks were most certainly far less accomplished. It is the same with much simpler jobs. Once a man has spent some weeks tightening nuts on the wheels of motor vehicles as they pass him on a production line his speed at performing this activity (and thus his productivity or output) increases.

(d) As the car worker mentioned above is performing the same operation all day long he needs only one tool. This represents a great saving. Where workers are working individually they need to be equipped with a full set of tools although most of these will stand idle all day. Once specialization occurs the need for such duplication disappears and one set of tools will meet the needs of all workers. This is a very important way of economizing on capital.

(e) While the man uses the same instrument each day to perform the same simple task he is likely to discover improvements that might be made to it, and the very fact that the tool is being used all day leads to enormous sophistication in handling. If I wish to change the wheel on my car it is a fairly laborious operation involving jacking up the car, loosening four, five or six nuts individually and manually with a wheel-brace and reversing the operation when the new wheel is on. In the car factories the operative usually stands still while the overhead production line brings the car to him, he then fits his specially developed wheel-brace to all four (five or six) nuts at the same time and a whiff of compressed air tightens them all. Obviously the cost of such sophistication is only justifiable if the equipment is going to be in constant use, but it is a very important labour-saving device for the worker.

(f) Not all the costs of production increase as output rises (see Unit Ten). It is true that the extra output caused by the division of labour will require greater expenditure on raw materials, but it is unlikely that rent and rates and other fixed costs will increase in proportion with output. Thus cost per unit is reduced.

(g) In Unit Two we saw that a high standard of living does not consist merely of producing and consuming more and more goods. It is one of the advantages of specialization that it allows more goods to be produced for the same expenditure of resources. It also allows workers to enjoy more leisure time as well as more goods.

(h) An often neglected advantage of specialization is that the greater productivity enables some units of labour to be released in order to develop great skills in a very confined field such as neuro-surgery, or to offer services in the form of entertainment to the rest of us. Neither case would be possible if we were each primarily concerned with producing our own food and drink.

5.6 The Disadvantages of Specialization

When specialization is carried to great lengths in advanced economies some disadvantages occur and must be set against the great increase in output that results, although there can be few people who feel that the disadvantages outweigh the advantages. Among the most important disadvantages are the following.

(a) Specialization involves a high degree of interdependence. Clearly no one can specialize in making a fractional part of a motor car unless others specialize in making the other parts. The car manufacturers, however, cannot afford to specialize in making cars unless thousands of others devote their energies to the production of food, furniture and clothing. When disruptions of production occur in one part of a plant this soon affects the other parts and some operatives cannot pass on goods that they have finished with before the breakdown; while those further on in the production line find that they are receiving no raw material. A prolonged breakdown or dispute in an important industry soon spreads to other sectors of the economy, causing temporary unemployment and inactivity.

(b) Minute division of labour often leads to inflexibility. Machinery is very likely to be made specifically for a particular job; for example, a machine developed for slicing and wrapping bread cannot conveniently be used for much else. People trained in one particular limited technique may need re-training before they can be absorbed elsewhere in the economy.

(c) A worker performing a simple operation several hundred times a day is likely to become so bored by monotonous repetition that his efficiency suffers as the day proceeds. It is certainly true that many of the tedious jobs forming part of a modern production line are not calculated to challenge the individual.

(d) As machinery becomes more sophisticated, and often replaces labour it may lead to unemployment. This of course is an argument that was used by the 'Luddites' who destroyed labour-saving machinery at the beginning of the nineteenth century. Since then, however, increasingly advanced techniques have not resulted in permanent increases in unemployment. The persistence of unemployment in the period 1970–2 has led many to believe that *automation*, or the use of machinery instead of men, has yielded an undesirable dividend.

(e) Specialization has been accompanied by a decline in craftsmanship in many areas of industry. It is true that some new crafts and skills are being developed but the majority of workers need relatively little skill. The corollary of this is that workers have no pride in their achievements and quality suffers. Such a decline in standards is wasteful not only in itself but also because it necessitates the employment of further staff to detect and correct mistakes.

(f) Machinery and labour have become so specialized that *standardization* of products is inevitable. The whole point of mass-production is that all

customers receive more or less the same good. It might cost a motor manufacturer anything up to £10 million in development costs and the purchase
of equipment before he would be able to sell the first car of a new design. In
such circumstances he is unlikely to be prepared to accommodate the whims
of individual consumers much beyond offering them a selection of colours
and engine sizes.

(g) The advancement of division of labour often aggravates social and
economic problems, as machinery replaces man and the unemployed have
to find new jobs, perhaps in an entirely different industry and in a different
part of the country. Also disputes about the distribution of the benefits of
specialization lead to increased hostility between capital and labour.

5.7 Division of Labour and the Extent of the Market

One advantage in primitive systems of direct production is that each family
just meets its own requirements and has no worry about the disposal of
surpluses. In more sophisticated systems where there is indirect production,
each producer has goods surplus to his requirements and these must be
exchanged with the surpluses available from other groups. In an isolated
community with a static population the opportunities for the exchange of
surpluses are limited and so the process of specialization slows down. There
is no point in producing a surplus of flour which cannot be exchanged for
beer or meat because the brewer's and butcher's requirements have already
been met. It would be more economic to produce less flour and devote the
available time to producing home-made beer, or meat. But as the extent of
the market widens, increased specialization becomes possible. The market
may increase in size in two ways:

(a) By means of improved communications, bringing more people within
range. While one view of the British industrial revolution emphasizes the
technological advances an equally important view stresses the great improvements in inland and sea transport which allowed the products of the new
technology to be marketed.

(b) By an increase in the size of the population of a given area, either
through natural changes or by immigration. Such changes are discussed in
Unit Six.

The extent of the market is the most important limiting factor to the
division of labour. The absence of an effective monetary system will also
greatly hamper specialization because, as we learn in Unit Sixteen, the
process of exchange is hopelessly complicated without money.

Finally, at any point in time technology itself sets limits to the extent of
specialization. These limits are only binding until some new technological
breakthrough allows further division of labour.

5.8 Questions and Exercises

1. Adam Smith described the advantages of division of labour in pin making. Consider the advantages for any industry known to you.
2. 'Specialization is limited by the extent of the market.' Consider the economic effects of a railway workers' strike in the light of this statement.
3. What do you understand by the term production? In what ways are the following productive: (*a*) a pop-singer? (*b*) a lorry driver?
4. One of the disadvantages of specialization is that it makes various sectors of the economy interdependent. Through the daily press trace the way in which the next major industrial dispute affects industries other than that in which it begins.
5. Describe and assess the relative importance of the advantages and disadvantages of the division of labour.
6. Distinguish between labour intensive and capital intensive industries, giving examples of each. Why is there a tendency for industries to become gradually more capital intensive?
7. By reference to the *Annual Abstract of Statistics*, analyse the changes that have occurred in the structure of employment in recent years.
8. 'Specialization in production may be wasted if there is not a well-developed market economy.' Explain and illustrate this statement.
9. Show how improvements in transport can facilitate the division of labour.
10. Use the *Annual Abstract of Statistics* to compile a list of Britain's main imports and exports. To what extent do they reflect international specialization?

The Factors of Production: Labour, Capital and Land

6.1 Introduction to Labour

Production of any kind requires a combination of the factors of production. In the following sections the supply of one of those factors, labour is examined against a background of population growth with its important dual role of providing labour for industry and commerce, and providing a market for the goods and services produced. First of all we will consider the historical growth of population, the causes of growth, the predicted changes and the economic problems created by a changing population structure.

6.2 The Growth of Population

Population and its structure is the backcloth against which all economic activity takes place and it has shown an almost continuous rise in the United Kingdom since the middle of the eighteenth century. This is not the place for a full-scale discussion of *demography* (the study of population) but the world population is putting increasing pressure on resources and the basic principles involved should be made clear.

Table 6.1 The growth of population in Great Britain

Year	Population (millions)
1801	10·5
1851	20·8
1901	36·9
1951	48·8
1961	51·2
1971	53·8

Source: The Annual Reports of the Registrar General

The first census was held in 1801 but the population was estimated to have been approximately 7 million in 1688. The fastest rate of growth occurred during the first half of the nineteenth century, but the growth rate has almost always given cause for concern and continues to do so. Population increases occur in two ways, by natural increases due to changes in the birth and death rates, or through the effects of migration.

6.3 Natural Changes in Population

(a) Birth rate

The *crude birth rate* is measured as the number of live births per 1000 population. For some purposes more precise measures are required, relating the number of births to the number of females in the population or to the number of women of child-bearing age, but we shall confine our attention to the crude rate. A glance at Fig. 6.1 shows that the growth of population cannot be ascribed to a rising birth rate. The birth rate shows a general reduction which may be attributed to many factors. Each of the following has its supporters:

   (i) The development and widening knowledge of methods of family limitation, and their increasing availability in the twentieth century.

   (ii) The decision by parents to enjoy a higher standard of living for themselves and a relatively small family.

   (iii) The decline in infant mortality rates (see Section 6.4), making the

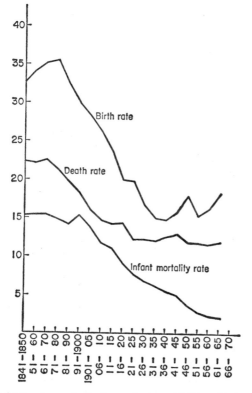

Fig. 6.1 Crude birth rate, crude death rate (no. per 1000) and infant mortality rate (no. per 100)

practice of having large families to ensure the survival of some members unnecessary.

(iv) The increase in material wealth, plus a growing social conscience, has long outlawed child labour. Children cannot now, therefore, be seen as an investment.

Short-run fluctuations may not fit into the general decline and Fig. 6.1 shows the way in which the birth rate accelerated after the Second World War.

There are signs that the size of families may once again be increasing, perhaps due partly to the fact that people often marry and have families at an earlier age. Increasing material wealth also allows the professional classes to support more children without reducing living standards. The introduction of family allowances and other welfare benefits, although not direct incentives for larger families, do at least remove the disincentive of absolute poverty, and may have encouraged the tendency towards larger families among the poor.

(b) Death rates

The *crude death rate* measures the number of deaths per 1000 of population and it is clear from Fig. 6.1 that its reduction during the nineteenth and twentieth centuries has contributed substantially to the increase in population. The most significant aspects of the decline are summarized as follows.

(i) The decline in the *infant mortality rate* (IMR) which measures the deaths of children before the age of one year per thousand related births has been of the greatest significance. The IMR has a double significance for demographers: not only does its reduction cause the population to increase initially, but the survival of more children to maturity and parenthood has cumulative effects. Despite advancing medical knowledge the rate remained stubbornly high in the nineteenth century but has fallen sharply since 1900, as general standards of living have improved.

(ii) Improvements in medicine, hygiene, and dietary standards are the main reasons for the decline in general mortality rates.

(c) Natural increase in population

A comparison of the crude birth rate and crude death rate gives the rate of natural increase (or decrease) of a population. It is clear from Fig. 6.1 that in the nineteenth century very high birth rates were partly offset by very high death rates and since then the subsequent decline in birth rates has not been as marked as the fall in death rates, so the population has increased.

6.4 An Observation on World Population

While the rate of growth of Britain's population has slowed down compared to the rates achieved in the nineteenth century it is still sufficient to cause concern to many authorities. The rates of expansion in other parts of the world, notably South East Asia and South America are very much greater.

This is largely due to the fact that medical advances have brought dramatic falls in mortality rates. It has, however, proved impossible to secure similar reductions in birth rates as the people have been reluctant or unable to adopt methods of family limitation. Table 6.2 gives an indication of the problem.

Table 6.2 The growth of world population (thousand millions)

Year	Population
1750	0·7
1800	1·0
1850	1·3
1860	1·3
1870	1·4
1880	1·4
1890	1·5
1900	1·6
1910	1·7
1920	1·8
1930	2·0
1940	2·2
1950	2·5
1960	3·0
1970	3·6

Estimates of world population are somewhat unreliable as the countries with the largest populations are frequently those with the least developed statistical techniques. Before this century, of course, such estimates were even less reliable. If we take the generally accepted figure of a world population of 550 million in 1650 we can see from the table that it took approximately 150 years for this number to double. In more recent times the population has doubled in 50 years between 1920 and 1970. The next doubling is unlikely to take so long, perhaps 30 to 35 years; and it is important to realize that each successive doubling represents a greater absolute increase in numbers and demands therefore a greater increase in food output. It is not surprising that many demographers adopt a view almost as gloomy as that of Malthus in the early nineteenth century (see Section 6.6).

It is not difficult to trace the causes of this rapid acceleration. Whereas in the United Kingdom and other European countries the benefits of medical science spread only slowly, the vast resources of modern medicine are now available even to the poorest countries through the agency of the United Nations and many other bodies. Scourges such as malaria and smallpox are now effectively contained and this has led to a dramatic fall in mortality rates (especially the IMR) and a consequent rise in life expectancy. Life expectancy is the number of years that a child at birth can expect to survive, and in many areas of the world it has more than doubled in 40 years.

6.5 Migration of Population

The size of a population can also be affected by emigration and immigration. To many countries such movements of population have been and still are of the greatest importance. The great economic advance of the United States could not have occurred without the influx of immigrants from Europe; Commonwealth countries such as Australia and New Zealand have been dependent on emigration from Europe, and the recent economic growth of West Germany has absorbed labour from East Germany, Southern Italy, Turkey and North Africa. In Britain the net effect of migration in the nineteenth and early twentieth centuries was an outward movement of perhaps 5 million people, which offset to some extent the natural increase in population. More recently net movements have been much smaller. Between 1931 and 1967 there was a net inflow into the United Kingdom, but the immigration boom of 1960-2 when the net immigration into the country was almost 1,100,000, persuaded the Government to take powers through the Commonwealth Immigrants Act (1962) to limit the numbers. Consequently there is now a general net loss by migration. In 1969 there was a net outward movement of 85,000 and in 1970, 65,000. Gains or losses to the economy cannot be measured simply by the numbers involved; the ages, intelligence and skills of emigrants and immigrants are important to any assessment of the effects of migration.

6.6 The Population and Other Resources

The fact that the British Government has to restrict immigration while other countries take steps to encourage it, draws attention to the most important economic aspect of population—its size in relation to other resources, and to the concept of an *optimum population*. Australia and Canada can encourage immigration since they have vast natural resources in relation to their existing population. Britain on the other hand has a population which exerts great pressure on very limited natural resources. Table 6.3 shows this in the very simplest way, comparing population size with land area.

Table 6.3 Population density: persons per square km, 1969

	Population (millions)	Area (thousand sq. km)	Persons per sq. km
Australia	11·5	7·686	2
Canada	20·0	9·976	2
Kenya	10·8	0·582	18
India	435·5	3·268	164
United Kingdom	52·7	0·244	228
West Germany	53·9	0·247	237
Hong Kong	3·1	0·001	3859

Source: *United Nations Statistical Yearbook*

A book by the Reverend Thomas Malthus entitled *An Essay on the Principle of Population as it affects the Future Improvement of Society* first brought population and resources to the public notice. Malthus was not concerned with the density of population but with what he regarded as the tendency of the population to grow at a rate faster than that of its means of support. He suggested that while the population grew in a geometric progression (1, 2, 4, 8, 16), agricultural output grew in an arithmetic progression (1, 2, 3, 4, 5). If such a hypothesis were right the result could only be disaster. Malthusian forecasts have not been upheld so far mainly because the means of supporting the rapidly growing population have increased more rapidly than Malthus foresaw. Improved technology has led to greater output, and improvements in transport have provided new sources of material and easier relief of shortages. It is not, however, too late for Malthus to be vindicated. As we have seen, population pressure is excessive in many regions and without continuing improvements in agricultural efficiency mass starvation will become inevitable in some parts of the world.

The concept of an *optimum population* has been developed in connexion with the problems of over-population. The optimum population of a country is that population which, given the availability of other factors of production and the state of technical knowledge, would generate the greatest national product. There are therefore three possibilities for the state of population in a country: over-population, optimum population and under-population. It may be clear that a state such as Hong Kong is suffering from over-population and could probably increase its living standards if it had a smaller population, and that a country like Canada is probably under-populated and needs more people to exploit its natural resources. But in most cases it is very difficult to categorize particular countries. There are several reasons for this:

(*a*) The state of technical knowledge is continually changing. Although it may be possible to estimate the optimum population for today, changes in technology rapidly necessitate new calculations.

(*b*) It may be insufficient to consider the size of the population without knowing its structure (see Section 6.7). The optimum will vary if, for example, different numbers of people are contained within the different age-groups; the numbers contained within the non-working groups, under 15 years and over 65 years, are particularly significant. The sex-distribution may affect the optimum in a similar way, especially if it is not the custom for females to undertake paid work.

(*c*) It may be difficult to forecast the effects on output of extra labour in the case of an individual firm. For an entire economy the difficulties are greatly multiplied and precision is impossible.

(*d*) The optimum for production may be different from the optimum for marketing. An advanced technology may require a small labour force, but the large output needs a large market.

(*e*) The level of production or national product depends on many things other than the combination of factors of production. A change in a Government's taxation policy may have beneficial or detrimental effects in this respect, and necessitate the recalculation of the optimum population.

6.7 The Age-Structure of the Population

The danger of generalizing about the population must be avoided especially when population is being considered as the source of labour. One of the main determinants of the size of the labour force at any time is the age structure of the population. It is normal practice to divide the population into three age-groups, those under 15, 15–64 and 65 years and over. Although these groupings do not correspond exactly with those in full-time education, those at work and those in retirement, changes in the proportions of the groupings are an important indication of the sizes of the dependent and working populations.

Table 6.4 The age distribution of the population of Great Britain

| Date | % in each age-group | | |
	0–14	15–64	65 and over
1911	30·8	63·9	5·3
1939	21·4	69·7	8·9
1947	21·5	68·1	10·5
1951	22·2	64·2	13·6
1961	23·1	62·2	14·7
1971	23·9	60·1	16·0*
1981	23·6	59·7	16·7*
1991	24·4	59·6	16·0*
2001	24·7	60·8	14·5*

* Based on data in *Population Projections 1970–2001*, HMSO

In Table 6.4 we can see that the proportion of the dependent population has been increasing since the war and is expected to go on rising. At the same time the population itself continues to expand. What implications do these facts hold for the economy?

(a) The first thing to remember is that changes occur gradually and are almost imperceptible as they arise, though if the allocation of resources in 1947 is compared with that in 1971, differences probably due to the changing age structure become apparent.

(b) As an increasing proportion of the population enters the over 65 years group the demand for old peoples' homes, geriatric and welfare services increases.

(c) Such changes require different patterns of production, but where the average age of the population is increasing geographical and occupational mobility are lowered.

(d) There is a greater burden on the working population. As the dependent population increases each worker has to produce a greater surplus if living standards are even to be maintained.

(e) Investment, particularly public sector investment, undertaken for a fairly youthful population may not meet the requirements of an older population without considerable modification. Thus infant and junior schools are not readily converted into old peoples' homes.

(f) As numbers increase the amount of land available to each person falls.

(g) An expanding market creates opportunities for producers, although this fact must be considered in relation to (c) above.

(h) The age structure of the population also affects the level of savings and investment. If the average age is rising and more people come in the higher age-group savings will tend to fall and investment will become more difficult to finance.

6.8 The Projected Population

Social and economic planning must be based on a fairly firm view of the future population and its structure, and the office of the Registrar-General provides such assessments for the Government. Population forecasts depend on applying current and anticipated birth and death rates to the present population. This sounds a fairly simple process but population projections have to be treated with the utmost care. The great difficulty is that the demographers must make an assessment on the future movements of birth and mortality rates, not just for the population as a whole, but also for each age-group within the population.

Table 6.5 Projected and actual population of the United Kingdom (000's)

Year of forecast	1960	1965	1970	1975	1980	1985	1990	1995	2000
1955	51796	52225	52624	52014	53270	53288	53091	52791	—
1960	—	53792	54931	56320	57719	59112	60599	62145	63822
1965	—	—	56606	58907	61223	—*	66767	—*	74594
1970	—	—	—	57136	58588	60139	61940	63916	66040
Actual popn.	52372	54361	55811	—	—	—	—	—	—

Sources: *Annual Abstract of Statistics* and *Population Projections 1970–2001*, HMSO
* Figure not available.

The figures given in Table 6.5 show how the projections of future population have had to be amended at different times. In three separate years the projections have been proved wrong by events. The assumptions on which the projections are based are more likely to be proved wrong if a long time period is involved. The development of the contraceptive pill may have led to a reduction in the birth rate, and the discovery of a cure for cancer would lead to a fall in mortality rates. It is impossible for demographers to predict these things or a thousand more subtle influences before the event, and so estimates are continually revised.

6.9 The Working Population

The supply of labour depends upon the size of the population, and in particular upon the numbers in the 16–64 age-group. We could say that other things being equal the larger the size of this group the larger will be the size of the working population. Other things are not, however, equal and many factors influence the size of the work force within a given population.

(a) Some people older than 65 years remain in employment even though they have reached retirement age and are regarded as part of the dependent population.

(b) The lower end of the 16–64 group contains many people in full-time education and not forming part of the working population. School attendance is now compulsory until the age of 16 years, so this fact is very important. Increasing numbers also remain in school or go to college or university well beyond the compulsory age.

(c) Within the working age-group there are also some unfortunate people chronically ill and unable to work. There may be a small number of others who find that Social Security benefits of various kinds enable them to survive without working. Wages may occasionally be so low that unskilled persons are actually better off when they are unemployed.

(d) A very important determinant of the size of the work force is the number of women at work. This is decided by many factors, many of them non-economic, including the prevailing attitude towards working mothers. In the post-war period the proportion of women in the labour force has risen from 32·8 per cent in 1948 to 37·7 per cent in 1970.

(e) One way of relating the supply of labour or the working population to the size of the 16–64 age-group is to assess the *activity rate*. This compares the number of employed persons and those seeking work with the total adult population between the ages of 16 and 64. The higher the activity rate the larger the supply of labour.

(f) While the number of people in employment is important, equally important is the number of hours they work.

In Fig. 6.2 we can see that the number of hours worked has hardly altered since 1948. This surprising finding, in view of successive declines in the negotiated normal week, has resulted from a great rise in overtime working. One argument in favour of reducing the working week shows that there are *diminishing returns* from labour; as the working day progresses the worker becomes less efficient and his productivity falls. In the nineteenth century the Ten Hour Act reduced the working day and increased output. Such an argument loses its force, however, if a reduction in the basic week is automatically followed by an increase in overtime.

The length of holidays allowed also affects the supply of labour. Here

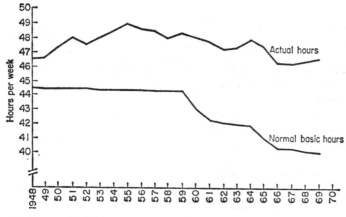

Fig. 6.2 The supply of labour: normal weekly hours and actual weekly hours (adult manual workers)

again, it may be argued that an increase in holidays does not necessarily reduce labour supply as it may lead to increased effort and efficiency.

(g) For an individual industry the quality of the labour force and its training probably outweigh the numerical factors. In Section 5.3 the alteration in the distribution of labour between sectors was shown. It would also be possible to examine the occupational distribution and find corresponding changes. The quality of labour includes many characteristics: it does not refer only to highly trained accountants or engineers, it also includes those able to plough a straight furrow or hew large quantities of coal. Training for some jobs may be more arduous and prolonged than for others, but almost all jobs require some training, and workers become more efficient with practice. The main problem with highly trained labour is that it tends to be very *specific*: a man may be highly skilled in one job but not very suitable for transfer to another.

The training of an adequate labour force is increasingly regarded as a Government function. This is suggested not only by the extension of compulsory schooling, but also by the expansion of Technical Colleges providing vocational training, by the passing of the 1964 Industrial Training Act, which set up a series of Industrial Training Boards, and by the establishment of Government Training Centres where redundant men are retrained.

The quality of labour is not something that can be measured in a vacuum without reference to other factors. To a great extent the efficiency of labour depends on the help that it gets from other factors of production. The intrinsic quality of a farm labourer may be greater now than was that of a similar man 50 years ago. The great difference in their productivity comes from the sophisticated equipment available to the modern worker. This other factor, *capital*, will soon be examined, but first the decline of the entrepreneur will be considered.

6.10 Entrepreneurism and Management

It was once common practice to regard the entrepreneur as a separate factor of production. He provided the capital necessary for his business, took all the decisions and risks, and then reaped all the profits at the end of the year. The growth of very large firms has led to a division of entrepreneurial functions as we shall see in Section 7.7. Thus the entrepreneur has gradually disappeared from the list of factors. The separate factor *management* is sometimes put in his place. Many regard management as a special kind of labour, but it may be argued that as management is responsible for the organization of production, rather than for production itself it may be treated as a separate factor. This view is reinforced by industrial conflicts over wage claims which often resolve into battles between labour and management, and not between labour and capital.

6.11 Introduction to Capital

Although we live within an economic system which is heavily dependent on capital, the idea of capital often causes some trouble to the newcomer to economics. In Sections 6.12–18 we define what economists mean by capital and consider its importance to the modern industrial economy.

6.12 A Definition of Capital

Capital may be defined as that portion of goods, physical assets or equipment created in the past but not consumed and available for use. As individuals, and as a community we have the choice of spending all of our current money income on goods and services for immediate consumption or reserving some of it for subsequent use (savings). The portion saved for future use may be referred to as capital, and it may for safety and profit be kept in one of the institutions discussed in Unit Nineteen. This money does not immediately become capital in the sense that we use the word in economics. The financially intermediate institutions with whom savings are deposited provide money for business enterprises to buy *productive assets*, such as plant and machinery to facilitate production. This kind of capital is important to economists. Self-denial on the part of the saver makes possible capital formation by the borrower as the new machinery will produce more goods in the future. Such capital may be regarded as the accumulated wealth of the community.

6.13 Types of Capital

As a factor of production capital is in the form of a stock of *producers' goods* which are available for use in production. These include machinery of all kinds, and with buildings represent the main investment of firms.

Stocks of raw materials and components or finished products also form part of the firm's capital, which is known as *circulating* or *working capital*, as the firm hopes that money spent on the items mentioned will be recovered rapidly from the sale of finished products.

Both kinds of capital existed even in the most primitive societies. Museums contain the originals or reproductions of tools and equipment which were the capital assets of some distant ancestor who realized that by devoting some of his time to shaping a flint or hollowing out a log he could make the task of 'producing' food easier in the future. We now have more sophisticated means of obtaining food, but the principle is the same—by devoting some current income to the development of equipment subsequent production of goods and services is made easier.

As time goes by and capital equipment becomes more sophisticated, it also tends to become more *specific*. The cave-man's axe no doubt served a variety of purposes, but as we saw in Section 5.6 (*b*) this is not necessarily the case with modern machinery, which may have been specially developed to perform one particular task in a particular industry. Nevertheless capital is the most mobile of the factors of production. The pool of savings accumulated by financial intermediaries is available for investment anywhere in the economy, and each completed project creates more wealth, making further savings possible. These savings may then be committed elsewhere to more profitable projects. It is the items of equipment that are specific, not the funds themselves.

It is not only business organizations which are involved in the accumulation of capital, many of the goods and services making up the national product are provided directly by the Government or Government agencies. The Government itself invests in many capital assets during the production of such goods. Sometimes, as with investment in the nationalized industries, such capital accumulation is indistinguishable from the investment of private firms. Government investment is often in the form of *social capital*, or capital provided for the benefit of society as a whole. The provision of roads, schools, hospitals, public libraries, museums, swimming pools and playing fields represents capital provided by the community as a whole. Such capital formation is made possible by the postponement of consumption. This is sometimes undertaken voluntarily and money is lent directly to the Government, sometimes less willingly when the Government reduces personal consumption through taxation.

6.14 The Importance of Capital

It is illuminating to imagine what would happen if for the next ten years the whole of the national product of the United Kingdom were spent on consumer goods, and none of it on buying or maintaining capital assets. For the first year or so almost everyone would be able to join in the bonanza and increase his consumption of goods and services and his standard of living. But it would not be very long before fewer goods came on to the

D

market as machinery failed, and road and rail services had breakdowns causing interruptions to the flow of production. In time few factories would remain in working order and the community would revert to direct methods of production, as the decaying machinery became unusable. Output and living standards would fall, and the benefits of thrift would become starkly apparent through their absence.

Thus it is demonstrable that a high level of output depends upon the accumulation of capital. Unit Five showed the interdependence of capital and the process of division of labour. The countries of Africa and Asia, often abundantly endowed with labour and other natural resources, find shortage of capital the greatest drawback in the exploitation of these resources. Capital deficiency derives from the vicious circle of poverty. Productivity is so low that almost all the national product is required for consumption purposes. Savings are therefore very low, little investment can occur and production remains low. Any increase in production is desperately needed for consumption by an expanding population. Such economies must try to obtain capital so that output may be sufficiently increased to generate enough savings to sustain investment. The task is difficult, but by no means impossible as can be seen from the rapid expansion of Japan. The problems are not as great as those overcome by Britain and other countries in the nineteenth century. Emerging nations have the benefit of modern technology. They do not have to develop it. They also have the assistance of a variety of international organizations in raising the capital they need for the initial vital expansion of their economies.

6.15 Capital Formation in Britain

The exact proportion of the national product which should be devoted to investment or *fixed capital formation* is difficult to determine and beyond the scope of this introductory book. Table 6.6 shows the amount that has recently been invested in relation to the gross national product. It has fluctuated between 18·1 per cent and 21·5 per cent in terms of current prices, and 17·7 per cent and 21·8 per cent in terms of constant 1963 prices which

Table 6.6 Gross domestic fixed capital formation in the United Kingdom 1960–70

Year	1960	1961	1962	1963	1964	1965	1966	1967	1968	1969	1970
Gross domestic fixed capital formation at current prices	4120	4619	4731	4912	5863	6315	6718	7261	7884	8121	8886
As a percentage of GNP	18·1	18·9	18·5	18·1	19·9	20·1	20·3	20·8	21·5	20·8	20·7
Gross domestic fixed capital formation at 1963 prices	4418	4847	4829	4912	5725	5960	6112	6524	6850	6777	6886
As a percentage of GNP at constant prices	17·7	18·7	18·3	18·1	20·0	20·2	20·4	21·6	21·8	21·2	21·1

Source: *National Income and Expenditure*, HMSO

are used to eliminate the effects of inflation. Many economists think this too low and accountable for the poor economic performance of the United Kingdom in relation to Japan and Germany. These countries invested respectively approximately 33 per cent and 24 per cent of their national products in the 1960s.

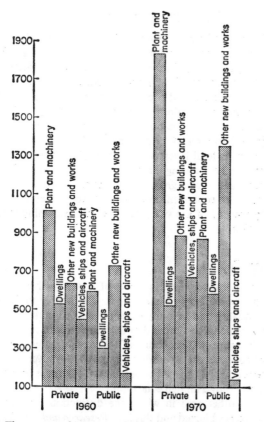

Fig. 6.3 The pattern of investment in the United Kingdom, 1960 and 1970

We cannot pursue the argument but Fig. 6.3 offers an interesting comparison between the types of investment that took place in Britain in 1960 and 1970.

6.16 The Finance of Investment

Although consumption must be foregone in order to finance investment it is not necessary for the sacrifice to be made by the individual or organization undertaking the investment. Indeed this is not usually the case. For example, the majority of people who buy a house (usually the major piece of individual capital accumulation) do so with the assistance of *other people's* savings obtained from a building society. In the same way companies rely heavily on the savings of private individuals to finance much of their capital accumulation.

6.17 Gross and Net Capital Formation

In Table 6.6 we used the term *Gross Domestic Fixed Capital Formation*, which is the terminology used for investment in the *Blue Book on National Income and Expenditure* published annually by Her Majesty's Stationery Office. This book gives details of the performance of the economy during the previous year. Domestic fixed capital formation simply refers to the investment in fixed assets, buildings, plant and machinery which took place within the economy. Fixed is not to be interpreted literally, for it means fixed as opposed to circulating and may include vehicles or aeroplanes. *Gross* investment is total investment including that needed to replace productive assets worn out during the year. If we wish to allow for this replacement, we must deduct from gross investment an allowance for depreciation (the extent to which equipment in existence at the beginning of the year has worn out by the end of the year) and the sum remaining is net domestic capital formation.

6.18 Human Capital

A society need not invest only in pieces of masonry or machinery, it may also invest in the skills of its manpower. The supply of labour depends partly on its quality and in many cases it needs to be highly trained. The proportion of national resources devoted to education may legitimately be regarded as investment. This is one of the great debates of economics for many regard education as a consumption good. However it is likely that a well-educated and highly trained population will be more productive than its counterpart with only a limited education.

There is one other area in which the distinction between labour and investment becomes blurred and this is in *Research and Development*. Many large companies employ a number of highly qualified people to study new ways of producing existing products, or develop new products for the company's future profits. These people are not producing anything of immediate use so, for example, the salaries of engineers working on prototype battery driven vehicles may be regarded as investment expenditure rather than labour costs.

6.19 Introduction to Land

The third factor of production to be considered is called land, but it includes more than the surface area of the five continents implied by the literal interpretation of the word. In this Section we examine land in the broadest sense, and some of the economic ideas associated with it.

6.20 Land: a Definition

Land is regarded in economics as including the various non-human natural resources found anywhere on earth. As such it would be far more accurate to refer to the factors of production as labour, capital and *natural resources*. In this sense the factor of production would include the following:

(*a*) The land surface of the world whether or not it is currently inhabited or habitable. In this very limited sense the supply of land to the United Kingdom is fixed at approximately 230,000 square kilometres. Early economists regarded land in this sense and were much concerned with the fixity of supply. This is an idea we shall consider later.

The land surface itself may have a variety of uses as a factor of production.

(i) Its most important use is probably as the basic factor in agricultural production, and it was in this respect that early economists developed the concept of *diminishing returns*. They observed the fact that if extra units of labour and capital are applied to a fixed amount of land there comes a point at which the extra output produced is less than proportional to the extra inputs of labour and capital. They also saw that extra units of land being brought into cultivation were frequently less productive than units in use. The early nineteenth-century writers attributed the characteristics of diminishing returns to land only, but they do also apply to other factors of production (see Unit Ten). The idea of diminishing returns is still fundamental to much of economic theory.

(ii) Land is also vital to secondary and tertiary activities. Manufacturing industry must have land, and one of the more important and difficult tasks of modern Governments is to confine the use of land for industrial purposes to certain clearly defined areas in order to maintain a balance between the various uses of land. Tertiary activities also require land.

(iii) Land is important for recreational purposes, and contemporary concern for the conservation of the environment has emphasized this aspect. Land can, however, only be used for such purposes if it has not been taken for agriculture or industry. It is difficult to calculate the opportunity costs involved.

(*b*) The rivers, seas and oceans of the world must be included as natural resources. In rare cases actual land is reclaimed from the sea, but even when this is not possible the seas and oceans are an important area of natural resources, and are likely to become more so as world food shortages enhance the value of farming the seas.

(*c*) The minerals and chemicals contained within the earth, the sea and the atmosphere are considered within this factor. Coal, oil, natural gas, iron ore and a thousand other items are included in these natural resources. This is one area in which the traditional classification of resources is arbitrary, since it is just as sensible to regard these deposits as wealth or capital rather than land.

6.21 The Supply of Land

Underlying Malthus's fears (see Section 6.6) was the fact that agricultural productivity fell as inferior land was brought into cultivation to meet the growing demand for food. This drew attention to the apparently fixed supply of land. The total acreage of land in a country is fixed, but the supply of land may be increased in other ways.

The scope for new geographical discoveries is almost exhausted, but the great advances in communications during the last century effectively increased the amount of agricultural land available to meet the demands for food in Europe. Modest amounts of land may be reclaimed from the seas: the airport at Foulness may be built on such land. Some land, previously useless for productive purposes may be cultivated through the intensive application of capital. Within the next ten years something approaching 100 million acres of artificially irrigated land will be cultivated in India. Such land would be useless without irrigation so this may be regarded as a method of increasing the supply of land to agriculture. The Volta Dam in Ghana, although designed primarily for the generation of hydro-electricity, has enabled the Plain of Accra to be cultivated. At the other extreme 'land' may be useless as it is waterlogged, and in this case extensive drainage works must precede agriculture.

The supply of land may be reduced by misuse or climatic factors. Intensive unscientific cultivation in India has left arid and unproductive areas which need very heavy investment to restore fertility. Wind action often causes serious erosion. Today Palmyra in Syria is surrounded by desert, yet in Roman times it was an important grain-producing area. The prevailing wind has changed it completely. In hilly or mountainous areas heavy rainfall can remove the fertile topsoil and slopes have to be protected or they will become barren.

If we think of land as natural resources it is even more clear that the supply is not fixed, for the removal of coal, oil, chalk, etc. has steadily reduced the known supply of natural resources. However, new discoveries of raw materials are made almost daily and existing supplies are now used more carefully and intensively so it is clear that the supply of land is not fixed. For example, modern furnaces allow much more energy to be harnessed from coal than was previously possible.

6.22 Work and Property

Our discussion on the supply of land may lead the reader to the conclusion that the border line between land and capital is difficult to define. For this reason some economists have abandoned the traditional divisions between the factors and distinguish only between *work* or *effort* and *property*. All kinds of labour, manual, clerical or managerial are gathered into the first category, while land and capital are linked together as property. This certainly overcomes the objections of those who would argue that much of the agricultural land of India is really capital since it would not be productive without large applications of capital. Since the division of factors into work and property does not help the description of payments to factors, we shall keep to the traditional classification.

6.23 The Payment of Factors of Production

The owners of factors of production do not usually supply them for productive purposes without reward. Exceptions occur when a factor is in such vast supply that anyone may enjoy its services. This applied to land in the early days of colonization; at that time such land commanded no rent.

All factors must receive payment today and as we saw in Unit Two the whole of the purchase price of a product (excluding tax) represents a payment to a particular factor of production: wages or salaries to labour; interest and profits to the owners of capital, and rent to the owner of the land. The proportions in which such payments are made depend partly on the proportions in which the factors of production are combined, but will be subject to many other influences as we shall discover later.

Meanwhile we will consider firms—those institutions within which the factors of production are organized.

6.24 Questions and Exercises

1. What is meant by the term 'population explosion'? To what extent does the problem exist in the United Kingdom?
2. 'The optimum population of a country is impossible to calculate.' To what extent do you agree with this view?
3. Explain the economic implications of (*a*) an aging population, (*b*) an expanding population.
4. Outline and assess the importance of the main factors affecting the supply of labour.
5. Explain carefully the meaning of the term 'capital' in economics.
6. What are the main determinants of the level of investment?
7. Describe the way in which land may be the subject of diminishing returns.
8. In what ways can the supply of land be increased?

The Unit of Control: the Firm

7.1 Introduction

The economy contains many *business units* or firms of all types and sizes. Their principal common characteristic is that they are the units whose decisions determine to a large extent the direction of the economy. It is within the firm that the factors of production are combined to create goods or services as decided by the owners. At this level the factors are organized, new products and developments are planned, and the many risks of economic and commercial life are taken. Firms show a progression from small-scale businesses often run by a single person to enormous enterprises of national or international standing. We will now look at these firms in more detail.

7.2 The Private Sector and the Public Sector

We have already distinguished between the public and private sectors of the economy. The main differences between them are those of ownership and motive. Business enterprises in the private sector have an obvious owner or group of owners and their principal objective is to make a profit. Businesses in the public sector have no clearly identifiable owners and belong to the community as a whole. Their pursuit of profit or surplus is likely to be tempered much more by a consideration for the public interest. In the United Kingdom economic activity is divided fairly evenly between the two sectors, but elsewhere things are different. In the United States of America the majority of enterprises are privately owned and decisions are taken by the owners of capital or their elected representatives within the firm. At the other extreme in Russia the majority of enterprises are controlled by the state. In this case the decision-making procedure is highly centralized and ultimately in the hands of the political representatives of the people.

In the private sector, pursuing efficiency and profits, are the following types of firms: sole traders, partnerships, private limited companies, public limited companies, and holding companies. Co-operative Societies may be included here though many regard them as non-profitmaking enterprises belonging neither to the private sector nor the public.

In the public sector are the public corporations and a variety of trading organizations linked directly to central or local government. In America the term 'public corporation' refers to what we call the public limited company, and we must take care not to confuse the two uses of the term.

7.3 The Sole Trader

The simplest and most widespread form of business organization is that of the sole trader. He starts business on his own account with his own capital and labour, assisted perhaps by one or two employees, and takes the profits as his reward. Such enterprise is not confined to the retail trade, although it is very widespread there. Any local newspaper will contain advertisements from builders, plumbers, hairdressers, printers and others in business on their own account.

It is an easy matter to establish such a business: there are no formal procedures, and the owner makes independent decisions, and has personal contact with employees and customers. There are, however, many difficulties associated with this form of business.

(*a*) The owner does not have the advantage of *limited liability*. He is liable for the debts of the business to the full extent of his private assets, not just to the extent of the capital he has committed to the business, as happens in other forms of business. Thus insolvency may mean the sale of his own home to pay creditors.

(*b*) There is no distinction made between the owner and the business, so not only is there unlimited liability but also a likely lack of continuity in the event of the owner's death. A business often has to be sold to meet estate duties.

(*c*) The sole proprietor usually has difficulty in raising capital. Initially he will probably use his own savings and later expand by ploughing back profits. He may borrow from a bank but in times of economic restraint small businessmen often find overdraft facilities withdrawn. Private loans are possible, but the sole proprietor is not allowed to appeal to the public for capital.

(*d*) Business these days is very competitive and success demands hard work, long hours and not inconsiderable worry. In most business firms these burdens can be shared, but the sole proprietor must bear them alone.

(*e*) Technological progress is often out of the question for the sole proprietor as he cannot afford the heavy capital outlay. But it is this progress that increases efficiency, so the gap between small and large firms widens and the markets of the small firms shrink as customers patronize the more efficient larger firms.

7.4 Partnerships

There are circumstances in which it might be appropriate for the sole proprietor to take a partner into the business, thus sharing control.

(*a*) It may be a convenient method of obtaining capital for expansion, although the number of partners is limited to 20.

(*b*) It may be a suitable way of acquiring expertise. Accountants specializing in different aspects of the profession: experts in personal taxation, company taxation, and bankruptcy for example, may form a partnership which would be more effective than three sole proprietorships. Similar specialization may occur in medical or legal partnerships.

(c) The burdens of prolonged hours of work, responsibility and organization of the business can be shared, so holidays are possible and illness is less of a nightmare.

So far as the general public are concerned there is normally no difference between the partnership and the sole proprietor. The liability of partners is still unlimited although the Limited Partnership Act 1907 does allow *sleeping partners* who contribute capital and have the benefit of limited liability but may take no part in the control of the business. Furthermore the action of one partner is binding on the others; he may buy or sell on behalf of the partnership and thus bind the others whether they agree or not. It is obviously very important for partners to have absolute confidence in each other.

It is most unlikely that even twenty partners will be able to raise enough capital to finance manufacturing processes on anything but a limited scale. Further expansion is necessary for large-scale manufacturing.

7.5 Limited Companies

The usual form of business organization which undertakes large-scale operations is called a Limited Company. The principal attraction of this business unit is that shareholder liability is limited to the nominal value of the shares held. This facility was conferred by Act of Parliament in 1856. In this way a large number of people can contribute funds to an enterprise without risking their entire personal possessions. Furthermore the company has its own legal existence quite separate from that of the shareholders, so its continuity is not threatened by the death of one of them. Such companies whose name always ends in the word 'limited', may be divided into two groups:

(a) The private limited company
This is a logical development from the partnership, and is frequently found where a family business has needed to expand beyond the resources of the partners, but the founders have needed to retain control. Such companies are restricted to fifty members, and must have at least two, shares cannot be sold by one shareholder without the agreement of the others, and the company cannot appeal to the public for funds.

(b) The public limited company
This is a form that embraces the very largest manufacturing units in the private sector. The mass production of such things as steel, motor cars and chemicals generally needs more capital than can be raised by fifty shareholders although there are some substantial and nationally known firms which have remained private companies. A public company must have at

least seven members, shareholders may dispose of their holdings at will (providing they can find a buyer!) and the company may appeal to the public for funds.

The directors of public companies are usually unknown personally to members of the share-buying public, so it is necessary for stringent control to be exercised over company formation and behaviour to protect shareholders from fraud. Such control is provided by the Department of Trade and Industry (formerly the Board of Trade) and the Registrar of Companies within the framework of the Companies Acts of 1948 and 1967.

At the end of 1970 there were well over 500,000 companies on the Register of Companies, 503,232 of them private and 15,425 public. There are three main reasons for forming a company.

(a) There is the advantage of limited liability, so investors may subscribe modest sums without endangering their whole capital should the enterprise fail and become bankrupt. In the first half of the nineteenth century there were many sad cases of investors losing all their belongings through the collapse of a company in which they had invested without the safeguard of limited liability.

(b) The limitation of liability allows the company to attract very large sums of money from a variety of sources to finance its undertakings. Capital raising is facilitated by the willingness of companies to accept funds on various terms to meet the requirements of different classes of investors. This is examined in Section 7.6.

(c) Companies, and especially public companies, enjoy the benefit of *economies of scale*; the cost per unit of production falls as the scale of operations increases. Economies of scale are fundamental to economic efficiency and are fully discussed in the next Unit.

7.6 The Share Capital of Companies

References to the capital of a company are often ambiguous as the term is applied to various aspects of the funds available.

(a) The *nominal* or *authorized capital* of a firm is the maximum amount of money the company is allowed to raise by issuing shares to the public.

(b) The company may not wish to issue the maximum amount of shares at the outset and the proportion they do issue is referred to as the *issued capital* of the company.

(c) When the shares are issued the company does not always need the full amount to be paid immediately. The proportion of issued capital that is actually paid for is the *paid up* capital of the firm. If a company fails the holders of shares which are not paid up are required to pay the difference between the nominal value and the paid up value of their shares.

(d) *Ordinary shares*. These represent the risk capital of the business. The holders of such shares are not guaranteed a dividend at the end of a trading year, but they do have the voting rights and elect the Board of Directors, responsible for the general policy of the company.

(e) *Preference shares.* While ordinary shares provide an attractive investment for those who do not mind the risk of getting no reward in some years, others find that preference shares offer a safer investment. Preference shares fall into several groups. There are basic preference shares which receive a fixed dividend out of profits before anything is paid to ordinary shareholders. With cumulative preference shares a dividend missed in one year is carried forward to the next. Most preference shareholders have no say in the control of the company as they have a privileged position with respect to dividends. Participating preference shares are an exception to this.

(f) *Debentures.* These are simply loans to the company which receive their fixed rate of interest before anything is paid to preference or ordinary shareholders. They are normally secured against property owned by the firm so that in the event of bankruptcy debenture holders are assured of getting their money.

These forms of capital are important to the investors, but none of them give a very accurate indication of the worth or size of the firm, which is determined by the net value of real assets employed.

7.7 The Entrepreneurial Function

The entrepreneur has been defined as the person who provides the capital, takes the risks, makes the decisions and reaps the profits or meets the losses of his business. There are still many thousands of such people in an advanced industrial economy, but they are not active in the large limited companies. Here the entrepreneurial function is divided between many groups.

(a) Ownership

This rests with the shareholders who have contributed capital by buying ordinary or preference shares.

(b) Control

This rests nominally with the owners of ordinary shares, as preference shareholders normally have no votes. There is thus an important division between ownership and control, and the extent of this division depends on the capital structure of the company. Consider the two companies in Table 7.1, each with a nominal capital of £1,000,000.

Table 7.1 Hypothetical capital structures

	Company A	Company B
Ordinary shares	100,000 × £1	900,000 × £1
Preference shares	900,000 × 6% × £1	100,000 × 6% × £1
Amount of capital controlled by each £1 ordinary share	£10	£1$\frac{1}{9}$

In company A where the majority of capital is in the form of preference shares the division between ownership and control is much greater than in company B. Companies such as A are said to be *high geared* and those like B, *low geared*. The *gearing ratio* is measured by the formula:

$$\text{Gearing ratio} = \frac{\text{Preference and loan capital}}{\text{Total nominal capital}} \times 100$$

This ratio is an important determinant of the amount of risk facing investors (see below).

The control vested in ordinary shareholders is in practice very limited. Each ordinary share carries one vote in the election of the board of directors at the annual general meeting. A large joint stock company may have upwards of 20,000 shareholders, many holding less than £100 worth of shares, so the influence and 'control' of any one of them is likely to be negligible. It is also impossible for all of them to attend the annual general meetings. Such meetings are usually quiet, orderly affairs, endorsing the policy of the existing board of directors, and not occasions for a close examination of policy by the owners of the business. It is only when a company has had a difficult year that the annual general meeting is likely to be the scene of strife. It is therefore more realistic to regard the directors as having control of the business especially as they alone can give orders to the staff.

(c) Decision-making

The entrepreneur of the economics text book has the clear and unobstructed task of maximizing profits. This in itself poses a number of questions that we will consider in Unit Fourteen, but the decisions that must be taken by a modern joint stock company cover a far wider field than profit maximization and would almost certainly be beyond the scope of the owners of the business. They therefore elect directors to assume responsibility for particular areas of activity and from these a chairman is elected to exercise overall control.

There is a further difficulty here, for while it is incumbent upon the directors to act in the interest of the shareholders, they need not, themselves, be major shareholders. It is therefore possible that the interests of the directors and shareholders may clash; for shareholders are looking for high dividends but directors may be more concerned with prestige, security or the long-term growth of the company. Many feel that this is one of the major weaknesses of the limited company.

The board of directors is responsible for establishing the general policy of the company, and senior, middle and junior management are appointed to implement that policy. Large companies are divided into many departments for managerial efficiency; so costing, production, safety, research, advertising, marketing and many other functions are the responsibility of particular departments which employ specialist staff. It is quite clear from this that the owners must leave almost all decisions in the hands of professional managers, and it seems that division of labour is just as important in management as it is in production. Indeed the management structure or the organization of decision-making is crucial to the success of a business.

(d) Risks
Before the Limited Liability Act of 1856 the owners of a business took a very real risk of losing their personal fortunes. Today the risks to shareholders are not so great and can be divided into two groups.

(i) **Capital losses.** The shareholder always faces the possibility that he may lose his entire investment through the failure of the company. A more likely event is that he will lose part of his investment as he may have to sell his shareholding at a price lower than that which he paid for it.

(ii) **Revenue losses.** Ordinary shareholders have no guarantee of an annual income from their shares. One of the risks they take is that in bad years they will receive no income as profits are too low. This is a more likely occurrence in high-geared companies than in low-geared companies. In Table 7.1 we see that the profits of company A must reach £54,000 if the demands of preference shareholders are to be met, and unless they are higher than this the ordinary shareholders gain nothing. In the case of company B, however, £6000 is enough to pay the preference shareholders, and any surplus left is then shared between the ordinary shareholders. Under the system of limited liability those who incur the greatest risks are often the creditors of a company—those to whom the company owes money. If one firm has supplied raw materials on credit to another which then goes into liquidation (goes bankrupt), there is little chance of the creditor company receiving full settlement of the debt. In this way risks are transferred from the owners of the business to its suppliers.

7.8 Public Enterprise
Some undertakings are not attractive to businessmen as they are non-profit-making, so the goods or services required by the community are undertaken by the State. In such cases the State may take over the supply of the product or service either directly or through the intermediary of the public corporation. Such activities as education, medical care and sanitation are amenities which are socially operated. Natural monopolies such as the supply of electricity, gas and water are also similarly run for the public benefit (see Section 7.10 (e)). Public enterprises may be divided into three groups:

(a) Trading activities are sometimes under the direct control of a government department. Her Majesty's Stationery Office for example, is under Treasury control, and a whole range of municipal trading activities are under the direct control of local authorities.

(b) The Government may (rarely in the United Kingdom) hold a majority shareholding in a public joint stock company such as British Petroleum Ltd. They may also have a 100 per cent holding in companies which have been established for a particular purpose. An example of this is Remploy Ltd which is wholly Government owned.

(c) Public corporations are the most prominent form of public enterprise, and they are of two types, those who sell a product directly to the public and those who do not. The latter group includes the BBC, the ITA and the Atomic Energy Authority; the former group includes the basic nationalized industries, coal, electricity, gas, air transport, the Post Office and the iron and steel industry.

7.9 The Organization of Public Corporations

Public corporations evolved comparatively slowly. An early example was the formation of the Port of London Authority in 1908, and the peak of the movement came with the nationalization of the transport, energy, and iron and steel industries during the post-war period 1945–51.

Each corporation has a legal identity separate from the Government, and has its own management board selected by the Government. The Government may lay down the general policy of the corporation, but the nationalized industries are normally free from day to day interference. Employees of nationalized industries are not civil servants, but are employed directly by the industry on terms which are, theoretically, a matter for negotiation between the board and the employees. Whereas the basic object of the limited company is to make a profit for its shareholders, the public corporation has for most of the post-war period been in an ambiguous position, torn between making a profit and providing a public service. Until 1961 the balance was largely in favour of providing a service irrespective of the losses incurred. In 1961 a Government White Paper, *The Financial and Economic Obligations of the Nationalized Industries*, called for a more commercial approach and since then specific economic targets have been set. The internal organization of the industries varies considerably. The National Coal Board has a highly centralized structure, despite the wide geographical dispersion of the industry. The electricity industry, however, is less highly centralized, supply is in the hands of a number of regional boards, though the generation of power is centralized in the Central Electricity Generating Board. Such constitutions are not, however, completely rigid. When the gas industry produced the majority of its gas from coal it was appropriate that the main unit of production and control should be the regional gas board. The advent of natural gas has left the gas industry with distribution as its principal task and a centralized structure has been thought more appropriate for this.

7.10 The Reasons for Nationalization

Nationalization has frequently been a matter of political belief, but economic arguments have always been advanced in support of any particular move towards public ownership.

(a) The interests of company shareholders and the general community may not always coincide. Nationalization allows the interests of the community to prevail. For example, shareholders in the steel industry might prefer to delay new investment in plant because of its high costs and because it may take years before profits benefit. The national interest may be better served by immediate investment.

(b) An essential industry may become so run down that it can be saved only by massive injections of capital by the State. If this happens state ownership normally follows. The nationalization of the coal industry, railways and, more recently, Rolls Royce all come partly within this category.

(c) The Government sometimes attempts to influence the general level of economic activity by way of the nationalized industries. It may attempt to off-set a rise in unemployment in the private sector by creating more jobs in the public corporations by encouraging expansion.

(d) In some cases there may be strategic arguments for nationalization. Coal was the only indigenous source of power when the industry was nationalized in 1946 and so was of extreme importance. Rolls Royce engines are similarly essential to Western defence so the company was nationalized when it faced difficulties in 1971.

(e) Efficient production often necessitates large units of control and production. In the relatively small United Kingdom market this may lead to a monopoly (a single producer in control of the market) which in turn may lead to the exploitation of consumers (see Unit Fifteen). Many nationalized industries are *natural monopolies*, where capital costs are so great that competition would be wasteful. In the early twentieth century, competition between local bus companies demonstrated the need for a monopoly situation to economize on capital costs and maintain a high quality of service to the consumer.

(f) The private sector if left to itself would not supply uneconomic services. Electricity supplies to outlying villages or single remote farms would be incredibly expensive if the consumer had to meet all the capital and installation costs, as would be the case with a private enterprise. Under public ownership such costs are shared by consumers as a whole. Many economists now urge a different approach, whereby charges would be made according to the costs incurred in providing the service.

7.11 The Difficulties of the Public Corporations

Section 7.10 summarizes the aims of the nationalized industries, but there are also many factors which have combined to prevent their fulfilment.

(a) *Dis-economies of scale* are often present in the public corporations. The cost of production per unit of output rises as the size of the enterprise increases. This is often due to difficulties of organization especially in the highly centralized corporations.

(b) Efficiency is often sacrificed to the requirements of the Government's general economic policy. It is much easier for the Government to exert pressure on prices or wages in the public sector and the corporations are in a very difficult position. On the one hand the corporation is set a certain profit target by the Government and on the other hand the Government prevents the price increases necessary to meet this target.

(c) In a bad year the limited company need not pay a dividend to its ordinary shareholders, and it can retain its resources for a greater effort in the future. Public corporations have to pay interest on the whole of their capital every year irrespective of performance.

(d) The knowledge that virtually unlimited money from taxation will meet any debts may induce a less vigorous attitude than prevails in the private sector. Much investment undertaken in the nationalized industries in the 1950s was made, irrespective of the likely returns.

(e) Pricing may also create difficulties for the public corporations. In general the policy has been to relate prices to the average cost of providing a service, rather than to the costs involved in the provision of a specific service. For example, I can post two identical letters in Westminster today, one to a resident of Westminster, the other to a friend in Glasgow. Each will cost 3p even though the latter involves the Post Office in greater expense than the former. The reason for such a policy is that it would be administratively complicated to charge each letter individually. This is a problem to which we return in Section 15.10 (a).

7.12 The Extent of Nationalization in the United Kingdom

The main acts of nationalization in the United Kingdom occurred during the period 1945–51. They included the nationalization of the Bank of England (an uncontroversial matter), the coal industry, British European Airways (the British Overseas Airways Corporation had already been established in 1939), the generation and supply of electricity (long subject to Government control), the production and sale of gas, inland transport, and the iron and steel industry. The latter was subsequently denationalized in 1951, but renationalized again in 1967. The Post Office was converted into a public corporation in 1970.

In 1971 the total investment of the public corporations was £1859 million of a United Kingdom total of £9923 million. They included in their number the four largest employers of labour in the country: the Post Office (408,000), the National Coal Board (376,000), British Rail (273,000), and the British Steel Corporation (250,000). In all they employed almost 1,900,000 people or about 8 per cent of the work force.

The significance of the industries is greater than this, for energy, transport and steel are at the very base of our economic life, and so the smooth running of the entire economy is dependent upon the efficiency of the public corporations.

7.13 Non-profitmaking Bodies

Between the private sector, motivated mainly by the desire for profit and the public sector with its twin objectives of profit and service to the public, are a number of non-profitmaking bodies such as co-operative and building societies. In this country the co-operative movement is restricted largely to the distributive trades. The co-operative movement began in 1844 to counter exploitation by capitalist shop owners. It spread rapidly on the basis of local retail societies buying goods in bulk, selling to members, and sharing any profit made between those members. In 1862 the retail societies established the Co-operative Wholesale Society which not only supplied goods but, increasingly produced them on behalf of the retail societies, who constituted the membership of the CWS.

In the post-war period the economic significance of the co-operative movement has declined in the face of severe competition from other forms of retail outlets. The movement now seems to be holding its own by adopting a more vigorous commercial approach.

Farmers' co-operatives have been established for a long while in other countries, but British farmers have only resorted to the co-operative purchase and use of expensive assets during the past few years. There are other important areas of co-operation between producers and these are the joint purchasing organizations which allow the bulk purchase of materials, and the joint marketing organizations which facilitate the sale of commodities. Thus producers gain some degree of market control.

7.14 The Size of Business Units

The size of enterprises varies enormously in each sector. We must now consider the measurement of firms, and the economic forces which encourage large firms, but allow small firms to survive.

7.15 Questions and Exercises

1. Distinguish between the main forms of business enterprise noting the main economic advantages of each.
2. Account for the growth of limited companies in the United Kingdom, showing how they have resulted in the decline of the entrepreneur.
3. Describe the main economic problems facing the public corporations, showing how they differ from those facing a private business organization.
4. Consider the economic implications of public ownership.
5. What are the principal justifications for nationalizing an industry?
6. What do you understand by the term 'natural monopoly'? Give examples of such monopolies.

Unit Eight

The Size of Firms and Plants

8.1 Introduction

Casual observation reveals a wide variety in the scale of economic activity. The greater efficiency of large-scale units has become part of the folklore of economics even though there is not much empirical evidence to support this view. In this Unit we will consider the ways in which the scale of economic activity may be measured. The reasons for growth and the way in which it occurs in firms are examined, and some attempt is made to see the effect of size on efficiency. The survival of small firms is considered against the dominance of large-scale firms in most industries.

8.2 The Measurement of Firms

The size of firms may be measured either in terms of their input of factors of production or else in terms of their output of finished products. If we use factors of production for measurement we merely count the number of employees or estimate the value of capital employed. If we measure in terms of the number of employees there is a danger of discriminating in favour of labour-intensive industries or firms. There is also the minor difficulty of dealing with part-time workers and the number of hours worked by the average employee. The use of capital employed as a standard of measurement is fraught with even greater difficulties and is a matter of much discussion among economists and accountants. We might consider measuring the size of companies by the current value of their issued capital in the Stock Exchange. As will be seen from Unit Nineteen this value is liable to fluctuate for reasons that may be quite unconnected with the company's size or performance. As an alternative the value of plant and machinery actually being employed may be used as an indicator of size. We may now turn to output to see whether similar difficulties are associated with its use as a standard of measurement. In a single industry producing a clearly identifiable product it might just be possible to count the number of units produced and compare the size of firms or plants on that basis, but a firm producing 1000 luxury limousines can hardly be equated with one producing 1000 mini-cars. Once we look beyond the confines of industry to compare the size of firms in general we have no alternative but to measure the value of their sales. This is known as the *turnover*, or gross sales.

This is the measure used by *The Times* in its annual publication, *The Times 1000*. This sets out details of the one thousand largest companies in the United Kingdom and of smaller numbers elsewhere in the world.

Table 8.1 The United Kingdom's largest firms (private sector) 1970

Company	Turnover	Capital employed	Employees	Equity market value	Net profit before tax
	£m	£m		£m	£m
British Petroleum	2,614·2	1,890·5	27,200	2,233·8	466·5
'Shell' Transport and Trading	2,608·2	2,043·5	—*	2,259·4	347·2
British American Tobacco	1,668·4	795·0	120,000	795·9	167·7
Imperial Chemical Industries	1,462·4	1,695·6	194,000	1,359·4	186·5
Unilever Ltd	1,292·0	570·8	100,553	567·5	71·1

Source: *The Times 1000*

* Figure not available.

Table 8.1 gives details of the five largest in the United Kingdom in terms of turnover. Only three other companies had a turnover of over a thousand million pounds. If we use other measures of size we get a different order. British Petroleum has only 27,200 employees (in the United Kingdom) and would not be placed in the first hundred firms on this basis, while the General Electric Company with 206,000 employees would be the largest firm though it is only ninth on the basis of turnover. When capital employed is considered only two firms which do not qualify for the top five on the basis of turnover would be admitted—Rio Tinto Zinc (£770·4 million capital employed) and Burmah Oil (£714·9 million capital employed). If the level of profits was used as an indicator of size then only Unilever would have to be removed from the top five, its place being taken by Rio Tinto Zinc with profits of £91·9 million (but only 13,320 employees in the United Kingdom).

It seems true that turnover, capital employed and profit levels all give a similar indication of size, but that the number of employees is often at variance with the other indicators and must be treated with care. At the lower end of the scale, among sole proprietors, partnerships and private companies, however, information on capital and turnover may be difficult to acquire and it is common practice to use the number of employees as an indicator of size.

8.3 The Growth of Firms

It is difficult to envisage the sizes of companies merely by looking at their turnover of assets, but an indication of the importance of the largest firms may be gained when we remember that the annual turnover of the ten largest United Kingdom companies amounts to over a third of the gross national product. These companies are dwarfed by General Motors and Standard Oil (New Jersey) of America whose combined annual sales amount to over £15,000 million, roughly equivalent to the expenditure of the central Government of the United Kingdom!

Although we are not especially concerned with such giants we must recognize that the economies of many countries are increasingly dominated by such large firms. It is interesting to see the ways in which small firms may grow in size. There are broadly speaking three ways in which a firm may expand:

(i) by joining in the general expansion of the market in which it finds itself;

(ii) by spontaneously producing new goods for different markets;

(iii) by acquiring other firms within or outside its own industry.

(a) An expanding market
If the economy is expanding many firms will grow simply because their market is growing. A company producing motor vehicles or television sets in 1950 can hardly have failed to increase its turnover since then. A firm may also actively seek new markets for its product, perhaps overseas, and so increase in size. In either case we should expect the turnover and capital employed to increase.

(b) Production of new goods
A further possibility is that the firm will grow by producing new goods, and in many cases these will be made from the same raw materials as its basic product. Thus the oil refining companies have benefited not only from the expanding demand for fuel but also from the development of a whole range of by-products.

(c) Acquisition of other companies
The most rapid expansion in company size is achieved by mergers or success-ful take-over bids. The terminology here is sometimes confusing and in this chapter we shall use the term *merger* which implies that two companies come together amicably and submerge their interests in a new third com-pany. An *acquisition* or successful *take-over bid* occurs when one company is absorbed into another company which retains its identity. Although the distinction between mergers and acquisitions is important during the nego-tiating period, the economic effects of them, once achieved, are likely to be similar.

Mergers may be classified in a number of ways though the student must recognize that few actual mergers will fit precisely into pre-determined categories. We may distinguish between four separate kinds of mergers or integrations:

(i) **Horizontal mergers** occur where the two companies concerned are engaged in the same stage of production. The merger of two brewing companies or two motor manufacturers would be horizontal. The aim of such mergers is to take advantage of economies of scale (discussed fully in Section 8.5) and reduce the costs of production. Such economies are undoubtedly available but it should be realized that lower costs do not necessarily mean lower prices; shareholders may demand higher profits. The extent of such mergers in brewing has been considerable: the number of brewers has fallen from approximately 4000 in 1914 to under 100 today with the bulk of production concentrated in the hands of seven major groups. Improvements in transport have been instrumental in allowing such concentration of production and they also account to some extent for the absorption of many small bakers into large groups. The dangers of such concentration are that it may leave the consumer at the mercy of a powerful monopolist; and it will almost certainly reduce the variety of goods available, and in the process of *rationalization* a large number of redundancies may occur. By rationalization economists mean the reorganization of production to achieve a given output with a different mixture of factors of production. It usually results in more capital-intensive techniques, and the concentration of production in fewer plants.

Horizontal mergers may occasionally be defensive in intent. The formation of the Metal Box Company from a number of producers was a reaction to intense American competition in the contracting markets of the 1930s. The British Aircraft Corporation came into being for similar reasons.

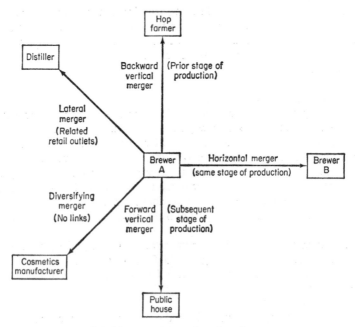

Fig. 8.1 Diagrammatic explanation of mergers

(ii) **Vertical mergers** occur between two companies engaged in different stages of the production of a good. In the brewing industry we may distinguish *backward vertical integration* when the brewer takes over the supplier of the hops or barley to safeguard his raw materials, or *forward vertical integration* when the brewer acquires a transport concern or public houses to facilitate the distribution of his product or safeguard his retail outlets. Backward integration allows the producer to by-pass the market to some extent and this may lead to economies; similarly forward integration and control of the market allows production to be planned with greater confidence. (Once again we should not overlook the likelihood that both types of vertical integration will lead to rising profits.) The outstanding example of economies from the vertical integration of production occurs in the iron and steel industry where continuous production from smelting to the formation of steel sheets saves enormously on fuel bills by eliminating the re-heating of materials. Other examples of vertical integration may be found in oil refining, where companies may control petrol stations as well as oil fields and in the building trade where, for example, one company, Redlands, not only produces sand and gravel (and other kinds of building materials) but has also acquired a waste disposal company which doubtless helps fill the exhausted gravel pits.

(iii) **Lateral mergers** involve two companies which produce related goods which do not compete directly with each other. The creation of Cadbury-Schweppes was in many ways a lateral merger, taking advantage of raw materials and marketing outlets common to both parent companies. The common link may be at only one end of the process: distilling and brewing are two distinct and separate techniques, but the lateral merger of brewers with distillers should lead to important marketing economies.

(iv) **Conglomerate or diversifying mergers** occur where the products of the companies involved are unrelated. The acquisition of a cosmetics or potato crisp manufacturer by a tobacco company constitutes a diversifying merger. Such mergers are frequently defensive, or in anticipation of a decline in the acquiring company's principal market.

Many diversifying mergers are of a more aggressive nature stemming from financial manipulation and they give rise to a *Holding Company* which may be defined as a company controlling a number of other companies by holding at least 51 per cent of their equity but not necessarily taking an active part in their management. In this way companies in a variety of industries may benefit from the financial skill and prestige of the holding company without losing their separate identities. By careful selection a holding company can establish a closely integrated group with each section complementary to others. On the other hand, there is more than a suspicion that many of today's diversifying mergers are inspired by speculative rather than economic motives and that loss making activities in one part of the group are subsidized by profitable activities elsewhere, which means that one set of consumers subsidizes another set.

It is worth noticing here that the establishment of holding companies may be facilitated by the existence of high-geared companies (see Section 7.7) for it is easier to acquire control of a company whose capital has been raised mainly in the form of preference shares, than one whose capital is largely in the form of ordinary shares.

8.4 Multi-national Corporations

It is appropriate at this stage to mention the multi-national corporations. These are a feature of economic life that has recently attracted a good deal of attention. The multi-national firm is one that operates in more than one country but under ownership based in only one of those countries. Many of the largest firms in the British economy come into this category; for example, Shell, Unilever and Ford, and they may be regarded as a natural development of the process of growth by internal development and acquisition. They merit separate treatment because of the special economic effects that are associated with them.

(a) National economies do not expand and contract in unison. The multi-national company by establishing manufacturing or other plant in two or three countries can often more easily weather an economic storm in one country.

(b) Transport economies are obviously possible if plant manufacturing bulky products can be located close to different national markets.

(c) Well-established firms are often the only ones with the gigantic sums needed for efficient mass production. By investing in a foreign country the firm not only extends its own market but also creates jobs in the host country. Moreover, to the extent that profits cannot be brought home they may be ploughed back to the benefit of the host country.

(d) The multi-national firm is of the greatest importance in the transfer of technical knowledge from one country to another. If ideas and knowledge did not cross national frontiers economic progress would be very slow.

(e) On the other hand the management and control of a large company operating exclusively in the United Kingdom is a very difficult problem. The complexity of the huge multi-nationals renders the problem of control even more difficult and many of the benefits deriving from these companies are likely to be offset by administrative problems. These can be very complex as company taxation is different in each country, and national Governments frequently and wisely place controls on the movement of capital and profits across their frontiers.

(f) The establishment of multi-national companies requires considerable international investment which has a dual effect on the balance of payments (see Unit Twenty-five for details). Initially large sums of money enter the recipient country, thus helping the balance of payments; but subsequently profits may be returned to the overseas holders of the business, thus adversely affecting the balance of payments.

(g) The clash between the interests of shareholders and the public may be even more marked with multi-national companies than it is with domestic companies. Consider a motor manufacturer with plants in half a dozen European countries; he may have less incentive to export goods from one of those countries to another than the firm based in one country and anxious to expand its market by exporting. It is also not inconceivable that groups of foreign-owned firms in a particular country might co-operate in their own interests rather than in those of the host economy.

(h) The most important of the multi-national firms are, of course, American, but many economists foresee a bright and important future for others in Europe as integration proceeds. The merger between Dunlop (United Kingdom) and Pirelli (Italy) in 1971 may be the first of many such ventures, and it emphasizes the need for the Governments concerned to come together to establish a framework within which such companies can operate efficiently without damaging national interests.

8.5 Economies of Scale

It is almost a truism of economics that large firms are able to take advantage of economies of scale. In this section we examine possible economies of scale available to firms or plants as they grow larger. The distinction between firms and plants is important as some economies of scale arise from the actual process of production while others are related to the organization of production. Both kinds of economies are known as *internal economies of scale* since they benefit a single firm. There are other economies available to many firms or industries which result from the general advancement of industrialization, and those are known as *external economies of scale*.

(a) Internal economies of scale
There are six basic types distinguished here.

(i) **Technical economies** are those primarily found in plants. Neither the capital costs nor the running costs of plants increase in proportion to their size. Thus a company that can justify the use of a 250,000 ton bulk oil carrier gains in two respects: the capital cost per 1000 tons of capacity is smaller than for a tanker of 100,000 tons; and neither the size of the crew nor the amount of fuel required by the vessel increases in proportion to its size. Similar considerations apply to the inland distribution of oil and account for the increasing size of road tankers. Chemical engineers have a rule of thumb used in connexion with the size of plant which they call the '0·6 rule'. This means that the doubling of the capacity of a machine or plant leads to a rise in costs of approximately 60 per cent.

Large firms may enjoy other technical economies by locating specialized plant on one site, by serving a particular market, or by using the same plant to make parts of different products.

An important source of technical economies in the mass production industries is the intensive use of equipment in order to spread fixed costs over as large an output as possible. Since a motor manufacturer might have to spend £10,000,000 on the development of a new model before the first car comes off the production line, he must sell a very large number in order to reduce the development costs per vehicle to a minimum.

(ii) **Managerial or administrative economies.** The cost of processing large orders is not likely to increase in proportion to the size of those orders. Salesmen, typists and accounts clerks can deal as comfortably with orders for 1000 units as they can with orders for 100 units. This kind of administrative economy is really an extension of the technical economy of spreading overhead costs over large outputs. In the multi-product firms many administrative costs may be shared between products but this is impossible in a single product firm. The principal managerial economies, however, are likely to be those derived from specialization: the large firm can employ specialist accountants, marketing managers, salesmen, production engineers who, by devoting all their attention to a relatively small part of the company's work, may do much to increase productivity.

(iii) **Research and development economies** are often available to the large firm. The design and development of new products may be a protracted and expensive undertaking, and small firms are unlikely to have the resources necessary to keep them ahead of competitors. Large firms develop new products and thus increase their competitive advantage over their smaller rivals, attract custom from them (or take the lion's share of any expansion in the market) and move further ahead of the field. The small firm may find it necessary to purchase new designs or patents from outside if it is to remain in business.

(iv) **Financial economies.** We have seen that large firms have not only a large turnover but also large assets. When it comes to raising capital for the purchase of new plant or for investment in stocks of raw materials the valuable assets of the large firm give it a further advantage. Investors in ordinary or preference shares or purchasers of debentures are likely to be more impressed by the status and achievements of a large nationally known company than by those of a small relatively unknown one. Moreover the actual administrative costs of raising money through the capital market will be proportionately lower for large firms as the costs do not increase in proportion to the size of the share issue. Large firms also have the advantage when it comes to short-term finance. An examination of the published accounts of many public companies will reveal bank overdrafts of hundreds of thousands of pounds. A small local business on the other hand may have difficulty in persuading the bank manager to part with even a few hundred pounds. The borrowing difficulties of small and medium sized firms have resulted in the formation of a number of Government-sponsored bodies whose main task is to channel funds towards deserving firms.

(v) **Marketing economies** may be available both in the purchase of raw materials and components and also in the sale of the finished products. A large firm may not only receive normal discounts for bulk purchases but may also be able to dictate very advantageous terms if it constitutes a large proportion of the supplier's market. To the extent that advertising increases the size of the market for a firm's products, and allows further division of labour in production it may lead to further economies of scale. The large firm is better able to support the advertising costs of launching new products and of keeping representatives in the field to maintain or expand its market share.

(vi) **Social economies** may be divided into two groups: those that build up the goodwill of the community in general and thus attract custom; and those that develop the loyalty of the firm's employees. The former may be regarded as an extension of the marketing effort and come within the public relations department. These involve, for example, the sponsorship of football or cricket competitions or the subscription of money to good causes and they may be expected to result in increased turnover. The latter consist of recreational facilities, housing, superannuation schemes, Christmas bonuses and any projects designed to make workers feel that they are an integral part of the firm and to win their loyalty. Such schemes are often seen as a means of compensating for the lack of personal contact between the owners and the operatives; and in other economies, notably Japan, such a paternalistic approach is regarded as one of the keys to business success.

(*b*) **External economies of scale**
These are available to all the firms in the industry and indeed may be of advantage to a number of firms in different industries. They may be divided into the following groups:

(i) **Economies related to a particular industry.** Many of these are derived from the concentration of an industry in one locality and are more fully discussed in Unit Nine. They include the provision of specialist maintenance services or training facilities in local technical colleges, or the development of a pool of labour with the skills appropriate to the industry. Other external economies may be realized through a *trade association*. This is an association of producers, corresponding to, though not normally parallel to, a trade union. Examples of such associations are the Society of Motor Manufacturers and Traders and the Brewers' Society. Economies obtained from membership of a trade association might include joint or *generic* advertising: 'Join the tea-set' is intended to increase the consumption of tea in general, not one particular brand of tea. Technical information and market trends may also become available through the association, who might be able to organize trade fairs or other marketing facilities beyond the scope of any single manufacturer.

(ii) **Economies related to industrialization.** Areas of high economic activity always include a number of firms dependent upon the major companies of the area, but providing an essential service to them. Thus the motor industry is served by a host of small firms who provide components and maintenance services to them and other industrial concerns. The activities of the services sector multiply, providing advantages to firms in the area compared with those in less developed regions.

(iii) **Economies related to society.** The provision of roads, railways, housing, schools, hospitals and other social services is largely the responsibility of the state. As industrialization proceeds the provision of these items of social capital increases and makes an area more attractive to firms and potential employees, in many cases also giving further advantages to firms already in the area. Better housing attracts better workers at all levels, and better communications facilitate purchases and sales.

It is not likely that any one firm will be able to enjoy all these economies of scale, both internal and external, at the same time, but the possibility of their taking advantage of them as they grow larger raises two important questions. Why is it that many industries are not dominated by one large firm? How do thousands of small firms manage to survive? We will now try to find answers to these questions.

8.6 The Limits to Growth—Diseconomies of Scale

As a firm expands its activities and takes advantage of economies of scale it is able to reduce the unit or average cost of production. Beyond a certain size the unit cost may begin to rise again owing to the effects of diseconomies of scale. These are of two types, internal and external.

(a) Internal diseconomies of scale

(i) **Technical diseconomies** occur as the size of plant increases. Bulk oil carriers, for example, need special berthing facilities the cost of which may go a long way towards offsetting the savings made by the use of larger units. In some cases part of the cargo of such tankers may be discharged into smaller vessels at sea thus reducing the draught of the ship (the depth under water) and allowing it to enter harbour. A further difficulty which applies to all kinds of vessels and carriers is that more expensive materials may have to be employed in construction as the capacity increases.

In the chemical industry enormously complicated plant is required for the production of various plastics. It may often be more economical to have two relatively small plants rather than one double the size, partly because the construction problems are greater with large plants and partly because the failure of one of two plants does not bring production to a halt.

(ii) **Administrative diseconomies** arise in the large organization through the minute division of managerial labour. It may be very expensive to inform staff of routine matters through internal memoranda; changing production techniques or market structures may render a given management structure obsolete and wasteful; and customers' inquiries may proceed through many managerial channels before a decision is reached. The greatest scope for future economies probably lies in the reduction of many top-heavy management structures.

(b) External diseconomies of scale
These result from the overcrowding of industrial areas and the consequent increase in the price of land, labour and services. An obvious and important example is provided by the congestion costs resulting from high traffic densities.

Besides diseconomies of scale there may be political reasons why firms do not expand until they completely dominate a market. There is a deep suspicion of the monopolist and as we shall see in Unit Fifteen the Government may take steps to prevent the establishment of monopolies. Even if this does not happen the firms themselves may consider it better for their image not to force all their rivals out of business by competition or take-over.

8.7 The Optimum Size and Returns to Scale

We may imagine a firm gradually increasing in size, at first enjoying the benefits of economies of scale with the average cost per unit of production falling, and later meeting diseconomies of scale and rising costs. At the point where the costs per unit of production are at a minimum the firm is said to be operating at its *optimum size* a concept to which we shall return in Unit Ten.

As a firm increases in size there are three possible effects on unit costs:

(a) They may fall: in which case the firm is said to be operating under *decreasing costs* or *increasing returns to scale*. For example, doubling the input of factors of production will more than double the output.

(b) They may rise: a case of *increasing costs* or *decreasing returns to scale*; a doubling of inputs will fail to double the output.

(c) They may be constant: a case of *constant costs* or *constant returns to scale*; a doubling of inputs exactly doubles the output.

Obviously the merits of expansion will depend partly on whether the firm is operating under increasing, decreasing or constant returns to scale.

8.8 How Do Small Firms Survive?

The Bolton Committee of Inquiry on Small Firms found it impossible to establish a definition of the small firm appropriate to all sectors of the economy. It therefore adopted different definitions for each sector; some of these appear in Table 8.2.

Table 8.2 The significance of small firms

Industry	Definition of small firm	Small firms as % of all firms in industry 1963	Propor- tion of total 1963 employ- ment in small firms (%)	Average employ- ment in small firms 1963
Manufacturing	200 employees or less	94	20	25
Retailing	Turnover £50,000 or less	96	49	3
Construction	25 employees or less	89	33	6

Source: *The Report of the Bolton Committee of Inquiry on Small Firms*, HMSO

Notice the difference between the percentage of small firms and the employment they offer. It is, however, their existence that is important in the context of this Unit, and not their significance as employers. There are many reasons for their survival.

(a) The entrepreneurial spirit of people who value independence and who may make a greater contribution to the economy by running their own businesses than they would if submerged in a larger organization. There are many cases of firms enjoying the possibilities of expansion, outlined in Section 8.3, who do not expand as the owner wishes to retain control or does not want the worry of a larger organization, or because he fails to spot the opportunities.

(b) Small and geographically dispersed markets are best served by small firms, as the optimum size is small. Moreover large firms are often not interested in the production of custom-made goods.

(c) Small firms frequently provide an important service to large firms not only by providing them with components, but by relieving them of the task of organizing the production of those components, and enabling them to concentrate on their main tasks.

(d) Despite what was said in Section 8.5 (a) (iii) small firms are often an important source of innovation; the management of small firms, being in many ways closer to the market than top management in large firms may be quicker to spot and appreciate the significance of new developments. Small firms are more flexible, and perhaps more amenable to the wishes of their clients.

(e) All the present industrial giants had small beginnings. Tomorrow's industrial giants will be different from yesterday's and are developing today. On most Monday mornings the leading national newspapers carry prospectuses of companies issuing shares to the public, often for the first time. A perusal of such a prospectus will show the reader how each firm has grown from small beginnings.

8.9 The Difficulties and Relative Decline of Small Firms

The Bolton Committee found a more or less continuous decline in the significance of the small firm in manufacturing since the 1930s and this is illustrated in Table 8.3.

Table 8.3 Employment in United Kingdom 1924–68

	All establishments (000's)	Small establishments (000's)	Small establishments as % of total
1924	5115	2257	44
1948	6871	2538	37
1968	7870*	2280*	29*

Source: *The Report of the Bolton Committee of Inquiry on Small Firms*, HMSO
* Estimates.

Similar trends are detectable in other sectors of the economy. Nevertheless, the small firm sector still generates between 20 and 25 per cent of the national product and it is certainly worth considering the special difficulties which confront such firms.

(a) Small firms suffer disproportionately in times of recession. They are frequently forced to rely on banks for finance and are the worst sufferers from a restriction in bank credit: they are often heavily dependent on one product so their turnover drops sharply if demand for that product falls.

(b) In many branches of manufacturing and distribution the optimum size of plant is increasing and this causes a reduction in the number of small firms.

(c) The development of new management techniques based mainly on computers may put the small firm, relying on traditional methods, at a disadvantage. This may, in some measure, be overcome by several firms sharing computer time.

(d) The activities of the State may contribute to the difficulties of the small firm by adding to its administrative burden (form-filling, complying with regulations), circumscribing its market by placing Government orders with large firms, encouraging industrial integration to promote efficiency, and taxing profits which could otherwise be re-invested.

8.10 Size and Efficiency

The motive for expansion and integration is normally stated to be the extra efficiency that is associated with large units. The measurement of economic efficiency is a matter of some debate among economists but the relationship between profits and capital is the most widely accepted measure.

Table 8.4 Size of firms and level of profits

Firms	Average turnover (£000)	Average capital employed (£000)	Average profits before tax (£000)	Profits as % of capital employed
*1– 50	590,767	355,619	50,339	14·8
601– 550	17,859	9,938	1,547	20·1
951–1000	7,877	3,966	510	18·1

Source: *The Times 1000*

* Column one refers to the ranking of firms in the top 1000

In Table 8.4 we compare the performance of the top fifty firms with two other groups. Profits as a percentage of capital employed are larger for the groups of smaller firms than for the largest firms, and this is a finding confirmed by other investigations. However there are many influences on the ratio of profit to capital, and it would be rash to conclude that the largest firms are the least efficient. A more valid comparison could be made if all the firms were in the same industry. Even then if the large firms were capital-intensive and the small firms labour-intensive the ratio of profits to capital would not be a good indication of efficiency. If, instead, we used as a measure the value added per unit of labour, we would be discriminating very heavily in favour of capital-intensive firms. A further possibility is to use the criterion of cost per unit of output as an indicator of efficiency. This ties up with the economic objective of using as few resources as possible to produce goods. Such an approach meets with difficulties over the prices of factors of production.

Clearly there are important problems in trying to measure the efficiency of firms. We can only conclude that the profit/capital ratio needs very careful handling, especially when it is remembered that profits may be entirely unrelated to efficiency: they may derive from monopoly power (see Unit Fifteen) or simply from speculative and unproductive buying and selling of property.

8.11   Questions and Exercises

1. Assess the difficulties involved in trying to measure the size of firms.
2. Explain, with current examples, the various ways in which firms may grow.
3. Describe the various kinds of mergers by which a firm may expand, showing the economic implications of each type.
4. 'Multi-national corporations may be so large as to rival Governments.' Consider the economic implications of such companies.
5. Explain the term 'economies of scale' outlining the main kinds of economy available to modern industry.

6. What factors prevent firms from growing until each industry is a monopoly?
7. What economic difficulties face the small firm? In what ways may they be overcome?
8. In what ways does the growth of a firm or industry lead to greater efficiency?
9. Consider what is meant by the efficiency of the firm. Analyse the difficulties involved in determining which of the following firms is the more efficient:

Business:	Firm A Oil refining	Firm B Manufacture of plastic toys	Firm C Motor car manufacture
Capital employed	£1500m	£3000	£900m
Employees	15,000	60	12,000
Profits (before tax)	£300m	£1200	£225m

10. The financial pages of the newspaper carry summaries of the annual reports of limited companies. Make a list of the results of companies and comment upon your findings, noting especially the levels of profits in different areas of the economy.
11. Make another list of mergers and explain the motives behind them and their likely effects.

R–E

Unit Nine

The Location of Economic Activity

9.1 Introduction

While the scale of economic activity may be an important determinant of economic efficiency, the geographical distribution of industry will also have an effect on performance. In this Unit we examine the principal influences affecting the location of industry, the extent to which they have changed, the problems arising from a changing pattern of location and the attempts made by the Government to ease the problems.

9.2 The Historical Factors

In the absence of Government interference there are a number of factors which influence the location decision. It would be convenient if we could assume that the firm in seeking to maximize its profits weighed up all the locational factors associated with various sites and then selected the best. Such rational locations, however, are rare if they exist at all; a lack of knowledge often prevents the perfect decision being reached. But it is safe to say that no firm will set up production or establish a branch factory without taking some account of most of the following factors.

(a) Power

Coal, the major source of power, was the most important single influence on nineteenth-century industrial location. Before that time the availability of swift-flowing water was a powerful inducement to industry. Coal is expensive to transport and is entirely consumed in the productive process so industry saved money by being situated on the coalfields. With the exception of the London area, a map showing the distribution of industry in the United Kingdom in 1850 would have shown a close correlation with the distribution of coalfields. The significance of coal as an attracting force declined first because transport became relatively cheaper but mainly because the generation of electricity from coal permitted the widespread distribution of energy. Similarly the commercial development of gas and the increasing availability of oil have helped to reduce the importance attached to the availability of fuel in the location decision in the United Kingdom and other industrial countries. Yet elsewhere proximity to fuel may still be of overriding importance. Hydro-electric power stations exercise a considerable pull on industrial location in many African countries.

(b) Raw materials

Like coal, raw materials may be expensive to transport, so they have often exerted an important influence on industrial location. Iron ore is a raw material whose cost of transport weighs heavily in location decisions. In the early days the British iron industry flourished as coal and iron were found in close proximity to each other. The exhaustion of the ore deposits and the discovery of new supplies in other areas led to changes in industrial location. Today a large proportion of ore is imported and coastal locations have assumed a new significance, as they had for oil refining whose basic raw material was almost all imported until recently. The cement industry in the United Kingdom is situated mainly in the South East so that the chalk of the Chiltern Hills and the North Downs is readily available. The availability of manufacturing components may be an important locational factor today.

(c) Labour
There is obviously no point in building factories where the supply of labour is inadequate. This is not merely a question of numbers but also of quality. There are vast reservoirs of labour in India, but it is not immediately suitable for the mass production techniques of American and European industry. Similar problems have occurred in the United Kingdom, as labour obtained through the decline of the coal mines has had to undergo considerable retraining before joining industrial firms. On the other hand an area with a population endowed with particular skills will be attractive to firms seeking those skills. Thus a prospective cutlery manufacturer would be ill-advised to ignore the skilled labour available in Sheffield. In more general terms many new towns have proved attractive locations for light electrical engineering companies and other firms whose processes allow them to employ married women, often on a part-time basis.

(d) Markets
While nineteenth-century locations were determined very largely by the factors outlined above, the availability and proximity of markets is likely to be more important to many industries today. As we have seen the structure of national output has gradually changed in the last 50 or 60 years with the result that industry is now much more consumer orientated. Accordingly there may be advantages in locating new plant near the principal markets rather than close to sources of raw materials. This tendency is reinforced by the fact that many raw materials have to be imported and can just as easily be imported near consumer markets as near traditional production sites. Furthermore consumer goods may be made from a variety of raw materials obtained from several sources, none of them exercising an overwhelming attraction. Market locations are again encouraged. The production of fragile or valuable articles is preferably undertaken near the market as this reduces the risk of damage or theft. In the Midlands many precision engineering companies have sprung up to serve the motor industry which provides a ready market for their products.

Perhaps the most important reason for market locations is that where there are markets there are people, and therefore there is a supply of labour readily available especially for mass production processes where little training is required.

(e) Transport

At any point in time the location of raw materials and markets will be fixed. The location decision may then be seen in terms of transport costs. Weber developed the theory that industries using 'weight-losing' raw materials (iron ore and many other minerals) would be situated near the materials to avoid the transport of waste, while industries using 'weight-gaining' materials would be expected to develop near the market. Bread making provides an example here, for although flour is bulky it gains considerable weight when water is added. Such a theory must of course take account of the value of materials or finished product in relation to its weight or bulk. It is important for pressed car bodies to be produced near the car assembly plant since they are difficult and relatively expensive to transport. It matters little, however, whether printed electrical circuits are produced in Basildon or Bolton for delivery to London, and the price of Swiss watches is not much affected by transport costs to the United States of America.

While transport costs may exercise an important influence on location when transport is inefficient and expensive, improvements in communications reduce the emphasis attached to this factor, and a large proportion of industry in the United Kingdom may be regarded as *foot-loose* and free to settle almost anywhere in the country without materially affecting costs. This is partly due to the compact nature of the British economy and its heavy dependence on imported raw materials. In the United States with its much greater area and natural resources transport costs are likely to be of greater significance.

(f) Industrial inertia

Although the original advantages of particular locations have disappeared, new firms or plants tend to become established in traditional areas. This is largely due to the availability of external economies of scale such as labour, training facilities, maintenance services, and marketing institutions. Such a tendency is called *industrial inertia.*

Once a firm has become established in an area it will be reluctant to move elsewhere when expansion is necessary because of the problems of co-ordinating the activities of different plants and of persuading key personnel to move from one area to another.

There are, therefore, many factors influencing the change of location though it is unlikely that most actual location decisions take all those factors into account. Indeed many locations may be haphazard in that a factory may be established in an area simply because its founder lives there; or a site may become available in one area before any others. There is one factor, however, which no firm can ignore nowadays when making its location decision, and that is Government policy.

9.3 The Need for Government Intervention

The British Government was forced to take a hand in the location of industry by the economic crises of the 1930s, although it was not until the end of the Second World War that any vigorous action occurred. In the thirties unemployment was high throughout the country and averaged over 10 per cent throughout the inter-war period, but in some areas the severity of the problem was much greater and unemployment reached 50 per cent in some places. The reason for this was that, in addition to the general unemployment of the period, some industries were experiencing their own special difficulties.

(a) Coalmining

The peak demand for coal occurred just before the First World War, and after this the use of oil for firing ships, and electricity for driving industrial machinery led to a contraction of, and extensive unemployment in, the coal industry.

(b) Textiles

Britain's traditional overseas markets were either producing their own textiles or importing them from India rather than from Lancashire.

(c) Shipbuilding

The general slump in world trade combined with a surplus of shipping reduced the demand for new vessels to a minimum. In addition the industry was slow to respond to the need for tankers. The decline in demand for new ships had adverse affects on the iron and steel industry and heavy engineering in general.

Unemployment of this kind, where there is a permanent fall in the demand for a product, is called *structural unemployment*, and in a developing economy this type of unemployment will be a continuous and not very serious feature, as some industries decline and others expand to absorb the unemployed. However, when the industries suffering from structural unemployment are situated close to one another and when there are no expanding industries nearby the result is often heavy *regional unemployment*. This was the case in the 1930s when the declining staple industries all shared coalfield locations. Lancashire, the North East, North West, Clydeside and South Wales all suffered severely when the demand for their staple products fell. At the same time the new locational factors, outlined in Section 9.2, were attracting light consumer-orientated industry to the South East and the Midlands. Government intervention was seen at first solely in terms of eliminating the high levels of regional unemployment. Large differences in employment levels are usually accompanied, however, by equally important social differences; health, education and housing, for example, all tend to vary with the level of economic activity and the case for Government intervention rests on the need to eliminate these social differences by restoring a proper level of economic activity.

9.4 Types of Intervention

One way to deal with regional unemployment is to encourage or compel the migration of labour to the areas offering work. This is not a popular policy in this country and the alternative method of moving work to the areas of high unemployment has been adopted. Four aspects of this policy may be identified:

(a) Incentives

Successive Governments have offered a variety of financial inducements to industry in attempts to reduce the level of unemployment in the regions. The legislation involved may be summarized as follows:

(i) **The Special Areas Acts** of 1934 and 1937 represented a largely unsuccessful attempt to bring stability to the areas of high unemployment by the introduction of a variety of light industry. Although little was achieved by this it is important as the beginning of a policy of *industrial diversification* designed to ensure that the areas concerned should never again be dependent on a few basic industries.

(ii) **The Distribution of Industry Acts** of 1945 and 1950 established Development Areas to replace the Special Areas. In these areas, centred on South Wales, Wrexham, Merseyside, South Lancashire, North Lancashire, West Cumberland, the North East Coast, the Scottish Lowlands and the Scottish Highlands, the Board of Trade provided factories at very low rents and special Treasury assistance was available for firms establishing plants there.

(iii) **The Local Employment Act** of 1960 was a piece of consolidating legislation seeking to channel Government help in the directions in which it was most urgently needed.

(iv) In 1966 the Government undertook to meet 40 per cent of investment costs in the Development Areas, but only 20 per cent outside.

(v) Tax privileges have been granted to firms in the Development Areas to encourage both investment and the creation of jobs. (See, for example, Section 29.11 which discusses Selective Employment Tax.)

(b) Physical controls

As a result of the Town and Country Planning Act of 1947 all factory building in excess of 5000 square feet needs Government approval in the form of an *Industrial Development Certificate* from the Department of Trade and Industry (originally the Board of Trade). It is therefore possible for the Government to exert a negative control over location. The establishment of the Location of Offices Bureau in 1963 represented an attempt to move offices out of London and to reduce (or stop the increase of) the pressure on road and rail services during rush hours. Although it was reinforced by the Office Development Act of 1965 the success of the policy has been somewhat limited, owing to a kind of 'office inertia'.

(c) Public investment and Government expenditure
Apart from its expenditure on the financial inducements outlined above, the Government has two ways of increasing the demand for goods and services in the Development Areas: it can be made public policy to divert Government orders for goods and equipment to those areas (providing prices are competitive); or pressure can be exerted to increase the expenditure of local authorities or nationalized industries in those areas. As the level of unemployment rose in 1971 the Government announced many schemes designed to create employment in the Development Areas and at the same time improve their amenities and social capital. Prior to this the Government decentralized some of its own services, the Post Office Giro, Inland Revenue Computer Centres, and other activities being taken away from the South East.

(d) Economic planning regions
Until 1965 regional economic problems were seen largely in terms of the Development Areas, with little attention being given to prosperous regions. In 1965, however, eleven Regional Economic Planning Councils were established, each responsible for devising a strategy for the future economic development of its region and advising the Government on its regional requirements.

9.5 Effects of Government Policy

In Table 9.1 we compare the levels of unemployment in the *New Standard Regions* of the United Kingdom and recent changes in them. (For the purposes of collation of statistics the country is divided into a number of Regions. Those used before 1966 were referred to as Standard Regions; those resulting from the boundary changes introduced in that year are the New Standard Regions.)

The South East had the lowest percentage unemployed in January 1971 and the increase that occurred in the following 12 months was below average, and much lower than that which occurred in the regions corresponding to the Development Areas thus indicating a widening of the gap despite Government efforts. In the year from January 1972 substantial reductions were achieved in all regions, but few of them had returned to the 1971 level.

Table 9.1 Regional changes in unemployment

New standard region	January 1971 Number (000's)	%	January 1972 Number (000's)	%	January 1973 Number (000's)	%
North Yorks	67·6	5·1	91·8	6·9	79·1	6·0
Humberside	67·3	3·3	97·1	4·8	75·4	3·8
East Midlands	37·8	2·7	51·6	3·6	36·6	2·8
East Anglia	18·4	2·9	23·6	3·6	16·8	2·6
South East	147·3	1·9	187·4	2·4	151·5	2·0
South West	45·0	3·4	56·9	4·2	45·4	3·4
West Midlands	56·9	2·5	112·5	5·0	68·1	3·0
North West	93·1	3·1	146·1	5·0	132·5	4·6
Wales	42·3	4·3	56·3	5·8	47·8	4·9
Scotland	115·1	5·3	154·4	7·1	129·8	6·1
Northern Ireland	40·9	7·9	46·0	8·9	38·8	7·5
Great Britain	669·0	3·0	997·6	4·3	705·0	3·1

9.6 Some Implications of Intervention

The effects of Government intervention in the location of industry may be considered from the point of view of the firm or of society as a whole. If it is known that firms locate new plant on a rational and profit-maximizing basis (which we have seen is somewhat doubtful), any interference with their decisions can be expected to increase their costs and reduce profits. Before accepting such an argument we would need to know a great deal more about the extra transport costs, the value of external economies of scale in the optimum site, the capital savings made by locating in a development area and many other factors. The greatest obstacle to a firm thinking of establishing new factories in a strange area is frequently the fear rather than the knowledge that there will be extra problems and costs.

If we accept the argument that persuading firms into development areas increases costs and reduces profits and leads to an inefficient use of resources, we must weigh these disadvantages against any advantages that might accrue to society in general from the move. The following points are important in assessing the implications of Government policy in this field.

(a) Private costs and social costs

Left to itself and market forces, the bulk of industry tends to locate itself in the Midlands and South East. It thus causes very heavy congestion in those regions, exerting great pressure on social capital, and imposing heavy costs on local ratepayers who have to finance local facilities to a large extent. In other words, the firms are subsidized by the community, for the market mechanism is not sufficiently sophisticated to allocate to the firm all the costs imposed by its existence. It is the divergence between private costs and social costs that is the basis of Government intervention.

The reverse of this particular coin is that areas of former prosperity become under-utilized, social capital goes to waste and is not replaced.

(b) Immobility of labour

Few people are eager to tear up their roots and move from one end of the country to another in search of work, owing to the domestic upheaval involved. The people who do leave are likely to be the younger, more vigorous, more enterprising members of the community whose departure renders the area even less desirable as a site for new enterprise.

(c) Economic growth

There are occasions when the South East and Midlands have more or less completely full employment while elsewhere there is considerable unemployment. In these cases the best way for the economy to produce more goods is to take up the slack in the development areas even if this does require Government pressure to force firms into particular locations.

(d) Social benefits

The gains to the community will be much greater than the value of goods and services produced in the new factories. People working in those factories will have far more money to spend than when they were dependent on Social Security benefits, and their expenditure will create further employment for other people. A successful regional policy will transform development areas so that eventually firms will not need special incentives to persuade them to locate there.

(e) Investment or employment incentives

One of the dilemmas clearly exposed by the policy of giving indiscriminate capital grants from 1966 to 1970 was the clash between investment and jobs. If the Government gives generous assistance to investment it may well find that it introduces capital-intensive industry to the regions, but does not create many jobs. Such industries are likely to be the most viable in the long run. An alternative approach is for the Government to subsidize the creation of jobs, but this may encourage relatively inefficient industry, not having long-term viability.

(f) Avoiding a repetition

The problems of the Development Areas arose from the concentration of basic industries on the coalfields and their simultaneous decline. This really has two implications for Government policy. A balanced pattern of industry must be established, and in particular industries with long-term prospects must be introduced. The Government must also notice and assist the contraction of declining industries. In the latter case it has had some success, for the problems of the Development Areas would have been much greater without Government assistance to the important declining industries, coal, textiles and shipbuilding.

9.7 Some Examples

(a) Iron and steel

The iron and steel industry in South Wales provides an example of the benefits of changing location, even though the industry is one normally exhibiting a high degree of industrial inertia, owing to the heavy capital investment involved. The nineteenth-century industry was located at the heads of the valleys where coal and iron ore were found together. The local ore became exhausted and the industry became dependent on supplies from North America and Africa. There was luckily a plentiful supply of the land necessary for today's vast plants so a gradual shift was made in the balance of location to the coast.

(b) Motor cars

The industry is based primarily on the London and Birmingham areas. Although the Midland plants such as those at Coventry have the considerable advantage of a central position and proximity to the steel producing areas, Dagenham was deliberately chosen by Fords as the centre of their British operations in 1925. The Thames offered a quick and easy export route to Europe; London County Council was in the process of turning the village of Dagenham into a vast housing estate which could supply the labour Ford needed, and the Essex marshes provided plenty of open, flat land, although the construction difficulties involved may be estimated from the fact that the plant was built on 22,000 concrete piles averaging forty feet in length!

In recent years the motor industry has spread further afield, establishing plants in South Wales, Merseyside and Scotland and taking advantage of Government incentives in those areas, but not without adverse effects. The Chrysler Company, for example, were persuaded by the Government to locate new plant at Linwood in Scotland, rather than expand their existing plant at Coventry where they had ample space. As a result, engines are built at Coventry, sent to Linwood, embodied in vehicles and then sent South again for sale. The company calculated that this would eventually involve them in an annual extra cost of £3 million for which the Government offered no compensation.

9.8 Questions and Exercises

1. Outline the main influences on industrial location, showing the extent to which they have changed in the twentieth century.
2. Consider the importance of transport costs to the entrepreneur hoping to build new plant.
3. 'The view that the Government intervenes in the location of industry to eliminate regional unemployment is too narrow.' Comment.
4. What are the main aims of the Government's regional policies?

5. 'Since firms need to be induced to move to the Development Areas it must be uneconomic for them to go there.' Discuss.

6. Examine the concepts of social costs and benefits in terms of the Government's regional economic policies.

7. Compare the various policies available to the Government in its attempts to eliminate regional unemployment.

8. What factors induce firms to locate in the South East of England?

9. Consider the advantages and disadvantages of the Government's policy for location of industry from the point of view of (a) the nation, (b) the firms involved.

10. Consider the likely effects of leaving industrial location to the discretion of individual firms, without Government interference.

Unit Ten

The Costs of Production

10.1 Introduction

In Units Five to Nine we have examined some of the more important aspects of the organization of production. At many points in the discussion we have mentioned the costs of production, and in this Unit we undertake a more formal analysis of those costs and the way in which they vary with output. After this we shall consider the output of the individual firm, and the supply of goods to the market.

10.2 Factors and Costs

We have already seen that production is the process by which raw materials are converted by a combination of the factors of production into a form acceptable to consumers. It will only rarely be necessary to employ factors of production in fixed proportions to secure a given output. For example, a given volume of wheat may be produced by *extensive* methods on the prairies of Canada or by *intensive* methods on the wheat farms of East Anglia, the proportion of land to the other factors of production being higher in the former case than in the latter. The proportions in which factors are combined will normally reflect their relative costs; if land is plentiful and therefore cheap extensive agriculture will prevail but if there is a shortage of land and its price or rent is high then more intensive cultivation will develop. The same principle applies to manufacturing industries, and also to services. As labour becomes relatively more expensive employers tend to substitute other, relatively cheaper factors for it.

However in any productive process at any one time there will be some factors of production that cannot easily be increased, while others may be varied on a day to day basis. In the refining of crude oil the amount of capital equipment employed is fixed over a fairly long period. (It is true that it can be reduced merely if the refining company refuses to use it but even in this unlikely event the cost of the equipment is still borne by the firm). An increase in capacity of 20 per cent by the installation of new equipment may take anything up to three years to achieve once the design has left the drawing board, and the time since the original decision will be even longer. Thus capital is regarded as a *fixed* factor of production. The fixed factors of production are those which cannot be changed in the *short run*. No precise time span can be allocated to the short run since this varies greatly between industries. It may be a number of years in an industry such as oil refining or the generation of electricity, but only a matter of weeks or even days in a car-hire business where the main units of capital are easily acquired. In this case, however, the fixed factor of production is more likely to be the land required for offices, garages, and maintenance services. In most industries the factor of production most easily altered in quantity is labour and we may regard this as a *variable* factor of production. Other important variable factors will be raw materials and fuel supplies which can usually be increased or decreased quite smoothly. There will, however, be degrees of variability: while some labour is hourly paid and perhaps subject to a week's notice at the most, some monthly paid staff may have contracts which entitle them to three months notice, or, in the case of some important employees, even longer. Thus managerial labour is somewhat less variable than manual labour.

In the long run all factors are variable, so that an industry may secure an increase in output by employing more units of each factor in their existing proportions or by taking on more of the variable factors and using the fixed factors more intensively.

If we categorize factors of production as fixed or variable we may also divide *costs* in the same way. (In this context we shall consider costs in the accounting sense of money spent on the various factors of production rather than in the broader economic sense discussed in Section 1.4.)

Fixed costs are those which do not vary directly with output and for this reason they are sometimes known as *indirect costs*. These costs have to be met whether or not the firm produces any goods. They include annual rent payable for land, interest on capital and the amortization of capital debts, standing charges for electrical equipment which need notice of cancellation, and the salaries of many employees which must be paid irrespective of output.

Variable or *direct costs* are those which do vary directly with output. If a firm is not producing anything it can quite easily reduce its expenditure on power, raw materials, and direct labour. If the break in production is prolonged then other costs initially regarded as fixed will become variable; the salaries of monthly paid staff provide the most obvious example. The distinction between fixed and variable costs is an important one to which we return in Section 10.5. First we must consider what happens when factors of production are combined together.

10.3 The Combination of Factors of Production in the Short Run

During the Napoleonic wars, British farmers were forced to cultivate as much land as possible to meet food shortages arising from the disruption of imports. It soon became apparent that the land brought into cultivation was less useful than the land already in use and a 10 per cent increase in land did not produce a 10 per cent rise in output. This is an example of diminishing returns, but it arises from an increased use of all factors of production and the decreasing suitability of one of them (see Section 10.7). The employment of the unsuitable land was itself necessary only because of diminishing returns resulting from the more intensive use of land already under cultivation. It was just not possible to meet the gap in supplies by a more intensive application of labour and capital to existing wheatland. The realization of this and further study led to a formal statement of the *principle of diminishing returns*: as increasing amounts of a variable factor of production are applied to fixed amounts of other factors of production, marginal product and average product at first increase but eventually decline.

This must be examined more carefully, but first certain terms must be defined.

(a) The average product
This is the total output divided by the number of units of the variable factor. If the variable factor is labour and a farmer employs ten men to produce 2000 kilos of wheat, the average product is obviously 200 kilos.

(b) The marginal product
This is the increase in total product resulting from the employment of an extra unit of the variable factor. If the above farmer employs a further eleventh man and output rises to 2100 kilos then the marginal product is 100 kilos. The concept of the margin is fundamental to a study of the theory of production, and it is essential that its characteristics at this stage should be clearly understood.

(i) As we have seen above marginal product is the amount to be added to total product from the efforts of one more man.

(ii) The sum of the marginal products will give the total product. If no one is employed output will be nil; if the employment of one man increases the output to ten, both the total product and the marginal product are ten; and if the employment of a second man increases total product by a further fifteen then total product rises to twenty-five which is the sum of the marginal products. If the marginal product were negative then total product would be reduced.

(iii) Marginal product exercises an important influence over average product. If the marginal product is greater than the previous average product, the average product rises, and if the marginal product is less than the previous average, the average product falls. It may be helpful to think of this in terms of a batsman's average. If he has had five innings and scored 150 runs he has an average of 30 runs (his average product); if on his next visit to the crease he scores 90 runs (his marginal product) his average rises to 40; if on the other hand he scores only 24 then his average falls to 29.

Let us now return to the principle of diminishing returns and imagine a farmer with a fixed amount of land experimenting to discover the effects of employing different amounts of labour for wheat production.

Table 10.1 Application of successive units of labour to a fixed amount of land

(1) No. of men	(2) Total product (kg)	(3) Marginal product (kg)	(4) Average product (kg)	(5) Fixed costs	(6) Variable costs	(7) Total costs	(8) Marginal costs	(9) Average fixed costs	(10) Average variable costs	(11) Average cost
1	5	5	5·0	50	10	60	2·00	10·00	2·00	12·00
2	12	7	6·0	50	20	70	1·45	4·16	1·67	5·83
3	23	11	7·6	50	30	80	0·90	2·17	1·30	3·47
4	36	13	9·0	50	40	90	0·76	1·39	1·11	2·50
5	48	12	9·6	50	50	100	0·83	1·04	1·04	2·08
6	58	10	9·7	50	60	110	1·00	0·86	1·03	1·89
7	65	7	9·3	50	70	120	1·42	0·77	1·07	1·84
8	68	3	8·5	50	80	130	3·33	0·73	1·17	1·90
9	70	2	7·7	50	90	140	5·00	0·71	1·29	2·00
10	71	1	7·1	50	100	150	10·00	0·71	1·40	2·11
11	71	0	6·4	50	110	160	—	0·71	1·54	2·25

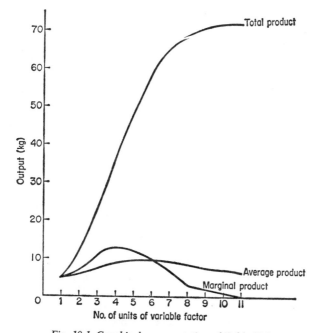

Fig. 10.1 Graphical representation of Table 10.1

When one man is employed total, marginal and average product are the same (assuming that nothing is produced with no employees), but as employment is increased the three quantities diverge. The least important from our point of view is total product, but we may notice that up to the employment of the tenth man total product rises but at a decreasing rate after the employment of the fifth man. This is what is meant by diminishing returns. There is an increase in output from successive applications of the same amount of a factor, but the increase eventually becomes smaller. If this were not so economic problems would not exist for it would be possible to produce the entire world's food requirements in one field!

Another way of expressing the growing difficulty in increasing total output is to say that marginal product eventually falls, marginal product being the increase in total product. We notice that when employment is doubled from one man to two men the marginal product rises from 5 kg to 7 kg. This is a case of increasing returns as a 100 per cent rise in the input of labour results in a more than proportionate rise in the total output—from 5 kg to 12 kg.

We must not make the mistake of thinking that the second worker is more productive than the first; the marginal product of 7 kg is partly attributable to the first worker as with two men working some specialization is possible and the fixed factor can be more fully exploited. Similar considerations arise from the employment of the third and fourth men: division of labour promotes greater efficiency in the use of the land.

We may conclude that the marginal product is not the output of the marginal unit of the factor, but the rise in total product obtained from the employment of the marginal unit of the factor and attributable to all units of the factor. When the fifth man is employed the marginal product declines and this shows that the benefits of division of labour are becoming exhausted. The employment of a tenth man gives an increase in output of only one kilo, not because the tenth unit of labour is any less hardworking than the previous nine, but because no amount of extra effort can squeeze further output from the fixed amount of land. On a more fertile piece of land marginal product might still be rising with ten employees, but diminishing returns would eventually set in.

We can see that average product does not rise as rapidly nor as far as marginal product. The reason is clear if we consider the rise in employment from one man to two. The extra man increases the total output to 12 kg; and the marginal product concept allocates the whole of the increase of 7 kg to the second unit even though we know that it is partly attributable to the first unit; when we measure average product we divide the total product equally between all units of the variable factor so that average product rises less rapidly than marginal product. When marginal product itself is less than average product, average product declines less rapidly than marginal product since the effects are again spread thinly over all the units of the variable factor rather than concentrated on the marginal unit.

The relationship between marginal product and average product is such that they are equal when average product is at its maximum. This follows from our discussion of average and marginal concepts in Section 10.3. So long as marginal product is greater than average product, the latter must be rising even if marginal product is falling. When marginal product is less than average product then the latter must be falling. When average product is neither rising nor falling, and is at its maximum, then it must be equal to marginal product.

The principle of diminishing returns does not apply only to agriculture. One may imagine the production line in a motor factory designed for a workforce of perhaps 300 men. If the owner tried to operate with only one employee, the total, marginal, and average products will be negligible. As the workforce is increased output will rise and average product may be expected to reach its maximum when 300 men are employed. Further increments of labour lead to a rise in total output but a fall in average and marginal product, until the production line is saturated with labour and the employment of perhaps the 400th man results in a marginal product of zero or even a fall in total product. Similarly the employment of three drivers each working an eight-hour shift will enable a lorry to be worked for 24 hours a day. The employment of a fourth driver will not increase total output (measured by journeys or mileage) very significantly. The point is that the fixed factor eventually becomes saturated with the variable factor. When this happens very quickly we say that diminishing returns set in early; when it takes much longer we may say that diminishing returns are delayed.

10.4 The Importance of Marginal Product

In view of the inevitability of diminishing returns we must discover the number of units of the variable factor to be employed. The answer depends upon the aims of the employer; if we assume that his objective is the maximization of production he will employ labour until the marginal product falls to zero. To proceed beyond this point would be to invite a negative marginal product which implies a fall in total product. If we assume that the employer's aim is to maximize his profits he will employ extra labour so long as it adds more to total revenue than it does to total costs. Profit maximization is the assumption underlying the theory of production and is more closely examined in Unit Fourteen. Here we may consider its implications in terms of Fig. 10.2.

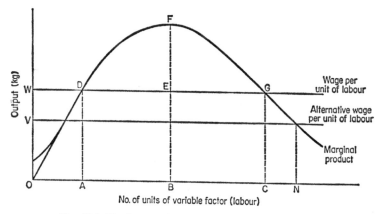

Fig. 10.2 The limit to the employment of a variable factor

Here we assume that extra units of labour are available for a wage of OW and that the employer has no other costs. If OA units are employed the total costs will therefore be OWDA (wages × number employed), but the total product is only ODA (the sum of the marginal products). If this is not clear, then re-read Section 10.3. There is a loss of OWD. When further jobs are offered the marginal product rises by more than the additional cost of employing an increment of labour. For example, when OB men are employed the wage of the marginal man is BE (=OW) but the marginal product is BF. Until OC men are employed the increase in total product (shown by the marginal product curve) is greater than the rise in the wage bill shown by OW, and there is a surplus of DFG on employing the units of labour A to C. (The total product derived from employing these men is ADFGC, while total wages are ADGC.) If there are more than OC men profits will be reduced, since the addition to output will be less than the wages payable to extra units of labour. It would, therefore, be inconsistent with the aim of profit maximization to employ more than OC men.

We may conclude that profits are maximized when the addition to total output falls to the point where it only just equals the cost of the last person, and in our example that profit is measured by the area DFG − OWD. Although we may regard OC as a profit-maximizing level of employment there may be circumstances in which it is a loss-minimizing situation, since the area OWD may be greater than DFG. It will be the best position for the firm if the level of wages and the productivity of workers are fixed. If wages change then a different amount of labour will be employed. At wage OV for example, it is worth employing ON units of labour, but if the wage rate rises above OW fewer people will be employed.

10.5 The Cost Curves of the Firm

The discussion so far has been conducted in terms of units of output and units of labour, but each firm will need precise estimates of its costs and revenues in monetary terms. In Table 10.1 the last seven columns deal with costs. We have assumed quite arbitrarily that the firm has to meet fixed costs of £50 per productive period and variable costs of £10 per unit of labour employed. Thus we can easily arrive at a figure for the total cost of producing any particular output. For example, 70 kg can be produced at a cost of £140 which can be broken down further into £50 fixed costs and £90 variable costs.

In Fig. 10.1 we isolated average product and marginal product, but we must now try to identify *average cost* and *marginal cost*.

Average cost is found by the formula:

$$AC = \frac{\text{Total costs}}{\text{Output}}$$

When output is 70 kg the average cost is £2 per kilo. Marginal cost strictly speaking, must be regarded as the cost of producing one more unit. In practice it is normally impossible to employ labour in sufficiently small

amounts to produce only one unit and then dismiss the workers. We therefore define marginal cost as:

$$MC = \frac{\text{Increase in total costs}}{\text{Increase in output}}$$

Consider an increase in the number employed from three to four. This
involves £10 in extra variable costs and a rise in output of 13 kgs.

$$MC = \frac{10}{13} = £0.77$$

It is also possible to calculate average fixed costs and average variable costs.
Average fixed costs are found by the formula:

$$AFC = \frac{\text{Total fixed costs}}{\text{Output}}$$

and average variable costs by the formula:

$$AVC = \frac{\text{Total variable costs}}{\text{Output}}$$

In Fig. 10.3 the relationship between the various curves is shown diagrammatically.

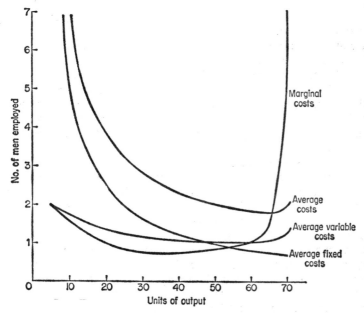

Fig. 10.3 The cost curves of the firm

Note especially the following points.

(*a*) **Average fixed costs.** As output rises the average fixed cost curve falls
since the fixed costs are spread more thinly over a large output. The AFC
curve is the same shape as a demand curve with an elasticity of one over its

entire length. All rectangles drawn from the curve to the axes will be equal in area.

(b) The marginal cost curve is U-shaped due to the operation of the Law of Diminishing Returns. When a second man is employed costs rise by £10 and output by 7 kg; the employment of a third man increases costs by a further £10 but gives rise to 11 kg of extra output. Thus marginal cost falls from $\frac{10}{7}$ to $\frac{10}{11}$. This reflects the more efficient deployment of labour and the more economic use of the fixed factor as more people are employed. The employment of a fourth person gives a further fall in marginal costs, but when the fifth employee is taken on there is a fall in the marginal product which causes the marginal cost curve to rise. The curve continues to rise until the marginal product falls to zero when the eleventh man is employed.

(c) The average cost curve is also U-shaped. The fall in its early stages is due partly to influences discussed in paragraph (b) which result in average variable costs falling. The fall is also partly due to the fact that average fixed costs fall as output expands. Average cost continues to fall after marginal cost begins to rise. The relationship between these two is exactly parallel to that between average product and marginal product discussed in Section 10.3, and the average cost curve is cut by the marginal cost curve when average cost is at a minimum. When average cost does begin to rise it is because the rise in average variable costs more than offsets the continuing fall in average fixed costs.

(d) It is essential to distinguish between marginal cost and average variable cost. Marginal cost relates an *increase* in output to an *increase* in variable costs, whereas average variable costs relates *total* output to *total* variable costs. Therefore, when we measure average variable cost we are interested in all the production that has taken place, while measurements of marginal costs relate only to increments of output.

(e) The cost curves and profit. An important element has been left out of our calculation of costs. The entrepreneur will not remain in business unless he is making a profit, and in the study of economics profits must be seen as a cost of production which must be met in the same way as wages. The amount of profit necessary to keep a firm in a given industry cannot be estimated precisely, but it is generally called a *normal rate of profit*. Anything less than a normal rate of profit will induce firms to leave the industry as equipment wears out. If profits are higher than normal, new firms will be attracted in and existing firms will be tempted to expand. Profit was omitted from Table 10.1 partly because of the uncertainty regarding the appropriate level and partly due to the difficulty of allocating it to fixed costs or variable costs. This is a matter of some debate. However the generalized cost curves that will be used subsequently must be regarded as incorporating a normal rate of profit.

The curves drawn in Fig. 10.3 are derived from the hypothetical figures in Table 10.1. It will be more convenient in our subsequent discussion to remove the kinks in the curves and to concentrate attention on two curves only, the average cost curve and the marginal cost curve. Of these the latter is by far the most important in fixing the output of the firm.

10.6 The Firm's Supply Curve

The firm's cost curves show how much it costs on average to produce a given output (average cost curve), or how much total costs increase when output is increased. The actual amount that the firm produces will depend on changes in total costs and changes in total revenue. We shall consider the case in Fig. 10.4.

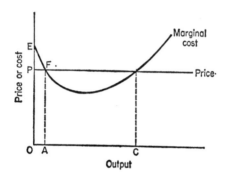

Fig. 10.4 The output of the individual firm

Here we are assuming that the price is fixed at OP and the firm (a profit maximizer) can sell as much as it wishes at that price. In these circumstances the firm knows that its marginal revenue is always constant. *Marginal revenue* is the increase in total revenue resulting from the sale of an extra unit of product.

In the light of our discussion in Section 10.4 it is a straightforward matter to establish the most profitable output, remembering that the cost curves include a normal rate of profit. If the entrepreneur decided to produce OA units his total variable costs would exceed his total revenue by PEF (OEFA − OPFA). On all the units of output between OA and OC the addition to total revenue (marginal revenue) is greater than the addition to total costs (marginal costs) and so the increase in output is profitable. The expansion of output beyond OC, however, pushes marginal cost (MC) above marginal revenue and therefore reduces profits. OC is thus the profit-maximizing output. Fig. 10.5 shows what will happen if price changes.

Let us suppose that the original price is OP. By equating MC with price (which is the same as MR) at *a*, the firm fixes output at OQ and maximizes profits. Now if price rises to OR, OQ is no longer the profit-maximizing output since MR is greater than MC at that price. It is therefore worth expanding output to OS, thereby adding more to total revenue than to total costs and attaining another profit-maximizing position at *b*. A further

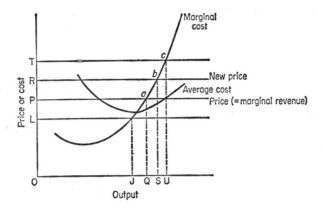

Fig. 10.5 The firm's supply curve

rise in price to OT gives yet another profit-maximizing output, OU, while a fall to OL gives an output of OJ.

In each case as the price changes the firm moves along the marginal cost curve in order to determine its output. We may conclude, therefore, that the *marginal cost curve is also the supply curve of this particular firm.* The supply curve shows the relationship between the quantity supplied and the market price. It shows the price that is necessary to induce the firm to put a particular quantity of its product on the market; or, looking at this a different way, it shows the amount that will be produced at any given price.

To be more precise we may say that the part of the marginal cost curve which is above the average cost curve forms the supply curve. If price fell to OL in the figure, the firm's best output would be OJ, but in this case average costs (including a normal rate of profit) are higher than the price received and the firm cannot afford to stay in business for very long. However, having established that the firm's marginal cost curve is also its supply curve we must defer a further discussion of supply to Unit Eleven.

10.7 Short-Run and Long-Run Costs

The cost curves discussed in the preceding sections are short-run cost curves which have been drawn up on the basis that at least one factor of production is fixed in supply. In this case the only way to increase output is by employing more units of the variable factor so that the fixed factor is used more intensively. This will be quite adequate to meet temporary fluctuations in demand. If the firm is faced by a permanent increase in demand, however, it may decide to increase its employment of all factors of production. If the employer does this then he may choose between a variety of short-run cost curves. We shall consider the possibilities shown in Fig. 10.6.

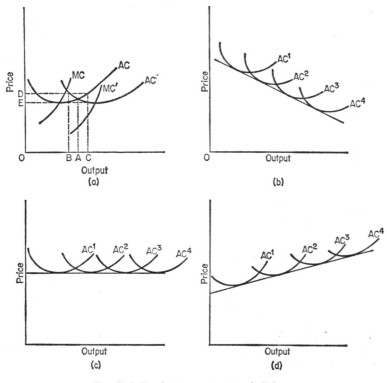

Fig. 10.6 Employing extra units of all factors

In Fig. 10.6 (a) we have a firm which has doubled its use of a factor that is fixed in the short run. (It may have installed a second production line.) We have therefore drawn two average cost curves and the two associated marginal cost curves. Up to an output of OA units average cost is lower if the capacity of the firm is restricted. It is wasteful to install the second production line to meet a demand of OB because a very large amount of *excess capacity* would remain. We use the phrase excess capacity to indicate a situation in which a firm is operating below the optimum output; the output where average cost is at a minimum. The output OC is more efficiently produced by installing new equipment than by trying to squeeze extra output from existing equipment. The average cost is OD using the original capacity, but only OE if extra equipment is brought into use. The installation of extra capacity cannot normally be achieved overnight: it may take months or years. It is, therefore, customary to regard the two combined short-run average cost curves as giving the long-run average cost curve AC′ showing costs when both old and new equipment are in use. Figs. 10.6 (b, c, d) indicate the possible trends of the long-run average cost curve. In Fig. 10.6 (c) as we increase capacity the optimum output increases but the average cost of that output is constant; this is a case of constant costs or constant returns to scale. In Fig. 10.6 (b) as optimum output increases, the cost falls, and this is a case of decreasing costs or increasing returns to scale, while in Fig. 10.6 (d) the rising trend indicates increasing costs, and would represent the case of diminishing returns in agriculture as capacity was expanded during the Napoleonic Wars.

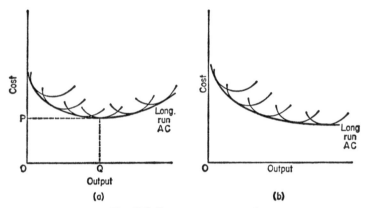

Fig. 10.7 Long-run average costs

The shape of the long-run average cost curve will depend very largely on the availability of technical and managerial economies of scale, and the rapidity with which diseconomies associated with the scarcity of labour or raw materials, or managerial deficiencies set in. In Fig. 10.7 (a) we find what many economists regard as a typical long-run average cost curve with falling costs in the early stages of expansion down to a long-run optimum level of OP, output OQ, then perhaps a period of constant costs followed by rising costs as diseconomies and bottlenecks in production offset earlier economies.

Others would argue that Fig. 10.7 (b) is more typical of real world situations, showing an L-shaped long-run average cost curve. In industries such as motor manufacturing and oil refining technological progress is so rapid and persistent as to allow average costs (measured in constant prices) to show a considerable fall over a long period.

With this we must leave long-run costs and move on to Unit Eleven and a further consideration of the short-run supply curve. Much more will be said about production in Units Fourteen and Fifteen.

10.8 Questions and Exercises

1. Distinguish between the following: fixed costs, variable costs, social costs, and opportunity costs.
2. Define and explain the following: average product, marginal product, average cost, and marginal costs.

3. Outline the operation of the law of diminishing returns, showing its importance to the cost curves of the firm.
4. Distinguish between the views of costs held by an accountant and an economist.
5. Explain briefly the relationship between total, average and marginal product.
6. Show how marginal product helps to determine how many units of a variable factor should be employed.
7. In what ways are marginal cost and average cost important to the firm?
8. Outline the relationship between short-run and long-run costs.
9. Explain with the help of diagrams the difference between increasing, decreasing and constant costs of production.

Unit Eleven

The Supply of Goods and Services

11.1 Introduction

In Unit Three we examined the demand for goods and services. The consumers who demand goods and services form one side of the market, while those people or enterprises who supply goods and services constitute the other side. We have seen that consumers try to maximize the utility they obtain from the goods they buy. We must now consider the factors governing supply before we see how demand and supply are combined. As we took golf balls as our example in Unit Three we shall now examine the supply of golf balls.

11.2 The Conditions of Supply

We saw in Section 10.6 that under certain assumed conditions, that part of a firm's marginal cost above the average cost curve may be regarded as the firm's supply curve. It slopes upwards from left to right showing that a greater quantity will be supplied as the price rises. The supply curve shows the relationship between price and the quantity supplied, but this quantity depends upon the influence of things other than price. In order to isolate the effects of price on the quantity supplied we must hold the *conditions of supply* constant. The most important conditions are as follows.

(a) *The techniques of production are fixed.* The short-run cost curves drawn up in Unit Ten were based on one factor of production being fixed. The fixed factor is normally capital and if an important scientific or engineering breakthrough allows that capital to be used more effectively, the number of goods supplied at any particular price will increase. If the improvement in technique coincides with a change in price it will be impossible to disentangle the effects of the change in technique and the change in price.

(b) *The scale of production.* Even if techniques do not change a company may double the amount of machinery employed so the level of supply will change irrespective of changes in price. This will be true of the firm's supply curve. In the case of the industry as a whole the introduction of new firms or the departure of existing firms will have an effect on the quantity supplied, thus the scale of production must be held constant.

(c) *The cost of the factors of production.* The cost curves discussed earlier were established on the assumption that extra units of labour were available to the producer at a fixed price per unit. If the price of labour changes then the cost curves and accordingly the supply curve must be redrawn. Thus an increase in wages means that any given output will only be forthcoming in response to a higher price.

(d) *Government action affects supply and must be held constant.* If the Government imposes a tax on a product then we may expect a price increase.

(e) *The price of other commodities should be constant.* As the aim of the firm is to maximize profits (see Section 10.4) the amount of a commodity that it supplies will depend partly on the prices of other goods that it could otherwise supply. If the prices of other goods rise, indicating an increase in the profitability of selling them then the firm may transfer some of its resources to producing them, thus affecting the supply of the original product. We will now consider the supply curve of the individual firm taking these conditions of supply as constant.

11.3 The Individual Supply Curve

Fig. 11.1 emphasizes one of the most important findings of Unit Ten.

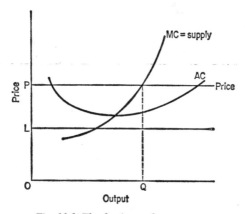

Fig. 11.1 The firm's supply curve again

When the price is fixed at, for example, OP the quantity supplied can be obtained from the marginal cost curve. If the price obtainable is at OL below the average cost of producing a given quantity then it will not be produced (except perhaps for a brief period). It is, therefore, the part of the marginal cost curve above AC (the heavily drawn section) that forms the supply curve.

Each firm will have its own supply schedule. Three possibilities are shown in Table 11.1, and the individual supply curves are drawn in Fig. 11.2.

Firm A is obviously an important supplier in the industry and it is prepared to supply a few balls even at very low prices. As prices rise, however, firm A can easily expand production to take advantage of the more profitable sales. We may infer that the price being paid is well above the average cost of production and therefore profits are high. Practical experience confirms the tendency of supply to rise as the price rises. During prolonged power crises the price of candles rises; for example, in the crisis of 1972 the price rose to such an extent that it was profitable to take the ferry to Europe, load a lorry with candles and bring them to the United Kingdom market. In normal times and at lower prices such a procedure is not worth while.

Table 11.1 Individual supply schedules for golf balls

Price (pence)	Quantity supplied per week by		
	Firm A	Firm B	Firm C
50	15,000	5,000	375
45	12,000	5,000	350
40	8,000	5,000	200
35	5,000	3,000	100
30	3,000	2,000	—
25	2,000	1,000	—
20	1,200	500	—
15	700	—	—
10	300	—	—
5	100	—	—

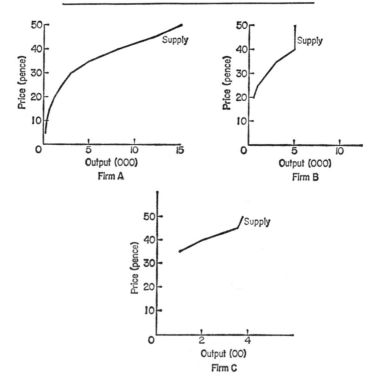

Fig. 11.2 Individual supply curves

In Table 11.1 we can see that firm B is a relatively high-cost producer, and finds it unprofitable to produce balls when the price is less than 20 pence. The capacity of the plant is well below that of A so whatever the price offered the supply cannot be more than 5000 balls per week with the existing equipment. (In the long run of course, the entrepreneur may buy more machinery, hire more labour and increase his supply.) Firm C is a much smaller concern altogether, perhaps a sole proprietor. Here costs are so high that the businessman will enter the market only when the price offered is 35p and later an increase in price from 45p to 50p can induce the production of only an extra 25 units so the plant is almost running at full capacity.

11.4 The Market Supply Curve

The number of firms in an industry may vary from one to several hundred, but however many there are the industry's supply curve may be calculated by adding the supply schedules of the individual firms. Thus if there were only three firms supplying golf balls we can find the quantity supplied at any price from Table 11.1. If the price is 25p then 3000 balls will be produced, while at a price of 50p 20,375 balls will be on offer. There are likely to be more than three firms supplying the market, and even if there are only three domestic manufacturers foreign producers are likely to compete in the market. When these factors are taken into account it is possible to construct the market supply schedule and curve as in Table 11.2 and Fig. 11.3.

Table 11.2 Market supply schedule for golf balls

Price	Quantity supplied per week (000's)
50	25
45	20
40	14
35	10
30	8
25	7
20	6
15	3
10	2
5	1

Remember that this is a short-run supply curve drawn up on the basis of the conditions of supply and deriving ultimately from the principle of diminishing returns. If the conditions of supply change we must reconsider the position of the supply curve. Let us first briefly consider the idea of producers' surplus.

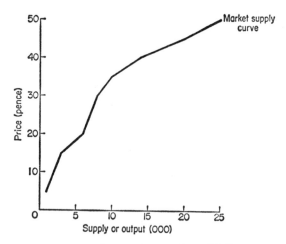

Fig. 11.3 Market supply curve for golf balls

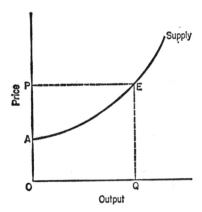

Fig. 11.4 Producers' surplus

11.5 Producers' Surplus

The supply curve in Fig. 11.4 shows that at a price fractionally above OA, one unit will be supplied, and it also shows the prices necessary to induce extra units of output. If producers received the amount of money required to induce the production of each unit of output their total revenue for an output OQ would be OAEQ. In actual fact they receive OPEQ by selling the whole output at OP. The difference between the two areas, APE is known as the producers' surplus and it is comparable with the concept of consumers' surplus discussed in Unit Three.

11.6 Changes in the Conditions of Supply

(a) The scale of operations

The size of an industry may increase as new firms enter or as existing firms expand and install more equipment thereby moving on to a new set of short-run cost curves like the firms in Fig. 10.6.

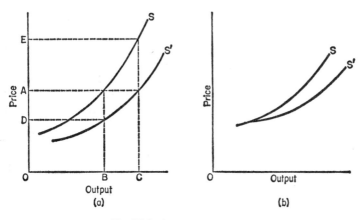

Fig. 11.5 An increase in supply

In either case more units are likely to be supplied at each price and the supply curve will, therefore, move to the right as in Fig. 11.5 (*a*). You will notice that the two supply curves are not parallel; there is no reason why they should be. If the increase in supply is attributable to the entry of new relatively high-cost firms the two supply curves might even merge at low prices where the new firms will produce nothing (Fig. 11.5 (*b*)). The move in Fig. 11.5 (*a*) may be interpreted in two ways. We may say that at price OA a greater quantity will be supplied than before; or that a quantity OB will now be supplied at a lower price than before (OD). If firms leave the industry or do not replace obsolete plant and thus reduce the scale of their operations then the supply curve will move to the left.

In order to be consistent with the terminology used in Unit Three we have referred to the move from OB to OC in Fig. 11.5 (*a*) as an *increase in supply* as it has been achieved by the firm shifting to a new supply curve. A supply of OC could also have been achieved if price had risen to OE with the capacity of the industry unchanged. Such a move gives rise to an *extension of supply*.

(b) Production techniques

Even if the size of an industry, as measured by the input of factors of production, does not change it may be possible to secure an increase in supply by reorganization or by using more efficient machinery. In such cases the supply curve will shift in the same way as in Fig. 11.5.

(c) The cost of factors of production

One of the most important influences on supply is the cost of the factors of production. If a trade union secures an increase in wages for its members, and the increase is not matched by extra output then its cost will almost certainly be embodied in the price of the product, causing the supply to move towards the left (from S' to S in terms of Fig. 11.5 (a)). The extent to which the curve moves towards the left depends on two things: the willingness of the producer to absorb the extra costs, perhaps by improving efficiency or by reducing profits; and on the size of labour costs as a proportion of total costs. An increase in wages in a capital-intensive industry, such as electricity generation, would not normally be expected to lead to a high increase in prices. But a similar percentage rise in salaries in an accountant's office where the cost of labour contributes much more to total costs would lead to a much larger increase in charges.

It is not only labour that can increase in price. The cost of raw materials may rise, shareholders may press for higher dividends, or the electricity board may charge more for power. In all these cases the supply curve of the industry will move to the left.

(d) Taxation

Another factor which has an important influence on supply is the taxation of goods and services. If the Government imposes a tax of 20p per unit on a commodity the producer will want to pass this on to the consumer. A quantity of goods previously supplied at £1·00 per unit will now be offered at £1·20. The whole supply curve will, therefore, move to the left by an amount representing 20p.

11.7 The Elasticity of Supply

The price elasticity of demand is a useful concept when we consider demand. In the same way it is convenient to measure the price elasticity of supply which measures the responsiveness of supply to a change in price:

$$\text{Elasticity of supply} = \frac{\% \text{ Change in quantity supplied}}{\% \text{ Change in price}}$$

If price falls from 25p to 24p and the quantity supplied falls from 100 units to 90, then

$$\text{Elasticity of supply} = \frac{10\%}{4\%} = 2\tfrac{1}{2}$$

In most cases the elasticity of supply will be positive since the changes in price and quantity are normally in the same direction. (Exceptional supply curves are discussed in Section 11.9.)

As with the elasticity of demand there are five broad categories of elasticity of supply each of which is illustrated in Fig. 11.6.

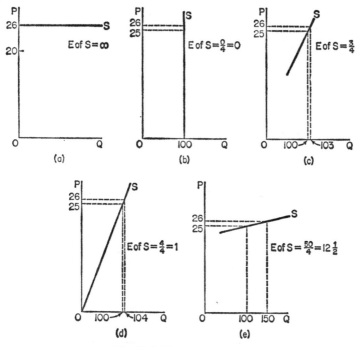

Fig. 11.6 Elasticity of supply

(a) *Elasticity of supply equals infinity.* In this case the industry is un-willing to supply anything at a price of 25p, but if the price rises to 26p then it is prepared to supply as much as anyone is prepared to buy. In this case an infinite change in quantity supplied is associated with a 4 per cent change in price. The elasticity equals infinity, and is an example of *perfect elasticity*. This sort of perfectly elastic supply is improbable in real life.

(b) *Elasticity of supply equals zero.* At the other extreme to infinite elasticity is zero elasticity where supply is perfectly inelastic. Here the increase in price has no effect on the quantity supplied. While neither of these two examples is likely to be found in practice they do provide limits towards which most real world supply curves will approximate. Example (a) is the extreme towards which manufacturing industry approximates, while example (b) is more representative of primary production where it is not so easy to transfer resources from one product to another.

(c) *Elasticity of supply is greater than zero but less than unity.* Here an increase in price of 4 per cent leads to an increase in supply of 3 per cent so that elasticity is ¾ or less than one. This is an example of an industry that has difficulty in increasing its output. Supply is *inelastic*.

(d) *Elasticity of supply equals one, a rather special case.* The price and the quantity supplied each rise by 4 per cent. In fact this curve has an elasticity of one along its entire length, as does any straight line passing through the origin. These are not the only circumstances in which the elasticity of supply will be equal to one. Indeed most supply curves, like most demand curves, will have an elasticity that varies considerably with prices and will include a range where elasticity equals one or is *unity*.

(*e*) *Elasticity of supply is greater than one but less than infinity.* In this case a 4 per cent rise in price induces a 50 per cent increase in output. This is an industry which can easily expand its output to meet an extra demand for its product. The supply is *elastic*.

11.8 The Determinants of Elasticity of Supply

The main determinants of the elasticity of supply include the following four main factors.

(*a*) The intensity of use of the fixed factor in the short run
We will consider the two firms, which are not necessarily in the same industry, shown in Fig. 11.7.

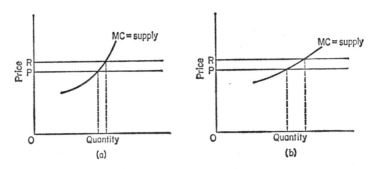

Fig. 11.7   Elasticity of supply and spare capacity

In Fig. 11.7 (*a*) an increase in price from OP to OR stimulates very little extra output because the plant (the fixed factor) is already being used very intensively and very little extra output can be squeezed from it even if an abundance of variable factors is available. In Fig. 11.7 (*b*) the supply can be increased fairly easily because the firm is a long way below full capacity. Whereas firm (*a*) may already be working a three-shift system and running the machines at nearly full speed, firm (*b*) may be working only eight hours a day and even then at a leisurely pace.

(*b*) The availability of other factors
If extra units of the variable factors are not available then the elasticity of supply will be low even if the plant is under-utilized. Even a considerable rise in the price offered for motor cars will have little effect on the quantity

supplied if a strike has prevented the motor manufacturer from obtaining the components he needs.

(c) Specific factors of production

If a productive process requires very specific factors of production, such as complex chemical equipment, then the elasticity of supply will be low when the existing plant is at full capacity. When a productive process depends mainly on non-specific factors, perhaps motor vehicles, then elasticity of supply will be much higher as extra units will be more easily obtainable. This influence is even more important in the primary sector of an economy. Once an African farmer has committed his land to tea production he cannot easily change to growing coffee; and a North Sea oil-well cannot be turned over to coal production when the price offered for the latter increases. It is one of the characteristics of many underdeveloped economies that they are heavily dependent on a single product which has an inelastic supply.

(d) The time factor

One of the most important influences on elasticity—the time span involved —has been left until last. Our earlier discussions have established the difference between the short-run and the long-run. It should be clear to the reader that in almost every industry supply will be more elastic in the long-run than in the short-run. In the example shown in Fig. 11.8 in the short-run the maximum supply is OM, but in the long-run extra units of the fixed factor can be employed enabling the industry to move on to a new short-run supply curve with a maximum supply of ON.

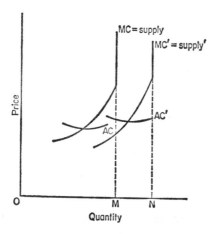

Fig. 11.8 Short-run and long-run elasticity of supply

It would be wrong to imagine that when we consider a short-run rise in supply, firms respond instantaneously to a rise in price. Let us imagine a company geared to the production of 100,000 cans of baked beans a week. The cans are produced one week and sold the next, and the firm will be employing sufficient quantities of their factors of production to achieve this output. If it becomes apparent that owing to a change in market conditions it would be profitable to sell 110,000 cans the firm cannot immediately put this amount on the market, for the present week's supply is fixed by the previous week's output. In the very short-run, supply (which is called *market supply*) is absolutely fixed. It will take at least a week to provide sufficient variable factors to increase supply by 10,000 units. The market supply period ends when this is achieved and the short-run supply period begins. If the rise in demand is permanent, however, it may be necessary to employ more fixed factors or new companies may enter the industry in the long-run. Thus we may conclude that market supply is less elastic than short-run supply which is again less elastic than long-run supply. As time goes by the supply curve gets flatter showing a larger percentage rise in supply for a given rise in price.

11.9 Regressive Supply Curves

In some circumstances the supply curve may be regressive over part of its length. Where labour is concerned (Fig. 11.9), an increase in wages may result in a fall in supply. When wages are OW, OM hours are worked; but when wages rise to OV the same total wage can be earned for less work and the supply of labour falls from OM to OL. In this situation absenteeism becomes a problem especially in industries with unpleasant work, for men can take greater amounts of leisure time without reducing their living standards.

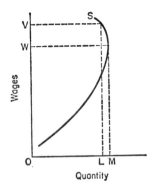

Fig. 11.9 The regressive supply curve of labour

A further example of a regressive supply curve occurs during periods of speculation. Producers may be supplying 1000 units a week at a price of £1·00. If price rises to £1·10 we should expect the supply to extend, but if producers believe that the rise to £1·10 is only the beginning of a series of price rises then they may prefer to reduce their supply temporarily, and hope to take advantage of higher rises later.

11.10 Questions and Exercises

1. What are the main conditions of supply? Show what happens to the supply curve if they change.
2. Explain the main determinants of the supply curve of the firm.
3. Define elasticity of supply, and examine its main determinants.
4. In what circumstances would you expect the supply curve to slope down from left to right?
5. Why does the supply curve normally slope up from left to right?
6. Why is the supply of primary products less elastic than the supply of manufactures? What are the economic implications of this for the producers of primary products?
7. Explain the effects of the following on the position of the supply curve of motor cars:
   (a) The trade unions secure a large rise in wages without a corresponding rise in output.
   (b) The Government removes indirect taxation from cars.
   (c) The price of imported iron ore increases.
   (d) The demand for cars increases.
   (e) An increase in the productivity of labour.
8. Calculate elasticity of supply in the following cases:

Good A		Good B		Good C	
Price	Quantity	Price	Quantity	Price	Quantity
10	150	20	150	15	1000
11	160	22	170	18	1200

9. To what extent does the elasticity of supply of a product depend upon the availability of labour?
10. A manufacturer has a plant that can produce 10,000 pairs of shoes per week. At present his weekly output is 5000 pairs. Consider the ways in which he can meet demand if it rises to (a) 6000 pairs per week; (b) 10,000 pairs; (c) 12,000 pairs.

Unit Twelve

Supply and Demand and Market Price

12.1 Introduction

It is now time to bring together the ideas discussed in Units Three and Eleven and investigate the establishment of a market price, and the forces that change it. Later in this Unit we will consider some of the criticisms levelled against the theory of supply and demand, but first we shall examine the theory itself. What do we mean by the term 'market'?

12.2 What is a Market?

In general terms a market is a place where people come together to buy and sell goods. There are market places near most homes; in this ordinary sense of the word all shopping centres are markets. Some markets are of national and even international importance—the Billingsgate fish market or Smithfield's meat market are examples of national markets while the London Metal Exchange is of international importance. Some markets have no precise location but are none the less easily accessible. The foreign exchange market in London exists in a variety of offices linked by telephone and telex with one another and with similar markets in other parts of the world. The housing market is also widely dispersed for there are thousands of Estate Agents specializing in the purchase and sale of property and there is also an equally important housing market in the columns of local newspapers.

The markets in which economists are interested include both highly localized and also widely dispersed markets, for the economist needs to know how many people contribute to the total demand for a product, and how many people or companies are responsible for the supply of the product. The market is therefore the number of people willing to buy or willing to sell a particular product at a particular price; they may or may not be in close contact with each other. In many of the large and highly organized commodity markets dealers in Britain will quite happily spend thousands of pounds on American produced items that they have not seen and which indeed might not yet have been produced!

Most of our discussion in this Unit will be in terms of *competitive markets* or *perfect markets*. A competitive market is one in which there are a large number of buyers and a large number of sellers. The buyers are all trying to obtain the maximum amount of satisfaction from their expenditure and will not pay 5p for an item if they can buy it elsewhere for 4p. This, of course, assumes perfect knowledge of the prices asked by all suppliers; it also assumes that there are no transport costs or other obstacles preventing would-be consumers from reaching all suppliers. There is no point in travelling several miles to save 1p if the transport costs 2p. Such obstacles in the market are called *imperfections* and although we know that they are of the utmost importance in practice we must disregard them for the time being. They will be re-introduced at a later stage. Market suppliers are assumed to maximize profits and so will not sell an item at 4p if they can obtain 5p for it.

12.3 The Establishment of Market Price

Table 12.1 Market demand and market supply schedules for golf balls

Price (pence)	Quantity demanded per week (000's)	Quantity supplied per week (000's)
60	2	40
55	3	30
50	5	25
45	6	20
40	8	14
35	10	10
30	12	8
25	15	7
20	20	6
15	30	3
10	50	2
5	90	1

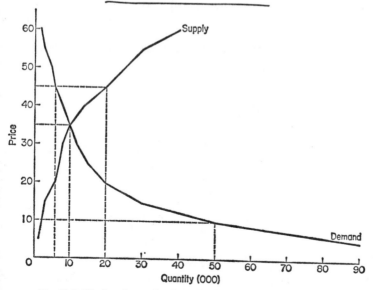

Fig. 12.1 Market demand and market supply curves for golf balls

Table 12.1 combines the information provided in Tables 3.3 and 11.2, and Fig. 12.1 shows the same information graphically. We will consider what would happen if price were set quite arbitrarily at 45p per ball. At this price, even relatively inefficient producers would find it worthwhile to enter the market, while efficient producers already in the market at lower prices will expand production in pursuit of further profits, so that 20,000 balls are on offer. Each consumer on the other hand will relate the price of 45p to the utility he expects to get from each ball. At a price of 45p some players prepared to buy new balls at a somewhat lower price decide to make do with their old ones while some may temporarily give up the game. Thus the demand is for 6000 balls in that week, and there is an *excess supply* of 14,000 balls, and at the end of the week suppliers will be left with this number on their hands. During the next week they will not produce so many and individually will move back down their marginal cost curves. At the same time, as they will be anxious to dispose of surplus stocks, suppliers will compete with each other by cutting prices. There is no need for them to reduce supply by 14,000, for as price falls more consumers find it worthwhile to enter the market and gradually demand extends until it is equal to supply at a price of 35p per ball. *Thus an excess supply leads to a downward pressure on price.* On the other hand if price is fixed temporarily at 10p, very few balls will be supplied as it is difficult to make a profit at this low price. On the other hand golf balls are such a bargain in relation to the utility gained at this price that demand reaches 50,000 and there is an *excess demand* of 48,000. In such circumstances suppliers rapidly increase prices to take advantage of the shortage. Output will expand, prices will tend to rise, and demand will fall until at a price level of 35p demand and supply are equal. *Thus excess demand results in an upward pressure on prices.*

When demand and supply are equated in this way an *equilibrium price* is

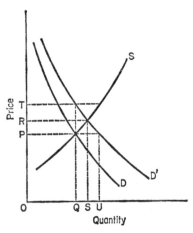

Fig. 12.2 An increase in demand

said to have been established. The equilibrium price is that which will remain in force in the absence of changes in the conditions of demand or supply. A change in those conditions, however, will necessitate a new equilibrium price. In perfect markets the adjustments towards an equilibrium are very swift—demand and supply are rapidly adjusted in the Stock Exchange, for example. In many real world imperfect markets, however, the process of adjustment may take a considerable time, and once again it is in primary production that the time lags are greatest. We will now see what happens when the conditions of demand change.

12.4 Changes in the Conditions of Demand

(a) An increase in demand

We will suppose that the market is in equilibrium at a price OP and quantity OQ with consumers just absorbing each week's supply. A favourable change in the conditions of demand will cause the demand curve to move to the right so that at the price OP, consumers now demand OU units. (See Fig. 12.2.)

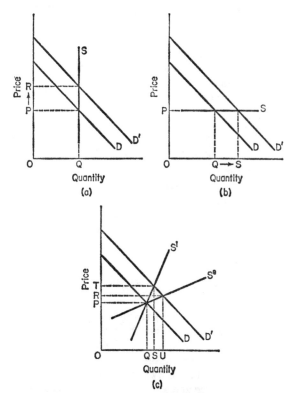

Fig. 12.3 The importance of elasticity of supply when conditions of demand change

(The change in the conditions of demand may have been a favourable change in tastes, an increase in incomes, or a rise in the price of a substitute good (see Section 3.2).) In the market period it is quite impossible for firms to supply OU units or indeed any more than OQ units since this week's supply is determined by last week's production. In a competitive market we may, therefore, expect the extra demand to be temporarily stopped by a rise in price to OT, which allows suppliers a very good profit. In response to such a high price producers will expand their output by moving individually along their marginal cost curves until a new equilibrium is established at price OR, quantity OS.

Can we predict then that the effect of the demand curve moving to the right will always be to increase both the equilibrium price and the equilibrium quantity? No. This is not always true. Let us consider the three diagrams in Fig. 12.3.

In Fig. 12.3 (a) supply is perfectly inelastic and the only effect of an increase in demand is a corresponding increase in price. In Fig. 12.3 (b) supply is perfectly elastic and as consumers can purchase as many units as they wish at the price OP the only effect is that the equilibrium quantity increases. In all other cases with an upward sloping supply curve the effect of an increase in demand is to increase both price and quantity. In Fig. 12.3 (c) there are two supply curves passing through the same point at the original equilibrium price. On the less elastic curve S^i a change in the conditions of demand causes a larger rise in price and a smaller rise in quantity than is the case with the more elastic supply curve S^e. Thus the effect of a given change in the conditions of demand depends on the elasticity of supply.

(b) A decrease in demand

The effects of a decrease in demand are much as one would expect in view of the preceding discussion. In Fig. 12.4 we have a market where production has been geared to the equilibrium output of OQ. When the demand curve shifts to D' the only way to dispose of the excess supply MQ is to allow price to fall to OK. This does not bring the market to equilibrium owing to excess demand and a new equilibrium is reached at OL with a quantity of ON units. The elasticity of supply will, of course, be as significant here as it is when the demand increases.

12.5 Changes in the Conditions of Supply

(a) An increase in supply

In Fig. 12.5 an increase in supply has resulted perhaps from the entry of new firms to the industry, which is now prepared to supply OS units rather than OQ at the original equilibrium price of OP. However, an excess supply of QS units exerts a downward pressure on prices until a new equilibrium is established at price ON and quantity OR. Thus the increase in supply leads to a lower equilibrium price and a higher equilibrium quantity. Whether it will always do so depends upon the elasticity of demand.

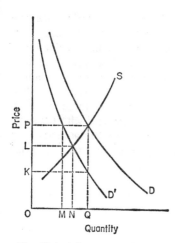

Fig. 12.4 A decrease in demand

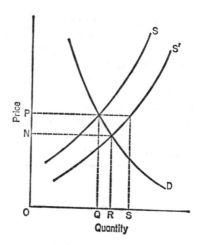

Fig. 12.5 An increase in supply

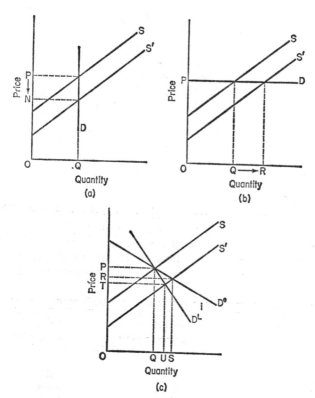

Fig. 12.6 The importance of elasticity of demand when conditions of supply change

In Fig. 12.6 (*a*) demand is perfectly inelastic and although the price may change consumers cannot be persuaded to buy more than OQ. Thus quantity remains the same and price falls. In Fig. 12.6 (*b*) consumers are prepared to buy as much as they can acquire at a price OP so there is no need for producers to reduce prices to sell the extra output, and only the equilibrium quantity changes. Neither of these cases is likely to occur very often in the real world. The majority of cases will be similar to those shown in Fig. 12.6 (*c*), where we have two demand curves associated with the same change in supply. When demand is relatively inelastic (D^i) the increase in supply leads to a smaller change in quantity but a larger change in price than when demand is relatively elastic (D^e).

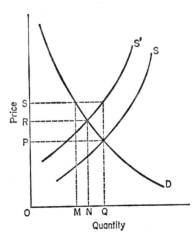

Fig. 12.7 A decrease in supply

(b) A decrease in supply

If there is a rise in the payment to factors of production then the supply curve moves to the left showing that a smaller quantity will be produced at each price. Thus the previous equilibrium quantity OQ will be put on the market for a price of OS rather than OP. But at this price there is excess supply of MQ and forces are set up which establish a new higher equilibrium

Table 12.2 Effects of changes in market conditions on equilibrium price and quantity

Change	Effect on equilibrium price	Effect on equilibrium quantity
Increase in demand	Rises	Rises
Decrease in demand	Falls	Falls
Increase in supply	Falls	Rises
Decrease in supply	Rises	Falls

price at OR and a smaller equilibrium quantity ON. There will not be a change in both price and quantity unless the demand curve slopes downwards for reasons explained in the discussion of Fig. 12.6.

The effects of a change in the conditions of demand or supply providing elasticities of between zero and infinity are shown in Table 12.2.

12.6 Changes in the Conditions of Supply and Demand

The changes in conditions described so far are short-run changes in that after the initial change in the conditions of demand or supply no subsequent changes occur. In fact when the demand curve shifts it is also likely that the supply curve will move. There are two possibilities to be considered.

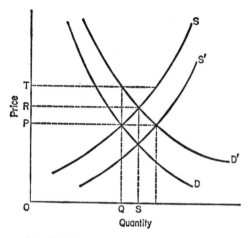

Fig. 12.8 An increase in demand and supply

(a) An increase in demand and an increase in supply

In this case an increase in demand after forcing price up to OT changes the equilibrium price and quantity from OP, OQ to OR, OS the adjustment being made by a movement along the supply curve. If the price OR is sufficiently attractive new firms will enter the industry causing the supply curve to move to the right and equilibrium price to fall below OR and quantity to rise above OS. We are unable to predict the extent of the change in supply yet, but in Fig. 12.8 the curve moves just far enough to offset the increase in demand and the original price is restored. In different circumstances the new equilibrium price may be above or below the original equilibrium price, depending upon whether there are decreasing or increasing returns to scale.

(b) A decrease in demand and a decrease in supply

In this case after forcing the price down to OT the decrease in demand leads to a lower price OR and a lower output OS. Some firms will not now be

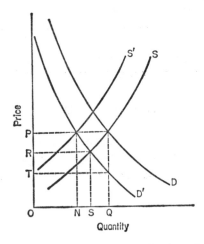

Fig. 12.9 A decrease in demand and supply

able to compete and will leave the industry. As they do this the supply curve moves gradually to the left, thus moving up D′ and establishing another equilibrium price. Once again we have made this equal to the original equilibrium price but there is no firm reason why it should not be elsewhere.

In the case of the original change being a change in the conditions of supply there is no economic reason why the demand should shift as well. If it does, however, happen to shift (by coincidence) it will obviously affect the long-run equilibrium price.

12.7 Demand Relationships

(a) Complementary demand

Many goods need to be consumed in fixed or nearly fixed proportions, and a change in the market conditions of one of them will almost certainly influence the market price of the others. Cups and saucers, bread and butter, gas fires and gas are examples. We normally refer to these as *complementary goods*. We will now consider the effects of changes in the conditions of demand or supply on the market for gas and gas fires.

In Fig. 12.10 (*a*) we assume that the cost of gas has fallen, perhaps as a result of the discovery of deposits of natural gas. The supply curve moves down to the right, reflecting the lower price at which any given quantity will now be supplied. The demand curve for gas does not move because the conditions of demand for gas have not changed. But we now have a lower equilibrium price for gas (OR). The conditions of demand for gas fires have changed and we may expect some consumers to switch from electricity to gas for cooking and heating. This will obviously not occur overnight, but there will be a gradual movement of the demand curve leading to a higher equilibrium price and quantity for gas fires. Now that a higher price is obtainable for producing gas fires it is likely that new firms

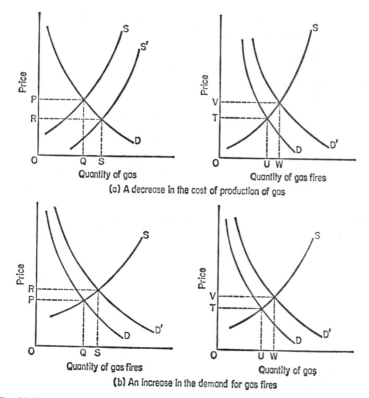

(a) A decrease in the cost of production of gas

(b) An increase in the demand for gas fires

Fig. 12.10 A change in market conditions with complementary goods (only equilibrium positions shown)

will be attracted into the industry thus causing the supply curve for gas fires to move to the right. We cannot pursue this possibility at this stage of our investigation. It is sufficient for us to examine the initial effects of a change in the conditions of demand or supply without exploring all the secondary effects.

In Fig. 12.10 (*b*) the demand for gas fires has increased perhaps because of a change in taste or fashion in favour of gas fires. The demand curve moves to the right giving a higher equilibrium price and quantity. If people buy more gas fires they will obviously need more gas, and the demand curve for gas moves to the right as well, giving a higher price and a larger quantity.

There are several other possibilities, and the student should work out for himself the effects on the market for gas and gas fires of the following:

(*a*) A decrease in demand for gas fires.
(*b*) An increase in supply of gas fires.
(*c*) A decrease in supply of gas.
(*d*) A decrease in the demand for gas.

It is essential to remember that a new demand curve is drawn only when the conditions of demand change and a new supply curve is drawn only when the conditions of supply change.

(b) Derived demand

Some goods are only needed when others are in demand. While such cases bear a close resemblance to complementary goods they are usually referred to as cases of derived demand. The best examples are among the factors of production. The demand for thatchers falls as the demand for thatched roofs falls; the demand for computer programmers increases as the demand for computers increases; and many people would argue that the demand for gas is similarly derived from the demand for gas fires. Once again, however, there is no final adjudicator of the term to be applied.

(c) Competitive demand

Many goods are perfect substitutes for each other and are therefore competing with each other. A change in the price of gas is likely to affect the market for electricity and the direction of the changes is indicated in Fig. 12.11.

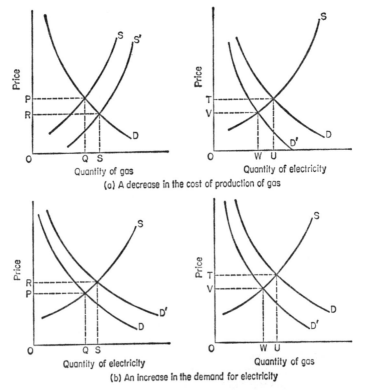

(a) A decrease in the cost of production of gas

(b) An increase in the demand for electricity

Fig. 12.11 A change in market conditions with substitute goods (only equilibrium positions shown)

In Fig. 12.11 (*a*) a fall in the cost of supplying gas leads to a decrease in the demand for electricity and a lower equilibrium price and quantity. In Fig. 12.11 (*b*) on the other hand a change in tastes in favour of electricity leads to a higher price for electricity and a decrease in the price of gas as the demand curve moves to the left.

(*d*) Cross elasticity of demand

In Section 3.11 (*a*), we noted that cross elasticity of demand measures the sensitivity of the demand for one commodity to a change in the price of another commodity.

$$\text{Cross elasticity of demand} = \frac{\% \text{ Change in quantity of good A demanded}}{\% \text{ Change in price of good B}}$$

We can confirm from our analysis of substitute and complementary goods that in the case of substitutes cross elasticity will be positive (an increase in the price of gas causing an increase in demand for electricity) and that the closer the substitutability the higher the cross elasticity. Thus a small rise in the price of one brand of cigarettes would result in a large increase in demand for a competitive brand, but in a smaller rise in the demand for pipe tobacco which is not so complete a substitute. With complementary goods the cross elasticity of demand will be negative (a rise in the price of gas fires leading to a fall in the demand for gas).

(*e*) Independent demand

There is no clear linkage between some commodities. We should not, for example, expect the market for bread to be very much affected by a change in demand for electric drills. However, if the community is spending a greater proportion of a fixed income on electric drills it has less money remaining to be spent on other things so the conditions of demand for bread have changed slightly. The link between the two, however, is so tenuous that we may ignore it and it is left to the common sense of the reader to distinguish between independent goods and substitutes.

12.8 Supply Relationships

There are many products which are of necessity produced together. Sometimes this is inevitable for natural reasons: for example, farmers cannot produce more mutton without producing more wool, although they can alter the proportions by breeding different kinds of sheep in the long-run; the production of extra beef increases the supply of hides. Sometimes joint supply occurs because the producer wants to avoid waste; during the process of refining crude oil into petroleum the oil refiners produce a multitude of by-products. In Fig. 12.12 we examine the markets for mutton and wool on the assumption that they are, at least in the short-run, produced in fixed proportions.

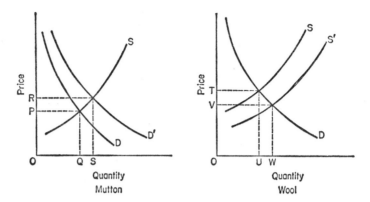

Fig. 12.12 Joint supply. The markets for mutton and wool (only equilibrium positions shown)

If the demand for mutton rises, perhaps because of successful advertising campaigns by the producers, the demand curve moves to the right causing a rise in price and quantity. As the supply of mutton rises from OQ to OS the supply of wool also increases, and as the conditions of supply for wool have changed a new supply curve is established to the right, giving a lower price OV and a higher quantity OW.

If the increase in the price of mutton had come about because of a decrease in the supply of mutton then the supply curve of wool would also have moved to the left causing a higher price and lower quantity.

12.9 Some Applications of Demand and Supply Theory

(a) The incidence of expenditure taxes

When the Government levies excise duties on goods it normally collects the tax from the producer of the goods involved and leaves it to the producer to recover the money from his customers. The consumers for their part may try to avoid paying the tax by forgoing the product altogether. In Fig. 12.13 we examine the general effect of imposing an expenditure tax before considering the importance of elasticity of demand and supply in determining the final effects of the tax.

In Fig. 12.13 (*a*) the free market equilibrium price is OP and quantity OQ. Suppose the Government imposes a tax of PT per unit on the product. This has the effect of moving the whole supply curve vertically upwards by PT as the producer tries to shift the burden of the tax on to consumers. As the tax is PT per unit whatever the price we know that it is a *specific tax*—charged as so much per unit, rather than an *ad valorem tax*—levied as a percentage of the value of the goods, and that the new supply curve S^t is parallel to the original supply curve. In the case of an ad valorem tax the new supply curve would rise more steeply than the original. The producers now try to sell the quantity OQ at a price OT, but considerable excess supply forces price down to OE and quantity to OF. For each unit sold the producer must remit to the Government the amount of the tax GE (= PT since S and S^t are parallel). Thus the final effects of the tax are

(i) The price to the consumers rises by PE per unit.

(ii) The receipts of the producers fall by GP per unit.

(iii) The Government receives revenue of value GEXY.

(iv) Resources employed in the industry fall from the level required to produce OQ units to the level required to produce OF units.

In Fig. 12.13 various possibilities are illustrated. In 12.13 (b) the original demand curve is combined with a much more elastic supply curve while the same tax is imposed; in 12.13 (c) a very inelastic supply curve accompanies the original demand curve. The student can see that other things being equal the greater the elasticity of supply, the greater the rise in unit price to the consumer, the smaller the fall in unit receipts to the producer and the greater the fall in output following the imposition of the tax.

Fig. 12.13 (d) and (e) shows that the greater the elasticity of demand

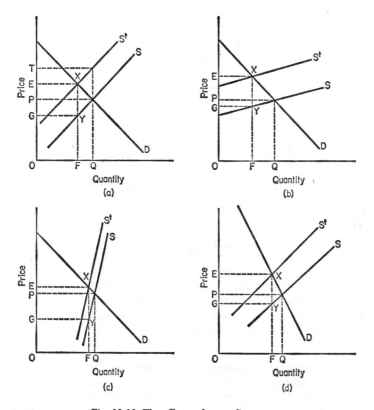

Fig. 12.13 The effects of expenditure taxes

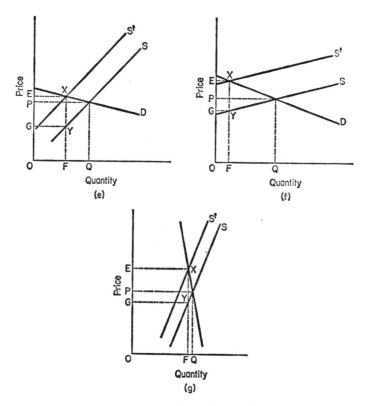

Fig. 12.13 cont. The effects of expenditure taxes

(other things being equal) the smaller the rise in consumer price, the greater the fall in producer price and the greater the fall in output following the imposition of the tax.

Finally in Fig. 12.13 (*f*) and (*g*) we have combined a very elastic demand curve with a very elastic supply curve; and a very inelastic demand curve with a very inelastic supply curve. Where the demand is elastic a tax has a great effect on the equilibrium quantity and therefore on the resources employed, but little revenue is raised for the Government. Where the market is inelastic there is very little change in the equilibrium quantity and the Government obtains more revenue. It is in the light of such reactions that the Government frames its taxation policy.

(*b*) Expenditure taxes and consumers' and producers' surplus

In Fig. 12.14 (*a*) with market price established at OP there is a consumers' surplus of PBE as the consumers obtain a total utility of OBEQ for which they spend only OPEQ (see Section 3.14). There is a producers' surplus of

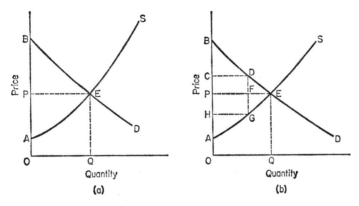

Fig. 12.14 Consumers' and producers' surpluses

APE showing the surplus revenue received by producers above that which is absolutely necessary to induce them to supply the amount OQ (see Section 11.5). Fig. 12.14 (*b*) shows what happens when the imposition of a tax establishes a new equilibrium price OC (we have omitted the second supply curve going through D to simplify the diagram). The consumers' surplus is reduced by PCDF to CBD; and the producers' surplus is reduced by PFGH to AHG. The triangle GDE represents the consumers' and producers' surplus which is wasted as a result of the tax; neither the parties concerned nor the Government are able to benefit from it. The higher the tax the greater this amount of waste. It is the extent of this waste that prompts many commentators to find income tax preferable to expenditure taxes.

(c) Subsidies

Governments generally impose taxes in order to raise revenue for public expenditure (but see Unit Twenty-nine). But the Government may give subsidies which may be regarded as negative taxes, and the main motives for these may be:

    (i) To reduce the price of essential goods to consumers.

    (ii) To increase the supply of these goods.

    (iii) To enable the producers of such goods to compete with overseas producers.

The general case is illustrated in Fig. 12.15. The free market price and quantity are OP and OQ. The Government believing the price to be too high, or that consumers are getting insufficient quantities of the product decides to subsidize the producers to the extent of PS per unit. The supply curve accordingly moves down by this amount indicating that any given quantity will be put on the market at a lower price than before. A new equilibrium is established at a price OR and quantity OT, and the effects are:

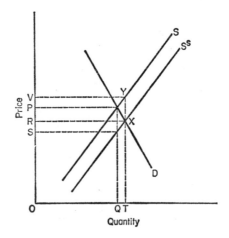

Fig. 12.15 The effects of a subsidy

(i) The consumers pay OR per unit instead of OP;

(ii) The producers receive OV per unit instead of OP. Of this OR is paid directly by the consumers and RV by the Government (or indirectly by the taxpayers).

(iii) The Government spends RVYX on subsidies.

(iv) The industry is larger than it would be without the subsidy.

The student should now draw diagrams similar to those in Fig. 12.13 to satisfy himself that the following points are true:

(i) If the elasticity of demand is taken as constant, then the reduction in consumers' price and the increase in output resulting from a given subsidy will increase as the elasticity of supply increases.

(ii) If the elasticity of supply is taken as fixed, then the reduction in consumers' price will be greater and the increase in output following the award of a subsidy will be smaller as the elasticity of demand decreases.

(d) Price controls

Sometimes the forces of supply and demand work erratically or lead to hardship in the case of essential goods or services which rise in price owing to shortages. At other times an exceptionally good harvest, for example, may depress prices to the detriment of producers. In such cases the Government may intervene to establish maximum or minimum prices. The schemes that may be operated by the Government are often exceptionally complicated and here we can only look at two simple cases and their implications.

(i) **Minimum prices.** In the free market the equilibrium position is reached at price OP and quantity OQ. If the Government feels that a price of OP gives producers insufficient return on their investment it may establish a minimum price at OS. The effect of this is to call forth a greater output of OR, but to reduce consumer demand to ON. It is clear that if the Govern-

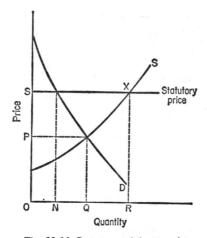

Fig. 12.16 Statutory minimum prices

ment interferes with the market by establishing a minimum price it must also intervene and establish an agency to buy up the surplus output. The line SX is in effect the demand curve for the industry's product. Of course, we have only considered the short-run supply curve and it may be that a price of OS is so attractive that even more producers will be drawn into the industry, causing an even greater excess supply. If the product is not perishable the buying agency may store it until the following year in the hope that supply will be lower, but if it is perishable, or if there is always an excess supply, then the agency must either destroy the surplus or sell it overseas. In many ways the fixing of artificially high statutory prices is characteristic of the EEC agricultural policy and the implications of this are discussed in Unit Twenty-four.

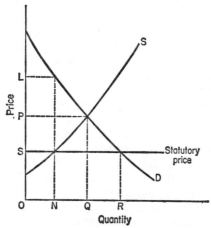

Fig. 12.17 Statutory maximum prices

(ii) **Maximum prices.** Statutory minimum prices lead to excess supply, and as we can see from Fig. 12.17 statutory maximum prices result in excess demand, in this case NR. The great danger of such a system, designed to prevent the exploitation of consumers, is that it will lead suppliers to attempt to direct their output on to the black market, leaving none for the open market. An extensive system of inspection becomes necessary to enforce the scheme and to prevent the black market price reaching OL. Moreover, once again interference in one area leads to an essential intervention elsewhere; unless customers are going to be served on a first come, first served basis a rationing scheme must be established to ensure equality of treatment of all those prepared to buy at price OS. One possible method would be to allow potential consumers a fraction $\frac{ON}{OR}$ of what they would buy at price OS; that is if the consumers and their demands can be identified.

12.10 Some Criticisms of Demand and Supply Theory

The 'laws' of demand and supply seem to indicate that in each market an equilibrium price will easily become established. There are several grounds on which the theory has been criticized and which deserve attention.

(a) The most fundamental criticism is that the theory doesn't work smoothly in practice. This criticism is frequently raised at the time of imposition of increased expenditure taxes. In Fig. 12.13 the burden of the expenditure tax is shared between producers and consumers, but if the Government announces an increase of 2p per gallon on petrol tax then it is all passed on to the consumer. This does not, however, invalidate the theory since it is a case of more or less perfectly inelastic demand.

(b) A more telling criticism is that the theory, in concentrating on the relationship between price and quantity, is emphasizing the wrong things. Fractional changes in price do not have much effect on the level of demand, but changes in the conditions of demand produce far more significant changes in the level of demand. Thus an increase in incomes or successful advertising (altering tastes) causes a larger change in sales than does a change in price. This is still an under-researched area of economics but it may well be that more attention should be given to these other factors, as they have considerable influence on consumer spending.

(c) An important criticism of the theory of demand, for the reader seeking to relate the theory to the world around him, is that there is a wide divergence between the theory and actuality. The reason for this is that while the theory predicts what would happen in competitive markets, real world markets are monopolistic; the consumer has little influence over price; he can buy bread at its existing price or not buy it at all. He cannot normally buy at a price lower than the one advertised.

On the other hand your own observations can show you areas in which the market mechanism does work. During times of power cuts, or threatened cuts, the rise in the price of candles is predictable; an increase in the price of public transport does induce more people to resort to private transport. With perishable commodities, such as fish or fresh vegetables, where the supply is fixed, the price depends on the level of demand.

However, it is true to say that prices are normally determined by competitive forces, and in Unit Fifteen we shall again examine the question of price determination.

12.11 The Price Mechanism

The operation of markets is often referred to as the *price mechanism*. In some ways this is misleading for it is really the mechanism through which the allocation of resources is influenced. Thus under competitive conditions an increase in demand for cars will lead to a rise in price which will attract extra resources to the industry. These extra resources may also be attracted through a reduction in costs, which in turn leads to lower prices and attracts more consumers. In the perfect market, demand will stop rising at the point where marginal utility is proportional to price, and production will cease when marginal cost equals price.

In practice the price mechanism does not work so smoothly. Frequently resources are dragged into an industry while their marginal cost to the community is in excess of the price paid for them. Thus road freight transport imposes costs in the form of noise pollution and congestion which are above the price that the users of such transport are asked to pay. In such cases the State may have to interfere, as we shall see in Unit Twenty-nine, to cause some re-allocation of resources.

Similarly in the capitalist system the supply of essential services may be inadequate because the price mechanism leads to a low price. Again the State may interfere to adjust the allocation of resources.

12.12 Questions and Exercises

1. Explain how supply and demand operate to establish an equilibrium price.
2. Show what is meant by (a) excess demand; (b) excess supply. Explain how they affect market price.
3. Show how a change of tastes in favour of beef affects the markets for (a) beef; (b) mutton; (c) beef hides.
4. 'The effects of a change in the conditions of demand depend upon the elasticity of supply.' Comment and illustrate.
5. Explain with the aid of diagrams the effects of imposing a specific tax on (a) holidays; (b) holidays at Seatown.
6. Show how a subsidy helps (a) consumers; (b) producers.
7. Examine the economic implications of the Government imposing (a) minimum prices; (b) maximum prices.
8. Outline the determination of market price. What are the functions of the price mechanism?

Unit Thirteen
Types of Markets

13.1 Introduction

Goods and services are normally sold in markets which have a fixed geographical location, but markets may occasionally exist in a number of widely dispersed offices linked only by telephone, and this is the case with the foreign exchange market. On the other hand retailers over a large area buy at a central wholesale produce market. Thus greengrocers, florists and fruiterers in the South East of England acquire their supplies largely from the wholesale market at Covent Garden. In this Unit we look briefly at the main characteristics of some important markets, which do not all work as smoothly as those discussed in Unit Twelve.

13.2 The Perfect Market

It is useful to begin with an abstraction, 'the perfect market'. As we have seen it is necessary in such a market that all potential buyers should have perfect knowledge of the prices asked by all potential sellers, and that all sellers should seek to maximize prices for their produce while all buyers should try to pay minimum prices in order to obtain the best value for money. All persons should also be dealing in readily identifiable products. While it is probably impossible to find a perfect market in practice, there are many highly organized markets which approach the ideal. Perhaps those coming nearest to the ideal are the Stock Exchange dealing in Government and company securities, Lloyds dealing in insurance, and the foreign exchange market. There are also many other commodity markets, which function very smoothly and have many of the characteristics of the perfect market.

13.3 Commodity Markets

Many of the most important commodity markets are in London; they include the following:

(i) The London Commodity Exchange. This is a group of markets, some of which have been established in London for centuries, and others of more recent origin. In this group are the markets for rubber, cocoa, vegetable oils, jute and sugar.

(ii) The London Wool Exchange.

(iii) The English Grains Market.

(iv) The Uncut Diamond Market, and the associated Hatton Garden Market.

(v) The Gold and Silver Bullion Market.

(vi) The London Metal Exchange.

(vii) The wholesale produce markets such as Covent Garden (vegetables), Billingsgate (fish), and Smithfield (meat).

(viii) The London Tea Auctions.

Outside London other specialist markets exist, the most notable being the Bradford Wool Exchange and the Manchester Royal Exchange which deals in cotton and rayon. Each of these markets has its own method of dealing, depending on the nature of the product and the conditions of supply. The methods of trading may be summarized as follows.

(a) Auction sales

These are appropriate where the quality of the commodity varies with each consignment, and cannot be standardized. Tea is an example of such a commodity. Brokers buy tea by auction only after their principals have examined the consignment and decided how much they are prepared to pay for it. Similarly wool is sold, after inspection, by auction to the highest bidder.

(b) Ring trading

This takes place in the London metal and sugar markets. The forty dealers sit in a ring in the metal market crying out their bids for consignments of metal. The metals are sold one at a time and five minutes is allocated for trading in each metal.

(c) Private treaties

These characterize the Stock Exchange, the Foreign Exchange Market, Lloyds, the Baltic Exchange and the Corn Exchange. In the Stock Exchange a broker will approach a number of jobbers (or wholesalers) on behalf of his client before concluding a deal on the most favourable terms available.

(d) The diamond market

This has its own method of dealing. There are only two suppliers, the Diamond Trading Company which deals in gem stones, and Industrial Distributors (sales) which controls industrial diamonds. The two suppliers quote a price to potential buyers on a take it or leave it basis, and in this way are able to maintain prices at what they regard as a desirable level. Markets such as these require a very broad specialist knowledge of the commodity being traded and dealings are in general restricted to specialist dealers acting on behalf of others wishing to buy and sell.

13.4 The Economic Importance of Markets

Outside these highly organized commodity markets the economy is full of much less formally organized markets concerned with the distribution of goods from the manufacturers via the wholesalers and retailers to the final consumers. These markets are vital to the whole process of production, but the proper study of them lies within the field of commerce. Here we will just briefly note the economic significance and functions of markets.

(*a*) Their most obvious function is to bring together buyers and sellers usually in the same place.

(*b*) They also reduce price fluctuations due to the seasonal nature of the product. One function of market specialists is to carry stocks of goods in order to prevent prices falling too rapidly in periods of high output, or rising too rapidly in periods of low output. They thus benefit producers in the first case and consumers in the second.

(*c*) In this connexion speculators, who are often condemned, contribute to stability by buying when prices are low (thus preventing prices falling further) and releasing their stocks as prices rise (thus preventing prices rising too far).

(*d*) Finally the establishment of centralized markets allows both producers and consumers to take advantage of the specialized services which can only be sustained where markets are large enough to lead to economies of scale.

13.5 Questions and Exercises

1. What are the main functions of markets?
2. Describe the operation of any market with which you are familiar either through your studies or through personal contact with the market.

Unit Fourteen

Perfect Competition

14.1 Introduction

In this Unit and the next production of the individual firm will be examined in greater detail. We begin with an analysis of perfect competition, a rarefied concept, but one that is invaluable as a guide to other kinds of competition. In Unit Fifteen we shall consider another extreme case, that of monopoly, before looking at situations of imperfect competition approximating more closely to the real world.

14.2 The Assumptions of Perfect Competition

We have already spoken of the construction of economic models, and the concept of perfect competition is one of the best-known models. Although some economic models are very sophisticated, requiring computers to accommodate them, they are all based on certain assumptions which allow the most important features of a situation to be isolated and examined. The main assumptions upon which the model of perfect competition rests are as follows:

(a) All the firms in an industry seek to maximize their profits. This is an assumption common to monopolistic and imperfectly competitive situations too. It is an assumption that has frequently been questioned as being too remote from the real world, but in the absence of an acceptable alternative it has stood the test of time. It might be argued that a businessman's main objective is to make his firm grow as large as possible by forcing his rivals out of business or by taking them over. Such a policy could not survive if it were not also profitable. If a firm is concerned with its public image it may reduce profits by spending money on public relations. But even allowing for these other motives it is reasonable to conclude that the main aim of firms is usually to maximize their profits.

(b) There is a very large number of producers but no single one is sufficiently large to be able to influence market price. Thus if one firm left the industry as depicted by an ordinary demand and supply curve, the leftward movement of the supply curve would be so slight as to leave price unchanged. Similarly a doubling of output by one firm would not cause the supply curve to move to the right. Only if several firms leave or enter the industry will the supply curve move sufficiently to affect price.

(c) There must be a large number of buyers, none of them large enough to influence market price. All consumers try to maximize their satisfaction and have no loyalty to particular suppliers.

(*d*) All firms must be selling a homogeneous product. A consumer will know that he will receive the same product from any supplier.

(*e*) New firms must be free to enter the market and there must be no obstacles to existing firms leaving the industry. Similar considerations apply to buyers.

(*f*) Perfect knowledge. All firms and all buyers are assumed to have perfect knowledge of the prices being asked and offered by other firms and buyers. In practice this is unlikely, but it does occur in some of the London commodity markets. We must now consider how much the firm will produce on the basis of these assumptions.

14.3 The Demand for the Firm's Product

In Unit Ten we established that the firm's short-run cost curves are likely to be U-shaped and this is true whether the firm is competitive or monopolistic, but the shape of the demand curve will differ according to market conditions. The shape of the demand curve for the firm's product is derived from assumption (*b*) in the previous section. As the firm is too small to influence market price it must accept the ruling price as determined by the forces of supply and demand. It can sell as much as it wishes at this price.

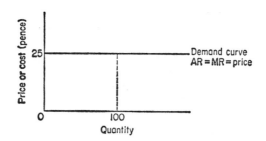

Fig. 14.1 The perfectly competitive firm's demand curve

In Fig. 14.1 we assume that the prevailing price is 25p. The firm can sell as many units as it wishes at this price and so the demand curve for its product is horizontal. If the firm decides to sell 100 units its total revenue will be £25·00, and its *average revenue* $\left(\dfrac{\text{total revenue}}{\text{number sold}}\right)$ will be equal to the price, 25p. The most important consideration for the firm is that its marginal revenue curve (showing the increase in total revenue derived from the sale of an extra unit of the product) coincides with the average revenue or demand curve. For example, if it sells 101 units its total revenue will be £25·25, an increase of 25p over total revenue from selling 100 units. This will be the case for all levels of output and accordingly: average revenue = marginal revenue = price in conditions of perfect competition.

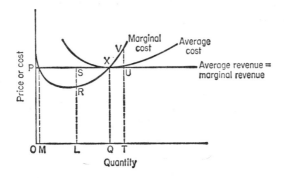

Fig. 14.2 The situation facing the perfectly competitive firm

In Fig. 14.2 we have combined the short-run cost curves and the revenue curve. Now we must decide how the firm fixes its output in order to maximize profits. This question has already been answered in a slightly different way in Unit Ten. There we said that the firm would go on employing labour so long as the addition to output was greater than the addition to costs. The same is true here: the firm is trying to maximize profits; therefore, so long as the revenue from selling an extra unit of production is greater than the cost of producing it (as long as MR is greater than MC), output will be increased. In terms of Fig. 14.2 this will apply at any stage between OM and OQ. (It is necessary to produce the unprofitable OM units in order to reach the profitable units beyond M.) Thus by producing the unit OL the firm adds LR to its costs and LS to its revenue and its profits increase. For any unit beyond OQ (e.g. unit OT) the increase in cost (TV) is greater than the increase in revenue (TU) and profits are therefore reduced. The firm will thus be justified in increasing output up to OQ but not beyond. Thus profits are at a maximum where the marginal cost curve cuts the marginal revenue curve from below. In this way the marginal cost curve can be used to show the entrepreneur when to cease expanding his business. This is the primary function of the marginal curves. The relationship between average costs and average revenue gives more information about the level of profits.

In the light of what we learnt in Unit Ten, namely that the average cost and marginal cost curves include a normal rate of profits, what can we say about the firm in Fig. 14.2? Total costs and total revenue are equal at OPXQ since average cost and average revenue are both equal at QX. The firm is thus making a normal rate of profit and is in a state of equilibrium, for given its cost and revenue curves it could not be in a more profitable position. If all the firms in an industry have the same cost curves then the industry is in equilibrium as all firms just make normal profits and are in a position where marginal cost = marginal revenue = average cost = average revenue = price.

Fig. 14.3 shows perfectly competitive firms whose short-run position is such that owing to a change in demand, average cost and average revenue are not equal. The firms are therefore in industries which are either expanding or contracting.

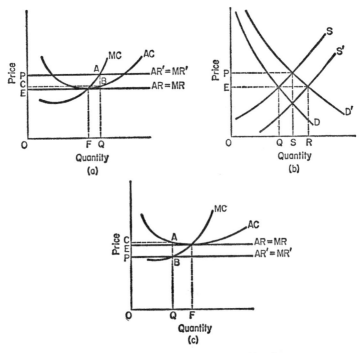

Fig. 14.3 Some more perfectly competitive firms

It is worth looking closely at what happens when the industry expands to meet an increase in demand for its product. Imagine that the firm in Fig. 14.3 (*a*) has been operating in equilibrium at price OE determined by the market forces illustrated in Fig. 14.3 (*b*). (The vertical scale is the same in each case, but the horizontal scales are very different.) An increase in demand shifts the market price to OP which the firm now accepts, and produces OQ rather than OF units and makes abnormal profits of CPAB. As new firms enter the industry the market supply curve gradually moves to the right, reducing market price until the abnormal profits are eliminated. If the supply curve moves to S' the old equilibrium price of OE is restored and the industry is experiencing constant costs. It would be unrealistic not to expect the firm in Fig. 14.3 (*a*) to join in the general expansion of the industry and we should therefore draw another set of cost curves to the right of the original pair and in the same horizontal plane (as in Fig. 10.6 (*a*)).

If abnormal profits were eliminated before the supply curve moved as far as S' the new equilibrium price would be between OE and OP. This would indicate that costs were increasing as the industry expanded and the new cost curves for the firm would be above and to the right of the old ones as in Fig. 10.6 (*d*).

The third possibility is that even when the supply curve reaches S′ there will still be abnormal profits because of the emergence of economies of scale. New firms will therefore continue to enter the industry until at a price below OE abnormal profits are finally eliminated and the industry reaches a new equilibrium. In this case the new cost curves of the firm will be to the right and below the original curves, as in Fig. 10.6 (*b*), indicating decreasing costs.

The firm in Fig. 14.3 (*c*) and firms like it will eventually have to leave the industry in circumstances discussed in Section 14.5. It may have been operating for some time in equilibrium at price OE and quantity OF when a fall in demand leaves it to face a price of OP. As firms leave the industry, the market supply curve moves to the left, causing price to rise until the remaining firms are again in equilibrium and making normal profits. This time the new equilibrium price will depend upon the extent of the leftward movement of the supply curve.

14.4 Pure Competition

If in addition to assumptions made in Section 14.2 we also assume that all the firms in the industry buy their factors of production in a perfect market, then their cost curves will be identical as the factors will be homogeneous. Assuming these conditions it would be impossible for the three firms depicted in Figs 14.2 and 14.3 to be in the same industry at the same time. If firms buy their factors in an imperfect market where some units will be more efficient than others then it would be quite possible for all three to be in the same industry at the same time, and indeed for many others with different cost patterns to be alongside them. In this case the firm in Fig. 14.2 would be a marginal firm, just in equilibrium, but likely to be in difficulty if price should fall. The firm in Fig. 14.3 (*a*) is an intra-marginal firm more than happy to remain in the industry at this level of profits; and the firm in Fig. 14.3 (*b*) is an extra-marginal firm which will be forced to leave the industry unless prices rise. If all these firms are in the industry simultaneously then it is unlikely that the industry will reach a position of equilibrium as there will always be some intra- or extra-marginal firms. Many economists prefer to call this situation (where firms are selling their product in a perfect market but buying their factors in imperfect markets) *pure competition* to distinguish it from perfect competition where both product and factor markets are perfect. In the same way impediments to perfect knowledge can reduce perfect competition to pure competition.

14.5 Leaving the Industry

We shall now investigate more closely the circumstances in which a firm will close down and leave a perfectly competitive industry.

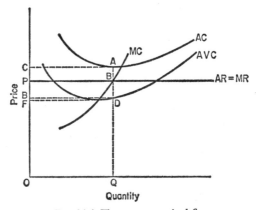

Fig. 14.4 The extra-marginal firm

If a firm cannot cover its average costs it will eventually have to leave the industry, but the timing of its exit will depend upon the extent of its deficit and the age of its fixed factors of production. In Fig. 14.4 we have drawn the firm's average cost curve and the average variable cost curve. The gap between the average cost curve and the average variable cost curve indicates average fixed costs and this is therefore smaller at high outputs than at low outputs. The firm has determined its output by equating marginal cost with marginal revenue but has to sell at price OP, below average costs OC. Profits are therefore below normal to the extent of PCAB, and if the firm continues in business for the time being this is the weekly or monthly deficit that must be met. If the owners of the business decide to cease production immediately and close down, they will still have to meet their fixed costs in the short-run, for by definition these are the same whether output is zero or a million. In this case the firm, producing nothing, would be spending BCAD (average fixed costs × previous output) on interest charges, rates, rent, standing charges and some salaries while the business was being wound up. It is obviously better to remain in production in the short-run and lose PCAB than to close down production and lose BCAD. If the short-run price falls below OF the firm would have to close down immediately for not only would it be failing to cover its fixed costs, but it could not even pay its variable costs. If the price remains at OP the best time for the firm to leave the industry would be when fixed factors of production had to be replaced. Of course if there are several other firms in the industry operating at a loss it may be that when they have to leave the industry, prices will rise sufficiently for other firms in similar positions to be able to survive.

14.6 The Justification for Perfect Competition

Since there is not a single case of perfect competition to be found in the real world the student may reasonably ask why time is devoted to its study in books such as this. Several arguments may be advanced in defence of the concept:

(a) The most important defence is that it is a standard against which other structures can be assessed. In its idealized world marginal costs and marginal revenue are equated in firms, and marginal utility and price are equated by consumers. Contemporary discussion about the pollution of the environment is very much a discussion of the inequality of marginal costs and prices: the social costs imposed by firms pumping pollutants into rivers are not reflected in the prices charged for their products. In other instances the consumer pays for more than the marginal cost of producing the goods he is buying. There may be good reasons for this, but the problem and many like it may be more rationally discussed with the aid of some knowledge of perfect competition.

(b) Although it is an idealized concept, perfect competition does allow us to discuss the happenings in an industry or market without being hampered by considerations peculiar to a special case. Modifications can be introduced as necessary but a provisional analysis can profitably be made in a world of perfect competition. In a similar way physical scientists ignore the effects of friction in their preliminary investigations; allowance can be made for it later.

(c) Perhaps most important from the student's point of view is that it is a relatively straightforward case to follow. From this beginning it is possible to introduce some important ideas which lead on to more advanced theory. The next Unit includes a study of monopoly which uses these basic concepts of competition in a different situation.

14.7 Questions and Exercises

1. What do you understand by the term 'perfect competition'? Of what use is this concept in economics?
2. Show how the perfectly competitive firm maximizes profits.
3. What are the main assumptions on which perfect competition rests?
4. Examine the relationship between total revenue, average revenue, and marginal revenue under conditions of perfect competition.
5. Show with the aid of diagrams how a perfectly competitive firm will react to (a) an increase; (b) a decrease in price.
6. Under what circumstances will it be necessary for a firm to leave the perfectly competitive industry?
7. What is meant by the term equilibrium? When will (a) the perfectly competitive firm; (b) the perfectly competitive industry be in equilibrium?

8.

Output (100)	Total fixed costs (£)	Total variable costs
1	100	20
2	100	50
3	100	90
4	100	115
5	100	144

Output (100)	Total fixed costs (£)	Total variable
6	100	175
7	100	210
8	100	300

(a) Complete the table above, by adding columns for average fixed cost, average variable cost, marginal cost, and average cost. (In each case assume that goods are produced in batches of 100.)

(b) What will be the profit-maximizing output if average revenue is 31p?

9. Explain carefully the importance of marginal decisions in economics. Show the difference between an intra-marginal firm, a marginal firm and an extra-marginal firm.

10. Outline the main differences between the concept of perfect competition and any market or industry with which you are familiar.

Monopoly, Imperfect Competition, Oligopoly

15.1 Introduction

Almost as idealized, but at the opposite extreme to the perfectly competitive industry, is the concept of the monopolist or single seller. In practice there is no perfect monopolist, but most real world situations do have monopolistic characteristics so it is worth examining the monopolist's behaviour before going on to examine more realistic situations.

15.2 The Characteristics of Monopoly

The monopolist is literally the single seller of a product. To some extent almost all producers are monopolists: for example only the Ford Motor Company manufactures Ford motor cars. The Ford Motor Company is not, however, the only manufacturer of motor cars and it is therefore not a monopolist; it is operating under conditions of monopolistic competition —it has a degree of monopoly power but must still consider the activities of its competitors who produce substitutes. Even if Ford were the world's only manufacturer of motor cars we might not regard the company as a monopolist for there would still be substitutes for motor cars—motor cycles, bicycles and public transport being the most obvious, but all goods and services ultimately competing with the motor manufacturer for the consumers' money. It is more accurate therefore to speak of degrees of monopoly rather than absolute monopoly.

15.3 The Sources of Monopoly Power

(a) Natural monopolies

Many monopolies are referred to as natural monopolies. They fall into two groups: literally natural monopolies deriving perhaps from the sole ownership of mineral deposits; Rhodesia thus has a virtual monopoly in the supply of chrome. The second group consists of those goods or services which are more efficiently supplied by a monopolist than under competitive conditions. Public transport comes into this category as does the supply of water, electricity and gas. Nineteenth-century experience in the field of public transport showed how wasteful it was to have two towns linked by perhaps three different railway lines owned and operated by three different companies. In this and many other public services amalgamations gradually established monopolies.

(b) Technological monopolies

A high degree of monopoly power is enjoyed by many large manufacturing companies because of the vast amounts of complex capital they employ. It would be impossible for any new firm to begin the mass production of motor cars, and hope to compete with existing producers. Thus, although they compete vigorously with each other the motor manufacturers enjoy monopoly power through the economies of scale.

(c) Statutory monopolies

These are established by the Government. They may coincide with the natural monopolies in (a) above. In this country the public corporations or nationalized industries are the most important statutory monopolies. We may also include in this category the patent rights enjoyed by inventors and the copyright of authors and composers who enjoy a legal, if temporary, monopoly.

(d) Cartels

These are perhaps the least desirable kind of monopoly. A cartel may be described as a group of firms acting together to reduce competition between themselves or from other firms. They may for example agree to charge the same prices for their goods or to limit their output on some quota basis. In the worst cases they may conspire to prevent the entry of new firms to the industry by exerting pressure on the suppliers of components or raw materials or on potential customers. It has usually been against this type of monopoly that Governments have taken actions and have framed controlling legislation. We will look more closely at the need for and nature of such legislation in section 15.7.

15.4 The Monopolist's Equilibrium

The monopolist like the perfectly competitive firm would be expected to have U-shaped average and marginal cost curves in the short-run since his operations are subject to the law of diminishing returns.

The demand or average revenue curve will be dissimilar. The demand for the monopolist's product is the same as the demand for the industry's product and the curve will therefore slope downwards from left to right. It cannot be vertical over its whole range for this would imply a willingness on the part of consumers to pay any price for the product, and a theoretical limit is set by the size of the national income as well as by common sense. The fact that the average revenue curve slopes down, showing that the monopolist can increase his sales by reducing his price, has an important implication for the position of the marginal revenue curve. Marginal revenue will always be below average revenue and slope down more steeply. We will suppose that the monopolist in Fig. 15.1 finds himself selling 50 units per week at £1·00 each. To sell one more unit per week he must reduce the price by one penny; he will therefore lose 50p on the 50 units that he could have sold at £1·00 and he will gain 99p on the 51st unit which could not have been sold at more than 99p. His total revenue increases by 49p and this is of course his marginal revenue, which is considerably below the average revenue of 99p.

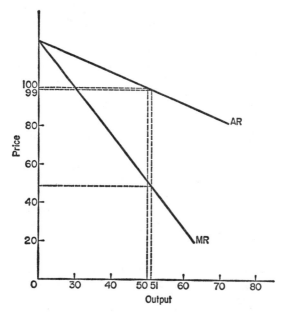

Fig. 15.1 Average revenue and marginal revenue for the monopolist

The student should establish for himself what happens to marginal revenue if price has to be reduced to 97p in order to sell 52 units. It can be shown that the MR curve will always bisect the horizontal line from the vertical axis to the AR curve. This is mentioned at this point more as an aid to drawing the diagrams than for its economic significance.

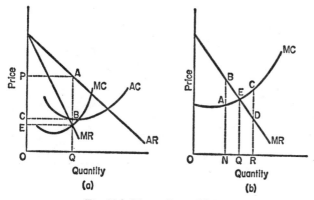

Fig. 15.2 Monopoly equilibrium

In Fig. 15.2 (*a*) the situation facing the monopolist is illustrated. In 15.2 (*b*) only the two marginal curves are drawn so that we can see that the monopolist like the perfectly competitive firm will fix his output at the point where MC = MR. If he produced less than the output OQ, for example ON, he would be depriving himself of the small area of profit ABE; if he produced more than OQ, perhaps OR, he would be reducing his profits by an amount ECD. OQ must be the profit maximizing output. The monopolist then charges what the market will bear for his product. In this case the average revenue curve shows that consumers are prepared to pay OP for a quantity OQ and this is what is charged. Notice that this is considerably above OE, the marginal cost of producing OQ units. Furthermore it is considerably above OC the average cost of producing OQ. The monopolist is thus able to make an abnormal profit of CPAB which is often referred to as *monopoly rent*. In economics 'rent' has the special meaning of being any payment over and above that required to keep factors of production employed in their present occupation. In this case the monopolist would be making a normal profit at a price of OC and would be prepared to remain in the industry if price fell to that level. He is thus receiving a rent of CPAB. We shall return to the concept of rent in Unit Twenty-one. Since the firm is by definition a monopolist, new firms are unable to enter the industry and the abnormal profits can continue indefinitely. We have only looked at one particular case of monopoly; the abnormal profits of other monopolists may be greater or smaller than those illustrated or in rare circumstances they may be non-existent.

15.5 Elasticity of Demand and Monopoly Equilibrium

There is an interesting relationship between the equilibrium output of the monopolist and the elasticity of demand for his product. In order to explore this relationship we must first examine a new method of measuring elasticity of demand. In Fig. 15.3 (*a*) the elasticity of demand at point A may be measured simply by measuring the distances *a* and *b* and dividing *a* by

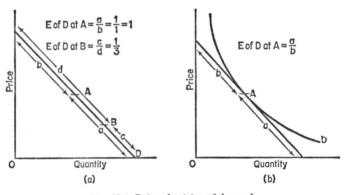

Fig. 15.3 Point elasticity of demand

b. Thus at a point half-way along the demand curve *a* and *b* will be equal and elasticity will be unity; three-quarters of the way along the curve at B elasticity will be one-third. If the demand curve is not a straight line point elasticity can still be measured by drawing a tangent to the curve at the relevant point and then assessing the ratio of the two parts of the tangents as in Fig. 15.3 (*b*).

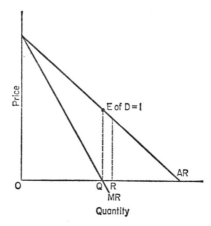

Fig. 15.4 Elasticity of demand and marginal revenue

We shall now consider the relationship between elasticity of demand and marginal revenue in Fig. 15.4. When output is OQ, marginal revenue is zero; therefore elasticity of demand is 1 (since total revenue must be constant if marginal revenue is zero). We know that elasticity of demand is less than one to the right of OQ and that marginal revenue is negative, causing a fall in total revenue. The monopolist could not be in a profit-maximizing position to the right of Q: he would be better off for example reducing output from OR and increasing total revenue. Accordingly we may conclude that the monopolist can only be in equilibrium to the left of Q, and that is when demand is elastic.

15.6 Discriminating Monopoly

We observed in Unit Three that consumers receive a surplus of satisfaction when they buy goods at the prevailing market price. If they had had to pay for the goods individually according to the utility afforded by each unit they would have spent more money on the first units purchased than on the marginal unit.

The bonus satisfaction is known as the *consumers' surplus* and it may be possible for the monopolist to reduce that bonus by charging different consumers different prices. A monopolist who can do this is known as a *discriminating monopolist*. The conditions necessary for price discrimination to be possible are:

(a) The monopolist must be able to separate his markets and keep them separate. Consumers must not be allowed to buy in the cheaper market and sell in the more expensive. The barrier between the markets may be geographical with transport costs preventing resale; or personal with the monopolist providing a personal service which cannot be transferred from one consumer to another; or temporal in that the same good may be sold at different prices at different times.

(b) The elasticity of demand in each market must be different. The object of discrimination is to raise price where demand is inelastic and reduce price where demand is elastic. If elasticity were the same in each market this would be impractical.

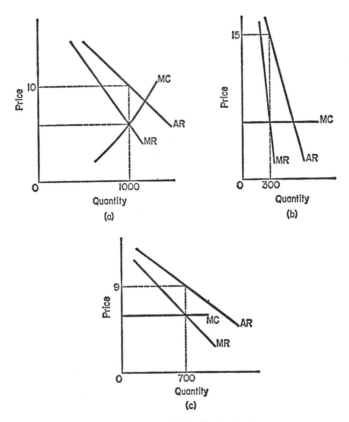

Fig. 15.5 Monopolistic discrimination

In Fig. 15.5 we can trace the effects of price discrimination. Actual prices and quantities are used as this clarifies the process. In Fig. 15.5 (a) the monopolist maximizes his profits by selling 1000 units at 10p each, giving him a total revenue of £100·00, but he knows that some consumers would not be put off by a higher price. If he can divide his market into two parts, one with inelastic demand (Fig. 15.5 (b)) and one with elastic demand (Fig. 15.5 (c)) then he will be able to divide the total output between them so as to increase his profits. He does this by equating the marginal cost of the whole output to marginal revenue in each market. Only a small reduction in price in the elastic market is necessary to increase sales. When elasticity of demand is greater than one a fall in price leads to an increase in total revenue. In this case 700 are sold for £63·00. In the inelastic market a considerable increase in price does not produce a great change in quantity and again total revenue increases, 300 being sold for £45·00, so that the total revenue is now £108·00 rather than £100·00.

Price discrimination is practised more widely than most people suspect.

(a) Nearly all the public corporations discriminate between their customers. The electricity boards, for example, differentiate between domestic and industrial consumers, selling electricity to industrial users on more favourable terms than to domestic users. In some cases they sell electricity at different prices at different times of the day. Since it is not possible to buy during the off-peak period for use during the peak period such discrimination is possible. Similar considerations lead British Rail to offer special off-peak fares to travellers, and airlines also offer preferential rates to bona fide students.

(b) The *dumping* of goods in international markets is an important example of price discrimination. Dumping is the term applied to the practice of selling goods overseas at prices below those in the home market and often below the cost of production. Surplus agricultural produce resulting from the European Economic Community's policies are often dumped in this way.

(c) The system of selling identical goods under the brand name of the manufacturer and also that of the retailer is a form of price discrimination.

(d) Where personal services are sold discrimination may be widespread. Barbers often cut old age pensioners' or children's hair at preferential rates; many people prefer to pay for private dental treatment even though treatment under the National Health Service may not be significantly different; doctors' prescriptions for the young, the old and the chronic sick do not incur charges in the United Kingdom.

15.7 Governments and Monopoly

We are now in a position to consider the effects of monopoly and the reasons why Governments seek to control them. Attitudes to monopoly have fluctuated over the past forty or fifty years. In the 1930s the Government openly encouraged the amalgamation of firms into much larger units as a defence against overseas competition. In the face of contracting markets domestic companies were crippling themselves in their efforts to force rivals out of business. The Government tried to encourage a more rational reduction of the number of firms in the hope that the emergent larger firms would be more efficient than the former small ones. It is of course one of the arguments in favour of monopolies that they enjoy the benefits of economies of scale; that their long-run average cost curve slopes downwards.

In the immediate post-war period the Government's attitude became more questioning and following a number of inquiries the first of a series of Acts of Parliament dealing with monopoly and restrictive practices was passed in 1948. This legislation is summarized below:

(a) Monopoly and Restrictive Practices (Inquiry and Control) Act 1948

This established the Monopolies Commission whose task was to investigate cases referred to it by the Government. A monopolist was defined as a firm responsible for 30 per cent of domestic production, but many inquiries were made into cartel type monopolies. The most important inquiry was one into Collective Discrimination by groups of firms against their suppliers, rivals or customers (1955). This, together with the considerable evidence thrown up by earlier reports, led to a hardening of attitudes in the form of new legislation.

(b) Monopoly and Restrictive Practices Act 1956

This was divided into three parts.

(i) The Act established a Registrar of Restrictive Practices who keeps a register of all collective agreements which restrict the freedom of the parties to decide the price of their product, the size or quality of their output or the destination of their product. It is the Registrar's duty to bring the parties to an agreement before the Restrictive Practices Court, especially established for the purpose, where the onus is on them to show that their agreement is not against the public interest. The Act provides seven ways by which their innocence may be established:

(i) the agreement is necessary to protect the public against injury;

(ii) the agreement ensures the public receive some specific and substantial benefit which would disappear if the agreement were ended;

(iii) the agreement is necessary to enable small firms to compete with large firms;

(iv) the agreement is necessary to prevent small firms being exploited by very large customers or suppliers;

(v) the agreement is necessary to maintain the level of employment in particular areas or industries;

(vi) the agreement is necessary to maintain exports;

(vii) the agreement is necessary to maintain another restriction already approved by the court.

Even if an agreement satisfies one of these points, it must still be shown that the benefit from it outweighs any general detriment to the public from the agreement as a whole. The success of the Court may be gauged from the fact that by the end of 1969, of 2660 agreements on the register 2370 had been abandoned or modified to make them acceptable.

(ii) The second provision of the Act was the abandonment of collective resale price maintenance. This is a practice whereby a group of manufacturers would fix prices between themselves and enforce these at the retail stage, if necessary by a collective boycott of retailers offering goods at cut prices. At the same time, to emphasize the Government's primary concern with *collective* discrimination the practice of an individual company stipulating resale prices for its goods was specifically upheld in the Act.

(iii) The Monopolies Commission was retained to investigate cases which didn't fall within the scope of the Registrar and the Court. The cases are referred to the Commission by the Department of Trade and Industry.

(c) Resale Prices Act 1964

Resale Price Maintenance was felt to be something of an anachronism in an increasingly competitive economy and was a practice more honoured in the breach than the observance. Co-operative society dividends, trading stamps, 'free gifts' and flagrant price-cutting combined to undermine the system, though the Courts on many occasions issued injunctions against price cutters at the request of manufacturers. After 1964 those manufacturers who wished to maintain their prices had to persuade the Restrictive Practices Court that this was in the public interest in the case of their product. This time there were five gateways available:

(*i*) the quality or variety of goods would be substantially reduced in the absence of RPM.

(*ii*) the number of retail outlets would be substantially reduced in the absence of RPM.

(*iii*) prices of goods would increase in the long-run;

(*iv*) goods would be retailed under conditions likely to endanger health;

(*v*) services provided with or after the sale of goods would be reduced.

Again there was the requirement that any detriment likely to occur to the public as a result of the abandonment of RPM should outweigh any detriment deriving from its retention.

As a result of the Act and the very narrow interpretation of escape clauses by the Court the practice of RPM has virtually disappeared in the United Kingdom, although it has been replaced in many industries by a system of *recommended prices*, which may be as effective as maintained prices.

(d) Monopolies and Mergers Act 1965

Previous legislation enabled the Government to deal with monopoly situations only after they had developed. This Act enabled them to order the postponement of proposed mergers for six months pending an inquiry into the likely effects by the Monopolies Commission. It is only mergers involving the transfer of more than £5 million of assets or which would result in one firm controlling one-third of domestic output which are subject to this delay. On the recommendation of the Commission some proposed mergers have been abandoned: in 1969 Unilever and Allied Brewers were prevented from merging, in 1968 Barclays and Lloyds were refused permission to merge even though (or perhaps because) Westminster and National Provincial had already amalgamated.

(e) The Industrial Reorganization Corporation 1966

This corporation was established to encourage mergers! It was provided with £150 million of public funds and was expected to be active among smaller firms. It became involved to varying degrees, however, in quite spectacular mergers, supplying funds in connexion with the take-over of Rootes Motor Company by Chryslers of America, and in connexion with the General Electric Company/Associated Electrical Industries take-over.

The existence of legislation to prevent mergers, and a publicly sponsored body to encourage them should not be regarded as inconsistent. Both can be seen as arms of the same policy designed to encourage efficiency and serve the public interest. It was perhaps unfortunate that the Industrial Reorganization Corporation was wound up by the new Conservative Government in 1971 before the fruits of its labours could be properly assessed.

15.8 Imperfect Competition

There are a variety of situations between the two extremes of perfect competition and monopoly. There may be four or five firms dominating a market or there may be hundreds competing for custom. It is the latter situation that concerns us here. In an imperfectly competitive industry each firm will be producing basically the same product but will seek to distinguish it from its rivals by product differentiation. The difference will often only be a matter of packaging but the manufacturer sets out to establish his product as something different from the others, even though they are in actual fact

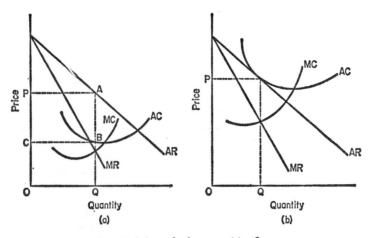

Fig. 15.6 Imperfectly competitive firms

very close substitutes. In this way the firm will be able to increase its sales by reducing prices, and it is faced by a demand curve similar to that of the monopolist, and depicted in each of the diagrams in Fig. 15.6. The short-run equilibrium position of the firm is identical to that of the monopolist and the firm is able to make abnormal profits of CPAB. Under conditions of imperfect competition new firms are free to enter the market, and we may expect them to attract some customers away from existing firms so that the average revenue curve of each existing firm moves to the left. The extent of the leftward movement depends upon the success of each firm in maintaining the loyalty of its customers but the limit of the movement is indicated in Fig. 15.6 (*b*) which shows the long-run equilibrium of the firm. In this case the AR curve has moved far enough to the left to become a tangent to AC and thus at the profit maximizing output of OQ the firm is just making a normal rate of profit, and is operating some way below the optimum size.

As there are similarities between this case and that of the monopolist many people refer to imperfect competition as monopolistic competition. It is worth noticing here that whereas under conditions of perfect competition we were able to use the firm's marginal cost curve as its supply curve, this is not possible for the imperfectly competitive firm. The only way to derive a supply curve for it is to observe its reaction to changes in demand. In Fig. 15.7, when AR and MR are the relevant demand and marginal revenue curves, we cannot read off the quantity supplied from the MC

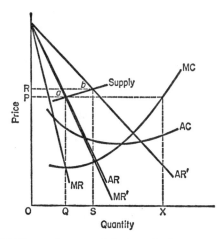

Fig. 15.7 The supply curve of the imperfectly competitive firm

curve, as this would give a supply of OX when in fact OQ is produced at price OP. If the demand curve shifts to the right, the MR curve now cuts the MC curve at a higher point and output rises to OS at price OR. A second move to the right would give a higher price and quantity and we may regard the line joining the points *a*, *b* and others on the successive AR curves as constituting the supply curve of the firm.

15.9 Oligopoly

A special and increasingly important case of imperfect competition is that of *oligopoly* or competition among the few. The motor car industry in the United Kingdom is dominated by four giant manufacturers and the supply of detergents is dominated by two firms. Such industries are *oligopolies*. The firms in these industries must consider far more than their own price and output when determining policy. There is a special case of oligopoly known as *duopoly* where only two firms compete in the same market. While the ordinary imperfectly competitive firm can concentrate on its customers' reactions to a change in price the duopolist must also try to assess the reaction of its competitor. Thus if the duopolist wants to increase his price he must assess the contraction in demand which will depend largely upon the closeness of other substitutes, and the brand loyalty that he has managed to establish among his customers. He must also make an assumption about the behaviour of his rival; will he increase his prices as well, or will he deliberately hold them down hoping to gain new customers? Only on the basis of a firm assumption can the duopolist plan his most profitable output.

In the circumstances it is not surprising that many economists feel that the outcome will be some form of collusion between the two firms. They may agree to inform one another in advance of price changes; or only change prices when important factors of production increase in price; or they may agree to limit advertising expenditure. The danger is that limited collusion may spread so that the firms form a rigid cartel perhaps fixing prices above a competitive level and restricting output. Indeed it may be argued that the firms will decide that their best interests will be served by trying to maximize their joint profits rather than their individul profits. In this case competition between them will be restricted largely to non-price competition—free offers, attractive wrappings, persuasive advertising and so on. It was to counteract such situations that the legislation outlined in Section 15.7 became necessary.

15.10 Criticisms of the Theory of Imperfect Competition

The traditional theory of the imperfectly competitive firm has been subjected to considerable criticism, partly because of the growth of oligopolistic industries. There are two connected criticisms.

(*a*) Modern industrial firms are not obsessed with marginal changes in costs and revenue; indeed it is very difficult to calculate marginal cost in many industries. Today's industrialist is more likely to try to establish the average variable cost of producing his output and then add a fairly arbitrary percentage in order to cover the fixed costs including profits. We cannot pursue this contention here but we must note it and also the fact that many firms employ a number of accountants to control the relationship between costs and revenue.

Pricing in the public sector provides particular difficulties. Many services are provided to consumers free of direct charges, being financed out of taxation. Others, like National Health Service prescriptions, are subject to arbitrary fixed charges, and others, like the charges made by the public corporations, are subject to many influences. But there are two broad approaches to this kind of pricing, based on average costs and marginal costs.

The first approach is to base prices on average costs. Total costs are estimated and divided by the anticipated sales. While this has the merit of simplicity it does lead to the problem of cross-subsidization of some consumers by others. Consider what happens with electricity. The peak demand for power occurs at about 8 a.m. and 9 p.m. and the Electricity Board must have sufficient generating equipment to meet that peak; much of this equipment will be idle for most of the time, and if the demand for electricity could be spread evenly over the day, less equipment would be required. It follows that the person who consumes his electricity at off-peak periods imposes less of a burden on the Board than the peak-period user. If both users are charged the same price per unit there is cross-subsidization of the on-peak user by the off-peak user, and this is what happens for the majority of consumers.

To overcome this, a system of marginal cost pricing would be necessary. This would result in peak-period users paying higher charges than off-peak users, since they are the consumers who necessitate the installation of extra generating plant. It is technically possible to record not only a consumer's intake of electricity but also the time of consumption. The consumer could then be charged retrospectively for his electricity, paying more for peak period consumption than for off-peak consumption.

There are other difficulties however. First it would lead to a great increase in the administrative costs of the Electricity Board. Furthermore, it is in practice very difficult to estimate marginal costs precisely. Accordingly the Electricity Boards compromise by charging a lower price to consumers taking electricity in off-peak periods for storage heaters. There is the further problem that it costs the Board more to supply electricity to a remote homestead than to one of a thousand terraced houses on a large estate. Here the introduction of marginal costing would lead to very heavy price rises for those living in rural areas.

(b) A further criticism of the theory of imperfect competition lies in the argument that the modern industrialist does not recognize the U-shaped cost curves familiar to economists. This is an argument concerning the long-term rather than the short-term: it is impossible for short-run cost curves not to be U-shaped owing to the operation of the law of diminishing returns. It is also quite possible that technical and managerial economies will lead to a falling or horizontal average cost curve up to very large outputs, in which case the long-run marginal cost curve will also be horizontal. But it is also extremely likely that managerial diseconomies, if nothing else, will eventually lead to rising costs so that the U-shaped long-run curve will be replaced by a saucer-shaped one.

15.11 Questions and Exercises

1. What do you understand by the term 'monopoly'? In what ways may a firm obtain monopoly power?
2. Show how the monopolist will maximize his profits.
3. How is a monopolist able to increase his profits by discriminating between his customers?
4. Examine the differences between monopolistic competition and perfect competition.
5. Outline the principal legislation by which United Kingdom Governments have sought to control monopoly and restrictive practices since the war.
6. What is meant by the term 'the public interest' in connexion with monopoly legislation.
7. The following charges were made by the Post Office for telephone calls in 1972:

Local calls: Time bought for 1p.

Mon—Fri	8 a.m.—6 p.m.	6 mins
All other times		12 mins

Calls over 50 miles: Time bought for 1p.

Mon—Fri	9 a.m.—Noon	8 secs
Mon—Fri	8 a.m.—9 a.m.	
	Noon—6 p.m.	10 secs
All other times		36 secs

Comment upon the economic motives for this kind of price discrimination.
8. For your own information keep a check on the prices of perhaps four different goods in two or more retail outlets over as long a period as possible. Comment on the changes that occur and the differences that exist between different retailers.
9. 'Trading stamps, free gifts and other promotional schemes lead to higher prices for the consumer.' What evidence can you find to support this claim? (You might start by discovering the price of petrol at different filling stations.)
10. If large firms are more efficient than small firms why is it necessary for the Government to establish a framework of monopoly legislation to control them?

Unit Sixteen
Money and Banking

16.1 Introduction

Unit Two showed that the wealth of a community exists in the goods
and services it produces, and that money is merely a convenient way of
measuring wealth. We must now investigate money more closely, and determine what it is, what it does, and what problems it creates. The principal
financial institutions, the banks and finance houses of various kinds that
have developed to handle and control money, must also be included here.

16.2 The Functions of Money

We all know what money is, but we are so familiar with it that we tend to
take it for granted. The very existence of our complex economic structure
depends upon the existence of money and we can best examine its functions
by imagining that from tomorrow we shall have to manage without money
in the economy. It would soon become clear that the main purposes of
money are as follows:

(a) To act as a means of exchange

In the absence of money goods would have to be bartered directly for each
other. Most children barter or swap comics, stamps, autographs or even
coins, with little difficulty; and a not inconsiderable amount of barter
occurs in international trade, particularly among the Communist bloc
countries. In the latter case, however, although money does not change
hands, it is lurking in the background in determining the amounts of commodities that should be swapped for each other. Imagine, however, going to
work on the production line in a car factory and being offered wages in the
form of a starter motor or a set of brake shoes! Even if you accepted payment in this form you would have the problem of exchanging the motor for
food or clothes. It would be simple to find a grocer anxious to dispose of
surplus food or a tailor selling clothes but unless either of them wanted a
starter motor in exchange you would be unable to deal with them. The
most important function of money is thus to eliminate the need for this
double coincidence of wants. Your employer pays you in cash which you can
then exchange for food and clothes without the supplier of food or clothes
needing anything that you have produced. They can then use the money to
buy commodities that they desire. In this way money serves as a *means of
exchange,* and for this reason it is sometimes called the oil of the economy.

(b) To act as a measure of value

If your employer tries to pay you with brake shoes you may well disagree with him over the number of brake shoes your labour is worth. Once agreement is reached you will then have to agree appropriate exchange rates of brake shoes against food and clothes. Meanwhile the grocer and tailor will be negotiating exchange rates with other customers offering books, pottery, leather goods and a thousand other things in exchange for food and clothes. Everyone would need to carry a host of exchange rates in his head in order to trade effectively. The second function of money is thus to obviate such chaos by serving as a *measure of value*. Comparisons of the value of goods is facilitated by reducing them all to a single standard. It thus becomes much easier to recognize a bargain and to avoid paying excessive prices.

(c) To act as a store of wealth
Workers may not wish to spend the whole of their income within a few days of receiving it, and if they are paid in some durable engineering product it is possible to keep it until a later date. If they were paid in the form of a perishable commodity such saving becomes more difficult. What would happen to a fisherman, for instance, if he were paid in fish! The third function of money is, therefore, to act as a *store of wealth* and to allow the occurrence of saving.

(d) To act as a standard for deferred payments
Most people do not in fact hold their savings in the form of money, largely because it yields them no interest. They place their money with one of a number of financial institutions, who promise to repay on demand or subject to specific notice, and to pay interest on the money. Again money is used to fix the repayments and the interest. Thousands of other agreements to settle debts at a future date also depend on money serving as a *standard for deferred payments;* wage agreements and mortgages are important examples.

By fulfilling each of these functions the commodity that serves as money allows the economy to develop; as money serves as a means of exchange it allows division of labour and specialization to occur. Each worker is confident that he is no longer dependent on the achievement of a double coincidence of wants to exchange his surplus output. Money as a standard of value permits the speedy settling of transactions and thereby acts as a stimulus to trade. In its roles of a store of wealth and standard for deferred payments money permits and even encourages savings, which, as we shall see later, in Section 22.6, are the very source of economic achievement.

16.3 What is Money?
Money may be defined as anything which is generally acceptable and accepted in exchange for goods and services. The commodities serving as money have varied enormously from time to time and place to place; cattle were once widely used in India and many other places as money, and the word

rupee is said to be derived from an ancient Sanskrit word for cattle. The Aztecs of Mexico had a complex economic structure which was dependent on cocoa beans for a currency, and the cultivation of these was carefully controlled. At other times maize was used in Mexico, beeswax was once used in Borneo, and pigs have been used in the New Hebrides; and throughout the islands of Oceania shells and animals' teeth have enjoyed a widespread use as a currency. There is considerable evidence to support the existence of a cattle standard in parts of Europe including Wales and Ireland.

Although livestock may serve as the basis of a currency system it must be supported by other commodities as it is indivisible. In most instances metals were used in this secondary role and have eventually taken over as the basis of the currency. Gold rings, iron swords, metal shells, copper crosses, and axeheads have all been used at various times. But before we consider the development of money in the United Kingdom we will consider the qualities necessary for an effective monetary commodity.

(a) Acceptability

Money needs to be generally acceptable; commodities were originally used as money because they had *intrinsic value* to the communities in which they were used: they were desirable for their own sake as well as being a means of exchange and acquiring other goods. Today money generally has no intrinsic value, but we accept it in the knowledge that other people will accept it from us in exchange for goods.

(b) Stability of value

We should be less eager to accept money if we expected its value to fall by 50 per cent by the end of the day. Indeed, when this happens, and people anticipate that rapid price rises will erode the value of money, the monetary system may collapse as it did in Germany in 1923. Prices normally rise by a small percentage each year so no modern money has complete stability of value.

(c) Durability, divisibility, portability

Money should be durable and not deteriorate over time—one of the disadvantages of livestock. It should also be readily divisible so that the smallest of payments can easily be made. It is also helpful if money is easily portable.

(d) Uniformity

A minor consideration is that money should be uniform in quality to prevent the bad money driving the good out of circulation. If one holds two coins, one of intrinsic value 10p and one of no intrinsic value, then one is tempted to spend the latter and keep the former. This is the principle known as Gresham's Law after an Elizabethan finance minister. In this connexion it is worth noting that money should be difficult for counterfeiters to copy; hence the ornate designs on banknotes.

16.4 The Development of British Money

Our understanding of the modern monetary system will be simplified if we know something of its background, although a detailed economic history of the United Kingdom cannot be given here. Coins were in widespread circulation by Norman times. They bore the stamp or effigy of the monarch and so the community accepted them as the general means of exchange. Control over the supply of money was negligible. One of the great weaknesses of the currency was the ease with which the king could debase it in order to increase his personal income. The monarch could call in existing gold or silver coins, melt them down, add perhaps 20 per cent of base metals, then reissue the same number of coins with the same face value and keep the balance of gold or silver for himself. The only merit of this policy was that it accustomed the community to accept money whose intrinsic value was less than its face value.

Paper money first developed in the seventeenth century. Wealthy traders began to leave their holdings of gold or silver coins at the local goldsmiths and accept a receipt or promise to pay in return. It was not long before the receipts came to be used as money. The depositor, rather than drawing coin from the goldsmith's vaults could simply endorse the receipt he had been given by the goldsmith, transferring his claim against the goldsmith to his own creditor. If the goldsmith was a trustworthy person the receipt could change hands many times and could properly be regarded as money, although it would circulate only on a local basis. At this time the notes or receipts were fully backed by gold: for every £100 worth of receipts issued the goldsmith held £100 in gold or silver coin.

During the eighteenth century the goldsmiths who already charged a fee for looking after gold deposits, developed a far more profitable technique. It had become obvious that most traders were willing to use notes issued by the goldsmiths in their everyday transactions, and would redeem them only on rare occasions. A banker (as we may now call them) would notice that on average only 20 per cent of the depositors' gold coin was withdrawn during a week, and that most of this was rapidly redeposited. If the banker always held gold coin equivalent to 20 per cent of the total number of notes, he could always meet the requirements of his customers. It therefore became the practice to issue notes in excess of deposits to customers wishing to borrow money. The balance sheet of such a banker might look like this:

Liabilities	£	Assets	£
Promises to pay gold on demand	5000	Gold coin	1000
		Customers' promises to repay	4000
	5000		5000

Thus customers holding notes issued without coin being deposited had exactly the same claim against the bank as those who had actually left coin with the bank. If all note holders tried to redeem their notes simultaneously then the banker would be unable to meet his liabilities, but this event was unlikely to occur if the banker kept a prudent reserve of gold. Some greedy bankers, however, issued notes far in excess of deposits, and on many occasions legitimate requests for cash led to bank closures and losses for depositors. Despite such setbacks the system of *fractionally backed* notes survived. The indiscriminate issue of such notes by what had become known as country banks was controlled, especially by the Bank Charter Act of 1844, which limited the issue of notes by such banks. This Act allowed the Bank of England an unlimited issue so long as the notes issued did not exceed the Banks' gold holding by more than £14 million. This £14 million was known as the *fiduciary issue,* and it was backed only by faith. When banks amalgamated their note issue was taken over by the Bank of England. As there were many amalgamations in the second half of the nineteenth century the Bank issue increased and that of the country banks gradually disappeared. But throughout the period notes were convertible into gold by the issuing bank on demand. This practice ceased in 1914 with the outbreak of war, and was re-established in 1925 for sums of £1700 or more, but was finally abandoned in 1931 when the gold standard finally collapsed (see Section 26.2). Since that time notes have been of token value only, although coins containing a high proportion of silver were minted until 1946.

Alongside these developments in the note issue even more fundamental changes occurred within the banking system. The early bankers lent to their customers by giving promises to pay which were in effect money, but in the nineteenth century their lending techniques became more sophisticated. When customers wished to borrow the bank would make entries in the ledger showing the sum available to each customer, who would be free to transfer his deposit in two ways; either by withdrawing cash or instructing the bank via a cheque to transfer funds to another's account. Banks will be described more fully in the next unit, but we must note here that the bulk of today's money supply consists of entries in bank ledgers and has no tangible existence.

16.5 The Value of Money

One important characteristic of money is that it should have stability of value. It will soon become apparent when money lacks this stability for the prices of other goods will be rising or falling significantly. Various Government bodies make their own assessment of changes in the value of money by examining changes in the prices of goods and services and constructing index numbers. Such numbers are widely used in economic discussions so it is important to understand their construction and deficiencies. First we shall consider the retail price index.

(a) The construction of index numbers

The object of a price index is to measure the average change in prices. It cannot contain every price change so the compilers take an imaginary basket of goods, representative of the expenditure of an average family, and trace changes in the prices of goods included in the basket. In the first year, the *base year,* each individual price is taken as a figure of 100 which is known as the *price relative.* We will suppose that the index contains three goods, food, shirts and petrol whose prices in year one are 25p per pound, £2·50 per shirt, and 35p per gallon. Each of these prices is regarded as 100 in year one. Subsequent changes are related to this 100. If the price of food rises to 30p in year two then its price relative becomes 120; when petrol falls to 28p its price relative becomes 80; and shirts at £2·75 give a price relative of 110. These changes are summarized in Table 16.1.

Table 16.1 A hypothetical price index

Com- modity	Actual prices Year 1	Year 2	Price relatives Year 1	Year 2	Weight	Expenditure relatives Year 1	Year 2
	£	£					
Food	0·25	0·30	100	120·0	6	600	720
Shirts	2·50	2·75	100	110·0	1	100	110
Petrol	0·35	0·28	100	80·0	3	300	240
Index number			100	103·3		100	107

A simple index may be obtained by averaging price relatives for year 2. In our example the index has risen from 100 to 103·3 which suggests that retail prices have risen by 3⅓ per cent between the two years and that the value of money has fallen by 3⅓ per cent. This procedure may be repeated to measure subsequent changes. Unfortunately it vastly over-simplifies the task of measuring changes in the value of money unless consumers divide their expenditure equally between the three commodities in the basket. In order to make the index more realistic we must give *weights* to each commodity according to their relative importance in household expenditure. The Government's retail price index has a total weighting of 1000 divided between various commodities on the basis of the pattern of expenditure of a previous year. If the investigation shows that consumers' expenditure is divided between food, shirts and petrol in the ratio 6:1:3, then the price relatives for each year are multiplied by the weights to give *expenditure relatives,* and the overall index is calculated by dividing the sum of the expenditure relatives by the sum of the weights. Thus the index moves from 100 in year 1 to 107 in year 2. When compared with the index calculated purely on the basis of price relatives more emphasis is given to the rise in the price of food, which is responsible for 60 per cent of expenditure.

(b) **Difficulties of constructing and using index numbers**

In practice the construction of an index is more complicated. The main difficulties include the following:

(i) The contents of the basket. The compilers must decide what goods to take into account. Their decision must largely depend upon the purpose of the index: there is no point in including mink coats if the index is meant to indicate changes in the value of money for the general public. The authorities do in fact construct a separate index with a different set of weights to gauge the effects of price changes on the budgets of old age pensioners, as they spend a large proportion of their income on food and fuel.

(ii) The weighting of the contents of the basket also poses problems. In the case of the retail price index the weights are frequently changed to reflect the growing importance of some goods and the relative decline of others in the average family's budget. Over a long period the usefulness of the index is reduced by such changes. The index also cannot take account of changes in quality or the introduction of new goods or services.

(iii) Measuring prices. It is unlikely that the same goods will be sold at the same prices all over the country. This can be overcome only to a limited extent by recording changes in a number of different centres.

(iv) The choice of a base year will affect the subsequent movements of the index. If prices are high in the base year then increases in the index in the following years may be damped down.

(c) **The current retail price indices**

In Table 16.2 the weighting of the current retail price indices is outlined. It can be seen that the weighting of the general index differs from that of the pensioners' index and that the weights have been changed to reflect changes in the pattern of expenditure.

Table 16.2 Weighting of the retail price indices

Commodity group	General retail price index		Single pensioner household index	
	January 1962	January 1972	January 1962	January 1972
Food	319	251	461	417
Alcoholic drink	64	66	16	17
Tobacco	79	53	34	29
Housing	102	121	Not included	—
Fuel and light	62	60	189	205
Consumer durables	64	58	39	46
Clothing, footwear	98	89	69	63
Transport, vehicles	92	139	26	34
Miscellaneous goods	64	65	77	80
Services	56	52	75	89
Meals out	—	46	14	20
All items	1000	1000	1000	1000

Source: *Department of Employment Gazette*, HMSO

Table 16.3 Movements in retail price indices 1962–72

Jan:	1962	1963	1964	1965	1966	1967	1968	1969	1970	1971	1972
General index	100	102·7	104·7	109·5	114·3	118·5	121·6	129·1	135·5	147·0	159·0
Single pensioner household index	100	103·9	107·0	111·5	116·3	119·0	124·5	131·1	140·2	154·4	162·5

Source: *Department of Employment Gazette*, HMSO

Table 16.3 gives changes in the indices since 1962. The general index shows that prices rose, or the value of money fell by 59 per cent between 1962 and 1972, but for the single pensioner household the decline has been somewhat greater.

16.6 Inflation: a Preliminary Word

The tendency of prices to rise, and the value of money to fall, is known as inflation. One of the main aims of Governments is to control the rate of inflation because of its undesirable effects on the economy. There is a full discussion of inflation in Unit Twenty-three, but it is important to realize the relationship between the supply of money and the rate at which prices are rising.

Let us imagine a primitive economy producing food as its only commodity. The economy has a supply of 1000 one pound notes each of which can be used once. In year 1 1000 units of food are sold at £1 each and consumed. If in year 2 the same amount of food is produced but the supply of money is increased to £2000, then those people with access to extra money will be prepared to offer more for each unit and prices will rise towards £2 a unit. If, on the other hand, the supply of food had increased by 100 per cent as well then we could expect prices to have been stabilized. The relationship is actually far more complex than this suggests, but the supply of money is an important consideration in the question of inflation. We must now explain the meaning of the supply of money.

16.7 The Supply of Money

For day to day transactions most people use the bank notes and coins issued by the authorities. But the most important source of money is in the form of bank deposits. In December 1972 notes and coins in circulation amounted to £3992 million, while bank deposits amounted to £21,221 million. It is clear why notes and coins have been referred to as the small change of the economy. It is now necessary to examine more closely the ways in which banks have built up their deposits to the point where they constitute over 85 per cent of the total money supply in the country.

16.8 Questions and Exercises

1. Examine and comment on the main functions and characteristics of money.
2. In what ways does the British monetary system meet the functions and characteristics discussed in Question 1?
3. Examine the system by which the goldsmiths were able to create money, bringing out the dangers of the system.
4. Explain how you would construct an index of retail prices. Consider the usefulness of this and other indices in economics.
5. What is meant by the term 'supply of money'? Consult the current *Financial Statistics* (HMSO) to discover how the supply is made up.
6. Trace the steps by which the United Kingdom's monetary system has evolved.
7. In what ways would each of the following be inadequate as the basis of our monetary system: (a) chalk; (b) cattle; (c) gold; (d) eggs; (e) lead.
8. What dangers are involved in using the *Retail Prices Index* as a cost of living index?
9. Trace the changes that have occurred in the *Retail Prices Index* over the last 12 months and comment on their significance.

Unit Seventeen

Financial Institutions: the Deposit Banks

17.1 Introduction

This and the two following Units will examine the role of the financial institutions. We shall begin with the Deposit or Clearing Banks, as these are the most important and most familiar. Unit Eighteen will describe the Bank of England which acts as a link between the Government and the deposit banks. Unit Nineteen explains the activities of some other financial institutions.

17.2 The Deposit Banks

The goldsmiths, who were the forerunners of the modern banks, were instrumental in increasing the money supply, initially through the issue of fully backed paper notes but eventually by the creation of deposits in their ledgers. It is the latter that constitute banks' essential function today and this will be investigated in the next section. The most important of the deposit banks are the London Clearing Banks of which the largest are Barclays, Lloyds, Midland and National Westminster. These have emerged as the result of a long period of amalgamations the most recent of which were the merger of National Provincial and Westminster and the absorption of Martins Bank by Barclays in 1968. The most important distinguishing feature of these banks is the facility they provide for customers of transferring deposits by cheque. The great power of the banks stems from this facility.

17.3 The Creation of Credit

(a) A one bank system

Let us imagine an economy in which there is only one bank. Soon after beginning business it finds that individuals and firms have placed £10,000 with it for safe-keeping. Its balance sheet (ignoring the shareholders' capital or property owned by the bank) would appear as follows:

Balance Sheet 1

Liabilities		Assets	
Customers' deposits	£10,000	Cash in hand	£10,000

The balance sheet is in effect a photograph of the bank's position at a particular point in time. The liabilities show the amounts that the bank may be called upon to provide to its customers and the assets show the cash and other resources available to the bank to meet its liabilities. At this stage it is quite clear that the bank has sufficient cash in its till to meet any demands made by its customers.

In practice customers prefer to settle their debts with each other by cheque, ordering the bank to transfer money from one account to another. Thus if Adam and Brown have each deposited £500 at the bank, and Adam owes Brown £100, he can settle his debt by instructing the bank to reduce his account by £100 and to increase Brown's by the same amount. No cash changes hands; the bank still has obligations to its customers of £10,000; there has simply been a slight readjustment to those obligations.

If all the bank's depositors were always prepared to settle their debts in this way the bank could forget all about its holdings of cash. Customers will, however, need to draw a certain amount of cash from the bank each week to make small payments (it is not usual to write cheques for very small amounts) and to pay those people who prefer not to use the banking system. If the bank discovers that, at the most, the weekly withdrawal of cash amounts to 10 per cent of total deposits, and that this is quickly re-deposited by traders accepting cash payments from customers, then the most cash the bank needs to meet demands from its customers with deposits of £10,000 is actually only £1000. Alternatively we may take the view that with cash in hand of £10,000 the bank can afford liabilities of £100,000. In this case let us imagine a customer, Mr Clark, who approaches the bank for a loan of £1000. The bank manager is agreeable and opens an account for him with a credit balance of £1000. Mr Clark can now write cheques to the value of £1000 although he has placed no money in the bank; he simply promises to repay the £1000 plus interest, having probably offered some security to the bank. The bank's balance sheet now shows a different picture.

Balance Sheet 2

Liabilities		Assets	
Customers' deposits	£11,000	Cash in hand	£10,000
		Loans to customers (or promises to repay by customers)	£1,000
Total	£11,000	Total	£11,000

There is now insufficient cash to supply all the customers if they wished to withdraw their deposits, but the bank knows that the most that is likely to be withdrawn is £1100. It will, therefore, be prepared to go on making loans (or creating credit, which is the same thing) until the cash that is held is equivalent to only 10 per cent of desposits (see Balance Sheet 3).

So far as customers are concerned the standing of their account is the same whether they have actually deposited cash to open the account or whether it has been created by a loan. When they spend their money the recipient has no means of knowing whether or not they originally deposited cash. Thus in creating credit the banks have added to the money supply.

Balance Sheet 3

Liabilities		Assets	
Customers' deposits	£100,000	Cash in hand	£10,000
		Loans to customers (promises to repay)	£90,000
Total	£100,000	Total	£100,000

In practice the banks will not create deposits for customers merely on the strength of promises to repay. Depending upon their creditworthiness the bank will insist upon its customers providing some kind of security which the bank can sell to raise cash in the case of default by the borrowers. Such security might be stocks and shares, the deeds of a house or an insurance policy with sufficient surrender value.

A more important refinement must be made in respect of the duration of loans, and the commitment of customers to repay. In the case mentioned above the bank could meet cash withdrawals of £10,000 because it has established a *cash ratio* of 10 per cent. The cash ratio is the ratio of cash to deposits. We must now consider what would happen if customers came and withdrew £1000 which is not then redeposited by other customers.

Balance Sheet 4

Liabilities		Assets	
Customers' deposits	£99,000	Cash	£9,000
		Loans to customers (promises to repay)	£90,000
Total	£99,000	Total	£99,000

Since £1000 has been withdrawn the banks assets have fallen by £1000 and their obligations to depositors have fallen by the same amount, but their holding of cash has fallen by a greater percentage than their obligations: the ratio of cash to deposits is now $\frac{9,000}{99,000} = 9.08\%$. If the bank wishes to maintain a ratio of 10 per cent, either because it is enforced by the Government or because it makes economic sense, then they must be able to replenish their cash holding quickly. When granting loans to customers the bank will, therefore, stipulate the period of the loan and will often include a clause requiring repayment at perhaps seven days or three months notice. Sometimes they will lend subject to repayment on demand. In these circumstances Balance Sheet 3 might be redrawn as Balance Sheet 5.

If customers draw out cash which is not quickly redeposited and the cash ratio becomes too low as in Balance Sheet 4, then the bank can demand repayment of some short-term loans and thus restore its holding of cash to the appropriate level.

Balance Sheet 5

Liabilities		Assets	
Customers' deposits	£100,000	Cash	£10,000
		Loans repayable within 3 months	£20,000
		Longer term loans	£70,000
Total	£100,000	Total	£100,000

It is now mainly of historical interest that until 1971 the deposit or clearing banks were required to keep a cash ratio of 8 per cent and a *liquidity ratio* of 28 per cent. The liquidity ratio is as follows:

$$\text{Liquidity ratio} = \frac{\text{Assets convertible into cash within 3 months}}{\text{Total deposits}}$$

Liquidity in this context refers to the nearness of an asset to cash. This stipulation is no longer made, but the banks have their own idea of the appropriate level of cash and other liquid assets. It is worth noticing here that the lower the cash ratio the greater will be the amount of credit that can be created on the basis of a given amount of cash. In our earlier examples a 10 per cent cash ratio resulted in total deposits of £100,000 on the basis of £10,000 of cash. The same amount of cash would support £125,000 of deposits with an 8 per cent cash ratio but only £50,000 with a 20 per cent cash ratio.

(b) A multi-bank system

The system outlined above is workable because, when Adam pays Brown or Clark by cheque they all use the same bank (X) and only a ledger adjustment is necessary. What would happen if Brown or Clark paid the cheque into an entirely different bank (Y) which insisted on X handing over the appropriate amount of cash so that Y could meet the requirements of its own customer? If the cheque was for £100, then bank X would lose this amount and Y would receive it. Bank X could not afford this to happen on a daily basis, as it would eventually run out of cash. But, of course some of Y's clients would write cheques in favour of X's customers and these would cancel out those going from X to Y. So long as the cheques going in each direction more or less balance, the banks have nothing to fear. If all the banks follow the same policy of credit creation, the required balance will be achieved, but if one bank operates with a cash ratio of 10 per cent and another with a cash ratio of 50 per cent then the former will gradually lose cash to the latter. An adjustment in lending policies will have to be made.

17.4 The London Clearing Banks

The actual direction and extent of bank lending and other activities may be demonstrated by an examination of the combined simplified balance sheet of the London Clearing Banks as reproduced in Table 17.1.

Table 17.1 London Clearing Banks' Balance Sheet (January 1973)

Liabilities £m	Assets £m	
(a) Customers' deposits 16,696	(b) Cash in hand and at the Bank of England	832
	(c) Cash with other UK banks	1,457
	(d) Money at call and short notice	1,130
	(e) United Kingdom treasury bills	136
	(f) Other bills of exchange	547
	(g) Special deposits with Bank of England	399
	(h) British Government stocks	1,222
	(i) Advances to customers	10,845

Source: *Financial Statistics*, HMSO

(*a*) Customers' deposits. The traditional function of the banks is the acceptance of money on deposit. Deposits are divided into two fairly equal groups, current accounts and deposit accounts. Money held in a current account is transferable on demand by cheque from one account holder to another and it earns no interest while it is at the bank. Money put into a deposit account is formally subject to at least seven days' notice of withdrawal, but it does earn interest whose rate was once fixed at 2 per cent below Bank Rate (see Unit Twenty-eight) by all the London Clearing Banks, but is now at the discretion of individual banks. If depositors are prepared to leave their funds with the banks for a longer period they can earn a higher rate of interest than if their money is subject to only seven days' notice.

An important liability omitted from Table 17.1 is shareholders' capital, money ultimately belonging to the shareholders and not to the bank. It is omitted because it is irrelevant to the creation of credit and to the methods whereby the Government tries to control bank lending. Nevertheless it is mentioned here to remind us that the overall function of a bank is to make a profit for its shareholders.

(*b*) Banks are obliged to keep a certain amount of cash in their tills in order to meet the daily demands of their customers. They also hold cash at the Bank of England, partly to replenish their own stock of cash should it fall too low, and partly to settle their debts with each other. If at the end of a day's business Barclays owe Midland £130,000 then the necessary adjustments are made to the banks' accounts at the Bank of England. These accounts are regarded as current accounts by the banks, but they cannot be overdrawn. The banks were previously required to maintain an 8 per cent cash ratio, but the ratio has fallen to only 5·00 per cent since the removal of limits in 1971. They are naturally anxious to keep their cash holding as low as possible as cash earns no interest, while all other assets contribute directly to profits.

(c) For a number of reasons the banks themselves advance money to other banks and the National Giro. The interest earned on these advances varies with their liquidity. The advance is made in the same way as advances mentioned in Section 17.3—simply as a book entry. If the National Giro wanted to borrow £200,000 from a London Clearing Bank, then the bank would simply open an account for National Giro and credit it with £200,000 which would be immediately available. There is no question of handing over 200,000 pound notes, although, of course, a borrower is perfectly entitled to insist on a cash payment. The same procedure applies to all the other assets: they are ledger transactions.

(d) In order to maintain a balanced spectrum of liquidity the banks ensure that some of the loans are very liquid. A most important liquid asset is money at call and short notice. The majority of this is advanced to the 11 members of the London Discount Market who then draw cheques to finance the purchase of Treasury Bills and Public Sector stocks of various kinds. Relatively small advances at call and short notice are made to other persons including stockbrokers and jobbers. As many of the advances are payable on demand in order to maintain the banks' liquidity, they earn a relatively low rate of interest. The Discount Houses who take the majority of money at call and short notice have an important role to play in the operation of the money market and we shall return to them in Section 19.2.

(e) Treasury Bills are another important liquid asset although they are of declining significance. (In 1960 the London Clearing Banks held over £1000 million in Treasury Bills.) The majority of the bills, which are sold by tender every Friday, are acquired by the Discount Houses, who until October 1971 submitted a joint tender but now bid individually. The number of bills issued each week depends mainly on the estimated excess of Government expenditure over receipts for the following week. A Treasury Bill is a promise by the Treasury to pay the holder £10,000, 91 days after the issue of the bill. Bills are allotted to the highest bidders, and holders make profits by paying the Government perhaps £9750 today, holding the bill for three months and then receiving £10,000. Thus by lending £9750 for three months the holder makes a profit of £250 or just over $2\frac{1}{2}$ per cent; this is equivalent to an annual rate of just over 10 per cent. The bills acquired by the Discount Houses are usually passed on to the Clearing Banks before they reach maturity. (There is a convention that the banks do not bid directly for the bills themselves.) In this way the banks can acquire bills of the exact maturity or liquidity they require. For example if a bank has cash surplus to its requirements for 26 days it can increase its liabilities, thus reducing its cash ratio, by buying Treasury Bills 65 days old (bills mature in 91 days). The price it pays must lie somewhere between the original purchase price and the redemption price and in the case of a bill issued for £9750 would be £9928·57, allowing the Discount House a profit of £178·57 $\left(£250 \times \frac{65}{91}\right)$ and the bank if it holds the bill for 26 days a profit of £71·43 $\left(£250 \times \frac{26}{91}\right)$. It is unlikely that the rate on Treasury Bills would be as high as 10 per cent (annual rate), but it will be higher than the rate paid for money at call and short notice, since the Discount Houses, having borrowed money at call at one rate will re-lend only at a higher rate.

(*f*) Equally liquid, though slightly less desirable since they do not carry a Government guarantee, are other Bills of Exchange used in the finance of commerce and trade. Where the purchaser of goods requires temporary credit which the seller cannot afford, the latter may draw a Bill of Exchange setting out the debt and demanding payment in perhaps three months' time. The purchaser acknowledges his debt and at the same time promises payment in three months by signing the bill. If the purchaser has a good name in the business world the bill may be accepted by an Accepting House (see below), or a Clearing Bank, who then endorse it. This means that in the event of the purchaser failing to meet his debt on the due date the seller can look to the Accepting House or Clearing Bank for payment. It is unlikely that the seller will hold the bill for three months; he will probably sell it to a bank at a discount in order to receive his money earlier. The bank then collects the money at the end of three months from the original purchaser and makes its profit.

The rate of discount on commercial bills of this kind will be slightly higher than on Treasury Bills. This reflects the slightly greater risk involved. Readers should notice the difference between a rate of discount and a rate of interest. If a three month £100 bill is bought for £97·50, that is at a discount of £2·50 and the discount rate is $2\frac{1}{2}\% \times 4 = 10\%$ per annum, the rate of interest is $\frac{2 \cdot 50}{97 \cdot 50} \times 100\% \times 4 = 10 \cdot 256\%$ per annum.

(*g*) The banks are likely to be instructed to make Special Deposits at the Bank of England in connexion with Government economic policy. Such deposits are completely frozen but do earn a rate of interest equivalent to that obtainable on Treasury Bills. They are fully discussed in Unit Twenty-eight.

(*h*) As an alternative to lending in the way outlined in section 17.3 the banks may lend to the Government by taking up bonds. In this case the banks' liabilities are increased as they credit the account of the Government with the value of the bonds bought. In general the bonds will be relatively liquid as the banks concern themselves primarily with the borrowing and lending of short-term funds. In January 1972 approximately two-thirds of the bonds had less than five years to run. It is not, of course, necessary for the banks to acquire bonds directly from the Government as they are always available in substantial numbers on the Stock Exchange. It is difficult to be precise about the earnings made from these Government securities since they pay a fixed rate of interest which means that the yield varies according to the price paid.

Let us suppose that in year one £100 of Government stock was issued at 5 per cent. This means that the long-term rate of interest was 5 per cent and that holders of the stock receive £5·00 per year. If the long-term rate of interest in the market has changed by year three to 10 per cent, the holders of the (fixed interest) stock will still receive £5·00 per year. But if they wished to sell the stock no one would offer them £100 for it, as they could buy new stock for £100 and earn £10 per year. The highest price payable for £100 of the original stock is that which will give the new owner of it a yield of 10 per cent. In this case the price is £100 $\times \frac{5}{10}$ = £50. This relationship is a most important feature of the gilt-edged market (the market in Government securities), and one to which the banks must pay careful attention.

(*i*) The most profitable assets of the banks are their advances to customers. These may be in the form of an *overdraft*, whereby the customer is allowed to draw cheques in excess of his deposit up to an agreed limit, and pays interest only on the amount overdrawn. The advance may be a *loan* whereby the customer's account is credited with the amount of the loan and interest is payable on the whole amount. This is a most important function of the banks and one without which industry and trade could not flourish.

Just as some important liabilities were omitted we have here made no reference to the least liquid asset held by the banks—their premises. The true value of these is immense but since they do not affect the basic banking operations we need not discuss them.

We may now summarize the functions of the Clearing Banks:

(i) The acceptance of deposits and their transfer via the cheque system.

(ii) The provision of short-term finance to the Government indirectly through money at call and by the acquisition of Treasury Bills.

(iii) Medium-term finance to the Government by the acquisition of Government Bonds.

(iv) Short-term finance to industry and trade through the discounting of Bills of Exchange.

(v) Medium-term finance to industry and trade by the system of overdrafts and loans.

(vi) There are also a whole range of non-banking functions: taxation advice, executorship and many others which do not concern us here.

(vii) Finally, banks exist to make profits for their shareholders.

17.5 Profitability versus Liquidity

The banks must strike a careful balance between their obligations to their customers and the requirements of their shareholders. Though customers will normally transfer their deposits by cheque they must be able to obtain cash on demand and the bank must retain some assets in a highly liquid form for this reason. If a bank were solely concerned with the possibility of customers withdrawing cash it would maintain a cash ratio of 100 per cent. But shareholders expect an annual dividend on their investment, and the bank has to make a profit by lending money at a higher rate than it pays for money left on deposit. If customers never wanted to withdraw cash then the bank could place much more emphasis on profits; and could happily operate on a cash ratio of 1 per cent or even less. In practice the policy of the banks has been to strike a balance, as we have seen, by maintaining a cash ratio of 8 per cent and a liquidity ratio of 28 per cent to ensure their ability to meet the requirements of their customers, but not contributing greatly to profits. The remaining 72 per cent of their assets are less liquid but contribute much more to profits.

In October 1971 the Government stipulation that banks should maintain a minimum cash and liquidity ratio was abandoned in favour of a new system of reserve ratios.

17.6 The Minimum Reserve Ratio

The new reserve ratio shows the relationship between two new concepts *eligible liabilities* and *eligible reserve assets.*

Eligible liabilities are the net sterling deposits of the banks excluding those deposits originally made for over two years, and which are not therefore characteristic of bank operations.

Eligible reserve assets include the following:

(a) Cash at the Bank of England (but not cash-in-hand, or Special Deposits at the Bank of England).

(b) United Kingdom Treasury Bills.

(c) Company tax reserve certificates.

(d) Money at Call with the London Money Markets, mainly the members of the London Discount Market Association.

(e) British Government Stocks with one year or less to maturity.

(f) Bills of Exchange issued by Local Authorities and eligible for rediscount at the Bank of England. (The Bank will only rediscount first class bills as will become clear in the next Unit.)

(g) Commercial Bills of Exchange eligible for rediscount at the Bank of England. Such bills must have been accepted in the way described in section 17.4. Commercial bills must not constitute more than 2 per cent of total reserve assets.

The banks must now keep a minimum reserve ratio of eligible assets to eligible liabilities of $12\frac{1}{2}$ per cent. Table 17.2 shows the position for various groups of banks in January 1972. The system does not yet apply to Northern Ireland Banks.

Other changes occurred at the same time, the most important of which so far as the Clearing Banks are concerned, was their decision to abandon their collective agreement on interest rates which had been heavily criticized by the Monopolies Commission and the National Board for Prices and Incomes. It tended to reduce competition between the banks.

Table 17.2 Reserve ratios of selected groups of banks (January 1973)

	Eligible Liabilities	Eligible Assets	Reserve Ratio
	£m	£m	
London Clearing Banks	13,829	2,063	14·9
Scottish Clearing Banks	1,393	187	13·4
Other Deposit Banks	332	48	14·4

The main significance of the changes lies in the field of monetary policy and is discussed in Unit Twenty-eight, but the Clearing Banks must now work from a base of 12½ per cent rather than the old 8 per cent or 28 per cent. Total deposits must now be related to total eligible assets: for every £100 of eligible assets a bank may have eligible liabilities (deposits) of £800. Previously one of two views could be taken: for every £100 of cash held the bank could have deposits of £1250 or, the view then adopted by the banks themselves, for every £100 of liquid assets held the bank could have deposits of £357.

In Unit Eighteen we will consider the work of the Bank of England which presides over the money market.

17.7 Questions and Exercises

1. Critically examine the system by which the banks create credit. In what ways is it different from the system operated by the goldsmiths in earlier times?
2. Examine the role of: (a) the cash ratio; (b) the liquidity ratio in the post-war British banking system.
3. By examining their balance sheets outline the functions of the London Clearing Banks.
4. Show how the banks achieve a balance between liquidity and profitability.
5. Explain how the changes introduced in 1971 affect the clearing banks.
6. Explain the importance of Treasury Bills to the London Clearing Banks.
7. Show how a compulsory cash ratio of 50 per cent would affect the London Clearing Banks.
8. Describe and assess the importance of the main profit-making activities of the London Clearing Banks.

Unit Eighteen

The Bank of England

18.1 Introduction

We have already seen that the Deposit or Clearing Banks are not entirely free to create credit as they wish. The Government has overall responsibility for the management of the economy and this includes the attempt to control the supply of money. The Bank of England is its agent in operating monetary policy and is the link between the Government and the banks. This Unit describes the activities of the Bank of England as a Central Bank: an analysis of monetary policy is given in Unit Twenty-eight.

18.2 The Weekly Return of the Bank

Table 18.1 The Bank of England's return (January 1973)

Issue Department			
Liabilities £m		Assets £m	
Notes in circulation	4160	Government securities	3638
Notes in banking department	15	Other securities	537
	4175		4175

Banking Department			
Capital	14	Government securities	1098
Public deposits	21	Advances and other accounts	24
Special deposits	692	Premises, equipment and other	
Bankers' deposits	196	securities	85
Reserves and other accounts	299	Notes and coin	15
	1222		1222

Source: *Bank of England Quarterly Bulletin*

The Bank of England publishes a weekly return in the form set out in Table 18.1. This type of statement dates from the Bank Charter Act 1844. It would be wrong to assume that the Bank is organized into just two departments: its work is far too extensive and complex to allow such a simple division. But just as the balance sheet of the London Clearing Banks reflects their main functions, so the weekly return (or balance sheet) of the Bank of England gives an indication of its main activities.

18.3 The Government's Bank

(a) Government accounts

One of the most important aspects of the work of a Central Bank is the administration of the Government's bank accounts, which appear under the heading Public Deposits in the above return. There are two principal accounts here: the *Exchequer Account* into which all taxation receipts are paid and from which all current expenditure is met; and the *National Loans Fund*, the Government's capital account for borrowing and lending. Although there are subsidiary accounts run by individual Government departments these two accounts between them are responsible for the majority of Government transactions. Since Government expenditure runs at approximately £16,000 million per year it is perhaps a little surprising that the Public Deposits are so small. This is a matter of deliberate policy on the part of the authorities (by which we mean in this instance the Treasury and the Bank): rather than allow a great balance to accumulate in the form of Public Deposits they would pay off part of the National Debt (see below). On the other hand the Bank takes steps to ensure that there is always enough in the accounts to meet the requirements of the Government and on rare occasions the Bank itself lends to the Government (overnight) in the form of *Ways and Means Advances*. The work of the Bank on behalf of the Government entails rather more than the management of the banking accounts.

(b) The note issue

The Government reserves the right to issue notes and coins, and these reach the public via the Bank of England. The size of the note issue is governed by the Currency and Bank Notes Act 1954, which set a limit to the fiduciary issue of £1500 million but allows it to be increased to meet public demand, subject to Treasury approval. These requirements vary according to seasonal factors (demand for notes reaches a peak at Christmas and the summer holiday periods) and according to the general level of prices and economic activity. The great increase in the issue since 1954 is mainly a reflection of rising prices. The notes actually reach the general public from the Clearing Banks. They acquire notes from the Banking Department of the Bank of England and thereby reduce the Bankers' deposits. The Banking Department receives notes from the Issue Department. The notes are backed by Government Securities and the securities issued by other Governments. Coins reach the public by a similar route but they originate at the Royal Mint rather than the Bank of England printing works.

(c) The National Debt

The Government can always increase the note issue to meet its own expenditure plans if it cannot raise sufficient funds through taxation or borrowing. Such a policy is likely to be dangerously inflationary and it is more likely that the Government will succeed in borrowing its requirements from the public. The accumulated borrowing is known as the *National Debt* and it represents money owed by the Government to its citizens and, to a much lesser extent, overseas Governments and residents. At the end of March 1972 the National Debt stood at £35,400 million and of this £4767 million was held overseas.

Table 18.2 Structure of the National Debt, March 1971

	£m
Government stocks	
Up to 5 years to maturity	7,387
Over 5 years and undated	17,955
Treasury Bills	3,586
Non-marketable debt	6,472
Total	35,400

Source: *Bank of England Quarterly Bulletin*

The Bank of England undertakes the administration of the debt on behalf of the Government. It is responsible for the issue of new securities, either to replace those maturing or to raise extra money for the Government. The Bank keeps a register of stockholders and pays them interest on the appropriate days and repays them as the debts mature.

Most of us own a part of the National Debt either directly or indirectly. The majority of the debt consists of fixed-interest Government Stocks of many kinds. In Table 18.2 they are divided into two broad groups, but there were eight separate issues of Gilt-Edged Securities totalling £4000 million in 1971 alone. Most Government Stocks are held by financial institutions of various kinds, and it is through such organizations as Insurance Companies, Pension Funds and Unit Trusts that most people have an indirect interest in the National Debt. Part of the debt—the floating debt—is in the form of Treasury Bills which we have discussed in Unit Seventeen. But it is by way of the non-marketable debt that most persons have a direct interest in the National Debt. Under this omnibus heading come National Savings Certificates, Defence Bonds, National Development Bonds, Premium Bonds and various contractual savings schemes, but not ordinary deposits with the National Savings Bank.

The National Debt is equal to about 80 per cent of the Gross National Product of the United Kingdom, and since the interest payable on the Debt in 1970–1 was £1302 million the Government feels that it has a duty not to allow the debt to increase without good cause. (Former hopes of repaying the debt have long since been discarded.) In fact only that part of National Debt interest which is payable to overseas holders is a burden to the economy, as interest paid to domestic holders is merely a transfer from taxpayers to debt holders who to some extent overlap. Even so efforts are made to keep both the interest charges and the administrative costs to a minimum, but there may be a conflict in this.

If interest rates are high, and are expected to fall during the next two or three years, then the Bank may feel that it would be unwise to issue for example 20 year stock and commit itself to the payment of high interest for the whole period. The Bank would rather issue two-year stock and hope to make a longer issue in two years' time at a lower rate of interest. But if it does this many of the administrative costs are doubled. The administrative costs may also be important in deciding whether the Government should raise perhaps £1000 million by one issue or by two separate issues. A further important administrative problem is the timing of the redemption of stocks. The Bank is not always happy at the prospect of redeeming say a £1000 million of stock on a given day (because unless a £1000 million of new stock is taken up the cash holdings of the Clearing Banks increase dramatically as those that are repaid deposit their cheques with the banks). This may be overcome in two ways: first by the Bank buying up stock for some months in advance of its redemption date, in the course of its ordinary market operations; and second, no fixed date for repayment will be given. For example one of the 1971 issues was £600 million of 9 per cent Treasury Stock 1992–6, giving the Bank a degree of flexibility in making repayments.

(d) Monetary policy

Perhaps the most important of the Bank's functions as the Government's bank is the implementation of monetary policy in the pursuit of the broad economic objectives of the Government. This involves controlling the supply of money, perhaps partly through the use of Special Deposits, and the price of money (the rate of interest), and is the most difficult of the Bank's jobs. We must defer our discussion of monetary policy until Unit Twenty-eight, as there is no point in talking of the nature of the policy and its effectiveness until we have considered the aims of economic policy in Units Twenty-two to Twenty-seven.

(e) Lender of last resort

We have seen that the Clearing Banks arrange their assets so that they always have access to liquid resources which can be converted into cash on demand. The most liquid of their assets other than actual cash consists of money at call with the Discount Houses. Normally the banks' management will be such that they do not have to insist on repayment of their 'call money' at times likely to embarrass the Discount Houses. There are, however, occasions when their holdings of cash are dangerously low and to ensure that they can meet the requirements of their customers who want cash they demand repayment of call money by the Discount Houses. The Discount Houses will have spent most of their call money in buying Treasury Bills, so they have to seek the necessary funds from other financial institutions. If no one else can oblige them they turn as a last resort to the Bank of England. The Bank is always prepared to advance money to the Discount Houses either by re-discounting Treasury Bills or by lending against acceptable security. The Bank of England will normally advance the money at the Minimum Lending Rate (see Unit Twenty-eight). In this case the Discount Houses are said to borrow 'at the front door', but sometimes they borrow at a lower rate 'at the back door'. The mechanics of the process are quite simple: the Bank of England in rediscounting Bills for the Discount Houses credits their accounts at the Bank (part of Bankers' Deposits) with the appropriate sum, which they are then free to transfer to the Clearing Banks in order to discharge their obligations. In this way the Bank ensures that the market never goes short of cash.

(f) External functions

The Bank also discharges a number of functions on behalf of the Government which may properly be regarded as belonging to the international economy rather than the domestic.

(i) The nation's gold and foreign exchange reserves are held at the Bank in the Exchange Equalization Account. It is from this account that the foreign currency needed by an importer to buy goods, or the spending money needed by a traveller abroad, ultimately comes.

(ii) Exchange control regulations are operated by the Bank. The Government sometimes wishes to limit the amount of foreign currency spent by residents and it is the Bank that supervises arrangements in this connexion. The Bank is also responsible for maintaining orderly exchange rates against other currencies, in the light of Government policy. Unit Twenty-six explains this in more detail.

(iii) The Bank's expertise is often called for in complex international monetary negotiations, particularly in connexion with the International Monetary Fund (see Unit Twenty-six).

18.4 The Bankers' Bank

As an essential aspect of its linking role between the Government and the banking system the Bank of England acts as banker to the banks. This function is reflected in the item Bankers' Deposits in Table 18.1. This item includes the accounts not only of the main Clearing Banks but also of the Discount Houses and a number of other institutions such as the Overseas Banks in London. The Clearing Banks have undertaken to keep funds to the extent of 1·5 per cent of their eligible liabilities in the form of cash at the Bank of England.

The role of the Bank as the Bankers' Bank is important in facilitating the settlement of inter-bank debts and contributing to the smooth running of the money market in general. The accounts that the banks hold at the Bank of England may be regarded as current accounts but they cannot be overdrawn.

18.5 Other Accounts

Although the Bank does not compete with the Clearing Banks it is worth noting that it does run a number of ordinary accounts for private customers, mainly long established accounts, or accounts for Bank employees. This is largely done to give Bank staff some training in the principles of Commercial Banking. Such accounts appear under the heading Other Accounts in Table 18.1 and also included here are the accounts of a number of international banks and banks under foreign control.

18.6 Questions and Exercises

1. What are the main functions of the Bank of England?
2. Assess the importance of the services of the Bank of England to the Government.
3. What is meant by the 'National Debt?' Examine the main economic implications of increasing the National Debt.
4. Describe the mechanism by which the Bank of England acts as the lender of last resort. Why is such a role important?
5. Examine the role of the Bank of England in relation to the National Debt.

Unit Nineteen
Other Financial Institutions

19.1 Introduction

While the Clearing Banks and the Bank of England are the most significant of the financial institutions there is a wide variety of others meeting the requirements of different groups of lenders and borrowers. Some of these, like the Discount Houses, are primarily concerned with short-term borrowing and lending in what we may regard as the money market. Others like the Building Societies and the Stock Exchange deal in the longer-term mobilization of funds and form part of the capital market, although it is impossible in practice to draw a firm boundary between the money market and the capital market.

19.2 The Discount Houses

We have already seen that the Discount Houses play an important part in the issue and distribution of Treasury Bills. A glance at their balance sheet as shown in Table 19.1 will give an indication of the extent of their activities.

Table 19.1 The Balance Sheet of the London discount houses (January 1973)

Liabilities £m		Assets £m	
Money borrowed from:		Invested in:	
Bank of England	—	British Government Stocks	400
London Clearing Banks	1058	British Government Treasury	
Scottish Clearing Banks	69	Bills	790
Other Deposit Banks	71	Other Sterling Bills	626
Accepting Houses and Other		Local Authority Bonds	383
Banks	1249	Sterling Certificates of	
Other sources	425	Deposit	421
		Dollar Certificates of Deposit	101
		Other investments	261

The balance sheet covers the activities of the London Discount Market Association, representing the 11 Discount Houses responsible for handling most of the business in short-term securities. The market became important in the nineteenth century in connexion with the financing of inland and international trade by means of short-term Bills of Exchange. Its function today involves gathering funds temporarily surplus to the requirements of large financial institutions, and channelling them towards other organizations temporarily in need of money. They thereby act as middlemen providing an important service to both sides of the market: those with surplus funds do not have to seek those wanting to borrow; they simply lend to the Discount Houses and those wanting to borrow go direct to the Discount Houses.

Most of the Discount Houses' funds come from the Clearing Banks although 'other sources' in Table 19.1 covers money from industrial and commercial firms with assets available for a short period. It can give us a good idea of the importance of the Discount Market if we imagine a situation where it has liquid cash of £1 million on its hands. What could it do with it?

(*a*) The historical outlet for such funds was in the discounting of bills of exchange ('other sterling bills' in Table 19.1). As we saw in Unit Seventeen such bills are used when a purchaser requires credit which the vendor cannot afford to grant. If the bills have been accepted in the way described in Section 17.4 then the Discount Houses will regard them as *Bank Bills* and will discount them at the minimum rate (i.e. will make very little profit on them). Bills not so accepted are called *Trade Bills* and as they carry a greater degree of risk they are discounted at a higher rate. Discount Houses frequently do not discount them and they are discounted directly by the drawer's bank. A further disadvantage of Trade Bills is that unlike Bank Bills they do not qualify as eligible security for a loan from the Bank of England should the Discount Houses run short of cash. A third category of Bills of Exchange is the *Finance Bill* which may be issued by any one of a number of financial institutions who need funds on a short-term basis.

(*b*) The Discount Houses may spend their £1 million on Treasury Bills for which they tender on a Friday, specifying the day on which they wish to take up the Bills. Such Bills, of course, carry no risk at all and the Discount Houses make a modest but certain profit. In most other countries Treasury Bills are sold directly to the banking system without the intermediary of a Discount Market. There are good reasons why the market survives here: it guarantees to take up the entire issue of Treasury Bills each week so that the Government knows that its borrowing requirements will always be met; it acts as a wholesaler to the banks releasing Bills to them at the precise time they are required, and with the exact maturity required.

(*c*) A Discount House may decide to place its funds in the recently developed Local Authority Market, where local government bodies raise money for a period of one to four years.

(*d*) Although it has been stressed that the Discount Houses are primarily concerned with short-term funds, it is perhaps surprising to find that they have a considerable holding of British Government Stocks. Most of such bonds, however, are relatively short-dated and would be unlikely to have more than two years to run to maturity. This is a market that the Discount Houses were forced into by a lack of other business during the inter-war period, but it is a field in which they coincidentally perform another important function for the authorities. We saw in Section 18.3 (*c*) that in its administration of the National Debt the Bank of England likes to buy up maturing stock in advance of its actual redemption date. As the Discount Market deals in short-dated bonds it accumulates a considerable amount of those bonds wanted by the Bank, and it therefore eases the problem for the Bank.

(e) The Discount Houses are also prepared to advance cash against *Sterling or Dollar Certificates of Deposit*. These are certificates issued for a fixed period by banks outside the clearing system, and are in effect a promise by the issuing body to pay a fixed sum on a specified day. Should the holder of a certificate need cash before the certificate matures the Discount Houses will normally buy them at a discount.

We may thus summarize the role of the Discount Market as it has emerged so far:

(i) the provision of short-term finance for industry and trade through Bills of Exchange;

(ii) the provision of short-term finance for the Government through the acquisition of Treasury Bills, in particular a guarantee to the Government that all Treasury Bills will be taken up each week;

(iii) the provision of safe, liquid assets to the banking system as and when they are wanted;

(iv) the market is the channel through which the Bank of England acts as the lender of last resort to the banking system.

(v) because of their high standing and integrity the Discount Houses have been able to branch out into markets other than the traditional ones, thereby providing finance to Local Authorities and other bodies.

19.3 The Merchant Banks

The term Merchant Bank is applied to many institutions today but refers specifically to the 16 members of the *Accepting Houses Committee* which include such famous financial names as Morgan Grenfell, Rothschilds, Hambros and Baring. Such banks grew to prominence through their skill in accepting the Bills of Exchange used in the finance of international trade. We have already seen that the Discount Houses will advance money against Bills of Exchange; but they are more willing to do so if the bill has been accepted by one of the major Accepting Houses thus rendering it a *fine bill*. As the acceptance of a bill is a guarantee that the acceptor will meet the debt in case of default, the Accepting House clearly needs a detailed knowledge of the traders involved. In the nineteenth century it was only the most skilful that survived. Today their important function is still to contribute to the smooth running of the money market by guaranteeing the value of paper securities.

Nowadays they have many other activities. In addition to their original trading or merchanting functions from which they derive their name, and which many maintain today, they are an important financial intermediary for industrial concerns. When a company wants to raise new long-term capital it will normally ask a Merchant Bank to arrange the issue of the shares. In general the Merchant Banks develop a role of general financial advisers to their industrial clients guiding them not only in connexion with new issues but also on the timing and scale of investment and on the merits and demerits of take-over bids and mergers. The bank itself might often make fixed-interest medium-term loans to its clients to finance the installation of new machinery which does not require the raising of extra long-term capital.

Recent expansion by the Merchant Banks in the long-term markets has included their involvement in the Unit Trust and Investment Trust movements, which are discussed later in the unit, and in the management of investment portfolios belonging to pension and superannuation funds.

Although they do not hold the same central position as the Discount Houses in the market, the Accepting Houses nevertheless have close connexions with the public sector, by holding modest amounts of Treasury Bills and other Government securities and making substantial advances to Local Authorities.

The close relationships between the various financial institutions noted earlier are further highlighted by the fact that the Accepting Houses frequently advance money at call to the Discount Market, and keep substantial balances with other British banks.

Such a range of activities requires considerable finances, and the Merchant Banks accept deposits on a current account basis and for a fixed period from both home and overseas residents. It is in respect of the fixed-term deposits, which are the more important, that the Certificates of Deposit mentioned earlier are issued. Approximately one-third of Merchant Bank deposits are from overseas residents and this reflects the traditional involvement with overseas trade, and their historical business of issuing bonds on the London Market for overseas Governments and institutions.

There are many other financial institutions primarily concerned with the collection and movement of short-term funds: the Overseas Banks in London and the Finance Houses are examples. But we must now examine the capital market which is at the other end of the financial spectrum.

19.4 The Capital Market

The terms 'money market' and 'capital market' need to be carefully distinguished. There is no clear boundary between them, but the money market is normally involved in the mobilization of funds on a short-term basis (perhaps for two years at the most) and the capital market caters for organizations who wish to raise money for much longer periods, perhaps indefinitely. Thus a manufacturer wishing to raise a million pounds to finance the import of raw materials which will be made up and sold within six months, enabling him to repay his debt, will be seeking the assistance of the money market. If the same manufacturer were seeking a million pounds to finance the installation of new machinery which would not pay for itself for ten years then he would look to the capital market. The fact that in each case he may turn to the same institution (the Merchant Bank) for assistance adds further confusion, but reminds us that there is no such place as either the money market or the capital market but each consists of the institutions dealing in money or capital. In the ordinary sense of the word all these institutions are dealing in money. It is however appropriate to refer to the transaction as one in capital where the loan or investment is for a period of greater than twelve months.

At the centre of the capital market is the Stock Exchange which is the place where stocks and shares are bought and sold. On one side of the market supplying funds to it are a number of institutions such as Insurance Companies, Unit Trust and Investment Trusts each of which gathers resources from a large number of subscribers and invests them on the Stock Exchange; there are also a large number of individuals who invest directly on the Stock Exchange, by-passing these institutions. On the other side of the market are those requiring funds to finance industrial and commercial development or social improvement. These include the Public Limited Companies (introduced in Unit Seven and frequently represented by the Merchant Banks), and the Government. The capital market exists to meet the requirements of these various components and its most important function is to channel funds towards the business community. We shall now examine the raising of business finance more closely.

All businesses require capital but only the Public Limited Companies are allowed to appeal directly to the public for their funds. When a private company is on the point of 'going public' it will normally consult an Issuing House about the best approach. The most important of the Issuing Houses are the Accepting Houses for this is another aspect of their work. They advise clients on the scale, structure and timing of capital issues, and help to decide whether the money should be raised mainly by the issue of debentures thereby saddling the company with heavy annual interest charges, or mainly by the issue of ordinary shares, which require subscribers to share in the risks of the enterprise. An Issuing House may be regarded as the best judge of such matters and of the best means of issue. (There are several different procedures for issuing capital, the technicalities of which come within a study of commerce and are therefore not pursued here.)

One of the tasks of the Issuing House is to find *underwriters* for the issue. These people guarantee to take up any part of the issue not sold in the market. The underwriters (there will normally be more than one) will be found among the institutional investors, who collect small sums from individuals and make a profit for themselves and their clients by buying shares. It is worth looking at the activities of some of these institutional investors as they perform a vital role in getting capital to industry.

(a) The insurance companies

Most people have a life insurance policy whereby the Insurance Company collects weekly, monthly or annual premiums, and undertakes to pay a lump sum either on a specific date or on the death of the insured. The company is able to repay more than it collects in premiums because it invests the bulk of the premiums in securities which yield interest or dividends. In October 1972 the Insurance Companies had investments of over £18,000 million, of which £7000 million were industrial securities of various kinds, and a further £4000 million British Government securities.

The majority of the funds that reach industry and commerce via the Insurance Companies would not be available for productive investment without the existence of the Insurance Companies as individually the contributions would be too small to be worth investing in securities. In recent years other organizations for the collection of small savings for investment purposes have sprung up and from the point of view of industry they have the same effect as the Insurance Companies. From the point of view of the individual saver they often offer a slightly higher return on the investment.

(b) Unit trusts
While part of most people's savings reach the Stock Exchange via the Insurance Companies, the number investing via a Unit Trust is much smaller. Small savers are able to buy units in the fund which the managers then use to buy securities in the particular sectors of industry to which the Trust is committed. (Unit Trusts deeds indicate the areas to which funds will be applied.) All earnings on the shares are payable to the Trust managers who then redistribute them to Unit Holders, who may, however, elect to convert them into further Units. The chief outlet for Unit Trust funds are ordinary shares and at the end of 1970, 97 per cent of Unit Trust assets were held in this form. The main advantage of a trust is its security, for it is able to spread its investment across a large number of companies and industries, thereby eliminating the danger of complete loss if one company collapses. But since the Trusts spread their assets precisely for this reason the Unit Holder cannot expect spectacular returns on his investment even though he has the advantage of expert professional advice in the handling of his money.

(c) Investment trusts
These are really Joint Stock Companies that use their capital to buy shares in other businesses in order to make a profit for their shareholders. Shares in Investment Trusts are bought in the same way as shares in other businesses and are therefore not obtained so easily as are Units in a Unit Trust. The latter can be acquired simply by answering newspaper advertisements.

Like the Unit Trusts they apply the funds mainly to industrial investments and approximately 80 per cent of their capital is invested in Ordinary Shares. The main advantage of the Investment Trust over the Unit Trust is that it is not so restricted in its activities by a Trust deed, and thus retains greater flexibility for switching funds from relatively unprofitable to more profitable uses.

These three kinds of institutions are among the most important in the capital market. Despite their differences, they are essential intermediaries in the movement of funds from those who temporarily do not require them to those who do. Between them they had investments totalling £21,000 million at the end of 1971, the bulk of it invested in the securities of industrial or commercial companies who were thus able to buy the plant and machinery they required for productive purposes.

But what happens if an investor, either an institution or an individual, wants to withdraw his money which the company has by now spent on machinery and therefore cannot repay? The Stock Exchange reconciles the opposing viewpoints for it is here that the shareholder can turn his asset into cash without damaging the company in which he has invested. If an investor has £100 of securities to sell, he has to contact a *stockbroker* who will go to the Stock Exchange to sell them on his behalf. Members of the public are not allowed into the Stock Exchange (except as observers), and must employ a broker as their agent. The Stock Exchange itself is divided into a number of markets and the broker will go to that area of the floor of the exchange where there is a market for the shares he has to sell. There he will obtain prices from *jobbers* (dealers in particular groups of shares) and he will sell the shares for the best price he can obtain. This price will depend on a number of factors including supply and demand. If the shares are those of a declining company there are likely to be many sellers and prices will be low; but if the shares are for expanding and profitable companies then demand and price will be high. Political movements, general elections, and military crises can also have an effect on the level of stock exchange prices. Whatever the price obtained, the Stock Exchange has allowed the original investor to obtain his money without embarrassing the company which regards the investment as permanent. The transaction affects the company only in so far as it will have to alter its register of shareholders.

Without the existence of a dependable market for securities individuals would be very reluctant to commit their savings to investments for fear that they would never be able to sell their securities. One of the most important aspects of launching a new public company is to obtain the Stock Exchange council's approval for dealing in the shares to take place. Without this approval, which is not lightly granted, the task of an Issuing House in raising money becomes monumental.

The provision of a market for securities is the most important function of the Stock Exchange but it does perform other tasks. As we have seen its very existence enables firms to raise capital. Then the mass of information it issues gives an important guide to the profitability of different industries and companies. Finally it serves as an important 'economic indicator', rising prices indicating optimism about the future of the economy, and falling prices showing pessimism.

To some extent price fluctuations may be the result of speculation. In the Stock Exchange special terms are applied to speculators. *Bulls* are those who buy shares today at £1·00 hoping to sell tomorrow at £1·05; if there are enough of them they will induce the rise in prices they are expecting. *Bears* are those who sell shares today at £1·00 hoping to buy them back tomorrow at £0·97; similarly if there are enough bears they can force prices down. Speculators incur a great deal of criticism but in general they actually bring stability to the market by anticipating fluctuations, acting on their anticipation and preventing the fluctuations from being excessive.

19.5 The Finance Corporations

Despite the variety of privately owned financial institutions, each developed to meet special requirements, it has proved necessary for Governments to sponsor special bodies from time to time to fill gaps in the availability of capital. The most notable are:

(a) The industrial and commercial finance corporation

This is jointly owned by the Bank of England and the London and Scottish Clearing Banks. It provides medium-term loans to industry up to a ceiling of £300,000, normally obtaining the funds that it disburses by the issue of debentures. Such loans are subject to very careful appraisal of the projects involved.

(b) The finance corporation for industry

This is owned by the Bank of England, and the Investment Trusts and Insurance Companies. It specializes in longer-term loans of over £200,000, in cases where companies do not want to go to the expense of raising money publicly.

19.6 Questions and Exercises

1. Describe and assess the role of the London Discount Houses.
2. Distinguish between the Clearing Banks and the Merchant Banks, showing the importance of the latter.
3. Examine the roles of the various institutions to be found in the capital market and comment on their economic importance.
4. Consider the main institutions of the capital market as vehicles for the savings of an individual.

Unit Twenty

Factor Incomes: Wages

20.1 Introduction

In Unit Two we saw that the size of the Gross National Product could be measured by adding together the incomes of all the factors of production, labour, land and capital. We did not discuss what determines the share of the national product received by each of the factors or what causes differences in the earnings of individual units of the factors. In this and the next Unit we shall examine the second of these two questions, the determination of the earnings of individual units of factors of production. Although we shall in general confine ourselves to the term wages in this unit the bulk of the argument applies equally well to other methods of paying labour—by salaries and fees. Salaries are becoming increasingly important as firms encourage their employees to accept monthly rather than weekly payment.

20.2 Types of Wage Payment

(a) Time rates

The simplest method of paying labour is in relation to the time worked; the employer may pay for labour at so much per hour or week, or in the case of salaries so much per year. In the last instance the employee will normally receive his salary in twelve equal instalments. Time rates are most appropriate where it is difficult to measure the work done or where quality is more important than quantity. The disadvantage is that without adequate supervision of labour, time wages may become mere attendance money, and the employer may not get full value for his expenditure.

(b) Piece rates

The main alternatives to time rates are piece-rate systems. The worker here is paid according to the number of pieces of work he completes; thus the more work he does the more money he earns. Such a system is appropriate where the work is easily measured and where quantity is more important than quality. There are many types of piece rate and one of the disadvantages of paying wages this way is the difficulty of establishing a system appropriate to a particular process, and of settling rates for particular jobs. The determination of rates for jobs has caused so much difficulty that it has given rise to the separate profession of 'work study'. Although quantity may be more important than quality, payment by the piece necessitates the employment of inspectors to ensure that workers are not paid for inefficient low quality work.

(c) Fees

Solicitors, accountants and many other professional workers charge fees for their labour. The methods of assessing such fees vary from profession to profession; sometimes they are related to the amount of work involved, and sometimes to the value of the transaction concerned.

20.3 The Marginal Productivity Theory of Wages

Before looking more closely at the major real world influences on wages, it is worth considering their determination in perfectly competitive conditions. In Table 20.1 part of Table 10.1 is reproduced, and once again we

Table 20.1 The law of diminishing returns again

No. of men	Total product (kg)	Marginal product (kg)	Wage per man (kg) (a)	(b)
1	5	5	10	12
2	12	7	10	12
3	23	11	10	12
4	36	13	10	12
5	48	12	10	12
6	58	10	10	12
7	65	7	10	12
8	68	3	10	12
9	70	2	10	12
10	71	1	10	12

imagine a fixed amount of land on which successive units of labour are sent to work. At first marginal product rises but with the introduction of the fifth man it begins to fall for the reasons discussed in Unit Ten. We already know that the employer will try to equate marginal cost to marginal product in order to maximize profits. In this case his costs increase by 10 kg for each extra man employed, and so long as the addition to total product is not smaller than 10 kg it is worth adding to the labour force. Accordingly the firm will employ six men. If wages rise to 12 kg it would be wasteful to employ the sixth man who adds only 10 kg to the total output. In this case only 5 men would be employed. If we present the same information graphically as in Fig. 20.1 we can see that the downward sloping part of the marginal product curve is the firm's demand curve for labour, and that the higher the price the smaller the quantity of labour employed. The industry's demand curve for labour could, of course, be obtained by adding together the individual demand curves.

We should notice here that the demand for labour and indeed for all the factors of production is a *derived demand*. Factors are not demanded for their own sake, but for what they can produce. If the demand for public transport is high the demand for bus drivers will be high; and as the demand for coal has fallen so has the demand for coalminers. The demand for a

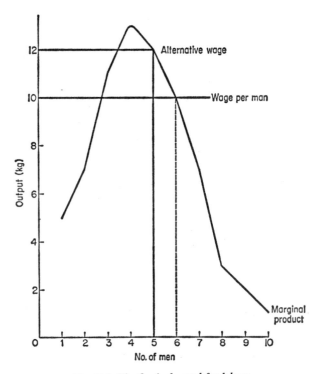

Fig. 20.1 The firm's demand for labour

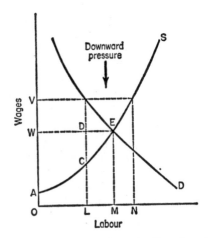

Fig. 20.2 Competitive determination of wages

factor does not depend only on the demand for the product. As techniques of production change so the proportions in which the factors of production are hired will alter. Modern navigational systems reduce the size of crew needed by ocean-going vessels even though they carry more cargo.

We have already seen in Unit Six that the supply of labour depends mainly upon the size and structure of the population. Unit Eleven showed that the supply curve of labour is likely to slope upwards from left to right as more labour is forthcoming at higher wage rates. In these circumstances the market for labour is comparable with other markets examined in Unit Thirteen and the pressures of demand and supply will establish an equilibrium wage as in Fig. 20.2.

Figure 20·2 is an outline of the marginal productivity theory of wages, and it goes some way towards explaining the real world situation. There are many examples of the forces of supply and demand influencing wages and it is clear that no employer will pay workers more than the value of their marginal product. But there are obvious weaknesses to the theory when postulated as an explanation of real world adjustments.

(a) We have dealt only with wages and a commodity called labour, the implication being that labour will move from low wage activities to high wage activities so that equilibrium will be established. It is really necessary to consider wages and other benefits and disadvantages of a particular job or the *net advantages*, including hours and conditions of work. Moreover we know that there is not a single commodity called labour but thousands of different working people, endowed with a huge variety of talents and skills, but not easily interchangeable between jobs.

The implication of the first of these considerations, the variety of benefits and disadvantages of a particular occupation, is that the net advantages of an occupation should be equated with the marginal product. Thus a university lecturer with generous holidays and congenial working conditions might be expected to accept a lower money income than a similarly qualified geologist working long and inconvenient hours in difficult terrain searching for oil. The implication of the second consideration, the different talents of the working force, is that those with very scarce talents may well earn more than the value of their marginal product, because of their scarcity (see Section 20.4).

(b) Most employers would find it very difficult to calculate the marginal product of a particular type of labour. Any theory based on the assumption that such calculations are possible is too precise to explain what actually happens in the labour market. There will be too many imperfections. The measurement of marginal productivity would be particularly difficult in the case of the introduction of new machinery, which results in an increase in output per worker. It would be quite erroneous to attribute the whole of the increase in output to labour, especially as there actually may be less work involved.

20.4 Economic Rent

In Fig. 20.2 a wage of OW would be established and OM people would be employed in the industry. Changes in the conditions of demand or supply would have effects similar to those explored in Unit Thirteen.

It is appropriate at this point to digress and discuss the concept of economic rent. We shall see in Unit Twenty-one that the early theory of rent was developed in relation to land, but it is now used for all factors of production, and is the term applied to any payment received by a factor of production over and above what is necessary to keep it in its present occupation. Thus if a pop-singer earns £1000 per week but would be prepared to work for £200 per week if he had to, then £800 of his earnings may be regarded as economic rent and the remaining £200 as his *transfer earnings*. If he is offered less than this for singing he will probably turn to some other occupation.

In the labour market illustrated in Fig. 20.2 no one is willing to work for less than OA. Let us consider the position of the man OL. The supply curve shows that he would be willing to work for a wage of LC; but the established wage which he receives is LD (=OW). Thus his transfer earnings are LC and he receives economic rent of CD. Only the marginal unit of the factor, OM, receives no rent, but the work force as a whole receives economic rent of AWE which represents the difference between what the employer actually pays for the labour he employs (OWEM) and what he would have paid had he been able to negotiate with each employee separately.

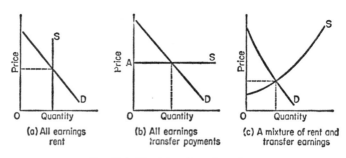

Fig. 20.3 Elasticity of supply and rent

It is clear from Fig. 20.3 that the proportions of economic rent and transfer earnings in a factor's income depend upon its elasticity of supply. In the case of Fig. 20.3 (*a*) the supply of the factor is completely inelastic; whatever the price offered for it the quantity supplied is constant. If income falls to zero the factor will still not leave its present occupation as there are no alternatives available. In this case all the earnings are economic rent and there are no transfer earnings. It is of course a theoretical limiting case and has little practical significance as there are usually alternative uses for most factors of production.

At the other extreme Fig. 20.3 (b) shows the position where the supply of the factor is perfectly elastic. Producers can employ as much as they wish at a price OA but could obtain none at a lower price. This might represent the position of a single firm employing labour at a fixed price under an agreement with a Trade Union. In this case as no one is willing to work for less than OA, total earnings are all transfer earnings and no economic rent is payable. In between these two extremes will be the majority of cases, where the elasticity of supply is greater than zero but less than infinity. There will be an element of both transfer earnings and economic rent in the factors' incomes.

Sometimes a factor of production may receive payments above its transfer earnings because of temporary shortage of supply. In this case at least part of the surplus can be expected to disappear as the supply of the factor increases. This kind of temporary surplus is known as *quasi-rent* or temporary rent and has been earned in the recent past by computer programmers. It is likely to occur wherever there is a lengthy period of training involved before extra workers become available to meet an increase in demand.

20.5 An Alternative Approach

A more realistic approach to wage determination is to recognize that market imperfections are so great as to render the marginal productivity theory almost meaningless, and then to consider the effect of those market imperfections which include the following:

(a) Monopolistic unions and employers

Approximately 45 per cent of the labour force belongs to a Trade Union and the majority of employers belong to a Trade Association, each group seeking to protect and promote the interests of their members. The functions of the Trade Unions may be divided into three groups:

(i) The traditional role of securing increased wages and improved conditions of work in terms of longer holidays and better welfare facilities.

(ii) General economic aims such as full employment, rapid economic growth and a high standard of living. Such aspirations are not confined to members of Trade Unions, but the leaders of the Union movement are frequently in the forefront of discussions on general economic policy.

(iii) Political aims, including the greater participation of labour in the running of industry may also be ascribed to Trade Unions. The Unions also have an increasingly important role in the formulation of economic policy, as is shown by their representation on numerous Government bodies such as Royal Commissions and the Boards of the Nationalized Industries.

The most important of these functions in the present context is the role of the Trade Unions in the negotiation of wages. Many people believe that the strength of the Unions is such that the only meaningful wages theory can be the *bargaining theory* relating the bargaining power of the Unions to that of the employers. This amounts to an examination of the circumstances in which the unions will be able to secure extra real wages for their members. It is important to stress the concept of real wages in this context since an increase in money wages which is offset by an increase in prices will leave real wages—the goods and services that the money wage will buy—unchanged. The circumstances likely to favour a union in seeking higher real wages include the following:

(i) Where wages constitute a small proportion of total cost an employer may be willing to concede a wage claim rather than risk a strike during which nothing is produced but heavy overhead costs still have to be met.

(ii) Where the demand for the product is buoyant and inelastic, a wage claim may be conceded in the knowledge that extra costs can be passed on to the public. These are the circumstances where an increase in money wages has an inflationary effect on the economy (see Unit Twenty-three). Although the workers concerned gain a real increase in wages it is at the expense of others.

(iii) Where workers are able to show that their productivity has increased, or will increase, a wage claim may be successful. In this connexion 'productivity bargains' are frequently associated with above average wage increases. The work force agrees to give up traditional and outmoded methods of working and adopt a more flexible attitude in return for high pay. At the same time surplus labour may be induced to leave the firm by the offer of redundancy payments.

(iv) Where wages for a job are below the wages offered for similar work it is often possible to bring pressure to bear on employers to secure a proper adjustment. Sometimes this may be merely a question of adjusting the balance between profits and wage payments, or of reducing the economic rent received by shareholders—the payment over and above that which they need to keep their money invested in its present projects.

(v) The nature of the product itself may be significant. If the workers supply an essential service or product, whose disruption causes immediate inconveniences to the community, then the pressure of public opinion may force the employer to a quick settlement.

(vi) A successful wage claim in one industry may set the scene for successful claims elsewhere.

(vii) The organization of the Union itself and the structure of the industry may be significant. Where the majority of the work force is concentrated in large plants, and belongs to the Union, the chances of successful wage claims are probably enhanced. But when the work force is thinly spread over a very large number of establishments, and is perhaps in close daily contact with the employer there may be greater reluctance to take industrial action and wage settlements are likely to be lower.

(viii) Finally, and of increasing importance, in prlonged wage disputes, are the increases in social security benefits. These benefits may favour labour, as they offer considerable support to workers' families during a strike.

This array of circumstances is helpful to Trade Unions in their pay negotiations, but there are fewer beneficial conditions for the employers, who tend to negotiate as individuals rather than as members of a Trade Association.

(i) In conditions of high unemployment, when the demand for goods and services is depressed (see Unit Twenty-two) the employer should be able to resist inflationary wage claims.

(ii) A monopolist who need not fear losing his market to competitors during a strike may be expected to resist wage increases vigorously. Though it is, of course, the monopolist who is more likely to be able to pass on wage increases to the general public in price rises.

(iii) In the same way a firm which is the sole buyer of a particular type of labour (technically called a *monopsonist*) will be able to hold down wages as the supply of labour to the firm is inelastic.

(iv) In a period of relatively constant prices wage demands are likely to be moderated and excessive claims will be easier to resist.

Although the determination of wages will depend heavily on the interaction of trade unions and employers there are several other factors that help to determine the pay for a particular job. We must now consider these.

(b) Traditional differentials

One of the most frequent complications during wage negotiations is the insistence of some groups of workers on the maintenance of the difference between themselves and some other group, even when changes in working conditions suggest that the differential is no longer justified. Related to this is the 'league table' approach to wage rates whereby each group of workers feels that it must at least maintain its position in the league even if other groups are increasing their productivity more rapidly.

(c) Non-monetary factors

These have already been mentioned briefly. In industries where workers receive considerable fringe benefits their money wage may be lower than it would otherwise be. Agricultural workers are often cited as an example as they frequently live in farm cottages let at low rents. In more lucrative occupations company houses, overseas holidays and company cars may be included in real wages, while subsidized travel constitutes part of the wage of railway workers, and free coal part of the wage of coalminers.

On the other hand some non-monetary factors have a negative impact and need to be compensated for by extra monetary payments. Steeple-jacks, miners, and others in exceptionally dangerous or arduous jobs might expect above average wages to balance the risks and hard work. Those on shift-work normally receive a premium to compensate for the inconvenience involved. People with heavy responsibilities may also expect to receive above average incomes.

(d) Training

A major cause of differences in wages is the length of training involved for the job. The time spent working through an apprenticeship or an engineering sandwich course may be regarded by the individual as an investment, for he will hope to see a return on his efforts in the form of wages higher than those of his less highly trained colleagues. He hopes in fact to receive an economic rent which we may regard as a rent of ability. In many cases this rent of ability may be protected by Professional Associations who limit entry to their profession and thus add a scarcity rent to the rent of ability.

(e) Female labour

The average earnings of men are often considerably higher than those of women in the same industry. This may be because the men do the most arduous jobs or because they work more overtime. In the past it was also because the basic wage of men was higher than that of women, although under the Equal Pay Act of 1970 such differences should be eliminated by 1975. Part of the difficulty in implementing equal pay is that for many women their job is often of secondary importance to their home commitments and an employer may feel that he cannot place so much reliance on them. Equal pay has, however, been established for a considerable time in many branches of the public sector.

20.6 Questions and Exercises

1. Consider the advantages and disadvantages of the various methods of wage payment.
2. In what ways is the marginal productivity theory of wages inadequate?
3. Assess the importance of the various influences determining the level of wages in any one occupation.
4. 'Differences in wages can largely be explained in terms of economic rent.'
5. In what circumstances will a Trade Union be able to secure an increase in real wages for its members?
6. Consider the possible economic effects of the introduction of equal pay for women.
7. Assess the role of Trade Unions in the modern industrial society.
8. Distinguish between the various systems of wage-payment, explaining the economic implications of each.
9. To what extent do wage increases depend upon the monopolistic power of Trade Unions?
10. 'Since labour is the only productive factor of production all wages should be increased.' Comment.

Unit Twenty-One

Factor Incomes: Interest, Profits and Rent

21.1 Introduction

We saw in Unit Five that the production of goods is made very much easier by the use of machinery. Machinery can be bought with income saved from one period and not spent on consumption during that period. The person who forgoes consumption in this way receives a reward for his savings in the form of interest. In addition to this the person who uses those savings (whether they are his own or someone else's) for productive purposes is taking a risk that his project may be a failure. The reward he receives for such risk-taking is known as profit. It is the nature, determination, and importance of interest and profits that we must examine in the first part of this unit. We shall also consider some further aspects of the theory of rent.

21.2 Rates of Interest

Interest is the payment made for the use of someone else's money or capital. It is common, but erroneous, to talk of *the* rate of interest. In fact there are many rates of interest and their relative levels vary according to the purpose for which funds are being advanced, and the ease with which they can be regained. When an individual lends money for a completely riskless project the return he earns may be regarded as *pure interest*—simply the reward of foregoing the current use of his money. The return on gilt-edged securities may be regarded as pure interest since their holders know that there is no risk of the Government failing to repay the stated sum. Many other 'rates of interest', however, embody elements of commission or other charges, and profit for taking the risk that repayment may not be made. Thus if a person borrows money from a Finance House the interest will normally include handling or service charges in addition to pure interest. In order to attract funds from savers the Finance House has to offer an interest rate higher than that offered by the Government as repayment of the loan is less certain. In practice the risk is remote but the differential in interest rates that it promotes is a clear feature of the capital market.

The return that depositors receive on their money or the interest that borrowers have to pay will also reflect the time for which the loan is arranged. The shorter the period of the loan or the shorter the notice required of repayment, the lower the rate of interest. Thus money lent to Discount Houses 'at call and short notice' will command a lower rate of interest than money advanced to the Government for 91 days against Treasury Bills. Local Authorities are frequent borrowers, and they offer depositors a range of different rates of interest according to the length of the loan. In March 1972 4¾ per cent was payable on deposits left for one month but 6 per cent was payable if the deposit was left for two years. The extra reward is compensation to the depositor for the loss of liquidity over a longer period.

A glance at the financial columns of a newspaper will indicate the prevailing rates of interest on different kinds of loan. It will normally be quite clear how those rates are related both to the risk involved and the loss of liquidity. We must now consider the factors determining the rate prevailing in a particular market at a particular time. There are two opposing but related theories of the rate of interest to consider. First there is the *loanable funds theory* which stresses the supply of funds and the demand for them in a particular market. Then there is the Keynesian *liquidity preference theory*, which concentrates on the demand or preference of individuals and institutions to hold their assets as money rather than in any other form.

21.3 The Loanable Funds Theory

We have seen that at any one time there are a large number of different rates of interest. For the remainder of this unit we shall assume that there is only one rate and we will consider the main determinants of that rate. The loanable funds theory examines the rate of interest in the same way as we analysed other markets in Unit Thirteen, by looking at the demand for funds and supply of funds.

(a) The demand for funds

Demand for funds for investment purposes comes from three main directions; business organizations, the Government, and individual persons. To some extent each group provides some of its own funds and need not enter the capital market to acquire them. A company may finance all of its current investment out of past profits or I may buy a new car out of my past savings. But frequently each of us will resort to the market. A company's demand for capital depends on its prospects of using the funds profitably. It will therefore estimate the likely trend of costs and revenue associated with the projected investment and establish the likely profit. If this is greater than the prevailing rate of interest then the investment is likely to go ahead; but if it is smaller the project will probably be deferred. Accordingly one of the main determinants of the demand for funds is the rate of interest, a reduction in the rate causing an extension of demand. However, the emphasis on the expectations of the company reminds us that the demand for capital is a derived demand, and while changes in the rate of interest may cause extensions or contractions in demand, changes in the underlying conditions of the economy will cause shifts in the demand schedule. In general a very active economy will raise entrepreneurs' expectations and move the demand curve for funds to the right, while a depressed economy with many thousands unemployed will cause the demand curve to move to the left. Indeed such influences are likely to be far more important in determining the level of demand for funds than the price of borrowing. This is particularly so where the anticipated level of profits is high in relation to the rate of interest, since a modest change in the interest rate will not make much difference to the overall prospects of the project.

Personal demands for capital resources may be divided into two categories, the demand for funds to finance house purchase, and the demand for funds to finance the purchase of consumer durable goods, especially cars. In the first case there is little indication that the rate of interest is a major determinant of the demand for funds. The level of income, the shortage of rented accommodation and the price of houses are far more important. In the second instance, the level of interest rates is more important but other conditions relating to hire purchase agreements such as the level of the initial deposit and the period allowed for repayments are equally significant.

The Government enters the capital market as a borrower when it cannot raise enough through taxation to meet its capital expenditure. The factors governing public investment are very complicated and may be only loosely related to economic conditions. The building of hospitals or schools for example, is not something that can be dictated by ordinary economic criteria, although it is safe to say that the Government endeavours to borrow most heavily when the rate of interest is low, a circumstance that they may themselves bring about (see Unit Twenty-eight).

(b) The supply of funds

The immediate sources of loanable funds are the institutions mentioned in Unit Nineteen, but as we have seen they derive resources in much smaller instalments from members of the public. One of the determinants of the supply of funds will be the reward offered for foregoing liquidity, the rate of interest itself. The higher the rate of interest the greater the supply of loanable funds. In fact it is again unlikely that slight changes in the rate of interest will cause much change in the supply of funds: there are very few savers who react to an increase in interest rates by saying 'now that the rate is 5 per cent instead of 4 per cent we will save more'. Much greater changes in the rate would be necessary to induce such a response. (Of course where the relative rates obtainable in different markets change sophisticated savers may switch their savings from one area to another.) Some people may even respond to an increase in interest rates by reducing their savings since they can earn the same amount from a smaller deposit.

More important than the rate of interest is the level of individual and national income. We have observed that the great difficulty of the poorest countries is that they spend all their income on consumption, and can afford to save nothing. This is true also of very poor individuals in comparatively rich communities. Wealthy societies and individuals on the other hand have more opportunity for saving and with any community it is the wealthy that are responsible for the bulk of saving.

The institutional framework can also influence the supply of funds. We may interpret this in its broadest sense to include the nature of the society itself. If it is prone to political upheaval savings will tend to be low since no one will risk leaving money in institutions that may be the subject of political dispute. Furthermore in a society like the United Kingdom with a large number of institutions specially geared to encourage savings, the supply of loanable funds is likely to be greater than in a society whose capital market and financial institutions are underdeveloped.

Other factors of importance will be the risks and liquidity factors discussed in Section 21.1 and the possibility of inflation. The latter is considered more fully in Unit Twenty-three, but here we may note that if the rate of interest is 5 per cent a person depositing £100 for one year will receive £105 at the end of the year, but he is worse off in real terms if prices have risen by 10 per cent during the year. His purchasing power is thus reduced. In such circumstances savers might seek alternative outlets for their funds.

(c) The equilibrium rate of interest

When the demand and supply curves for loanable funds have been established, we can then find the equilibrium rate of interest at the point of intersection of the two curves. At this point savings (the supply of funds) and investment (the demand for funds) will be equal and the market will be in equilibrium.

Like many economic theories the loanable funds theory contains more than a grain of truth. Rates of interest do respond to changes in the demand and supply of funds. In backward economies where savings are low and the demand for funds high the rate of interest is high. In advanced economies rates also respond to the pressures of demand and in the London money market they change on a day to day basis or even hour by hour as market conditions change. The question of interest rates must now be approached from a somewhat different angle. The traditional theory places too much emphasis on the rate of interest itself in determining the demand for and supply of funds, and there are probably other factors of more importance in determining the level of savings and the level of investment.

21.4 Liquidity Preference Theory

While the loanable funds theory emphasizes the importance of supply and demand it says nothing about the overall supply of money in the economy. John Maynard Keynes was primarily interested in the control of the chronic cycles of economic activity which were a feature of nineteenth- and early twentieth-century capitalist economies. In the course of his work, which is more closely examined in Unit Twenty-two, he developed his own theory of interest which relates the supply of money to the wish to hold it. Unlike the loanable funds theory this does not bring savings and investment into equilibrium. How this occurs is shown in Section 22.7.

(a) The supply of money
This may be regarded as fixed at any one time. It is not a quantity that can be easily controlled by the authorities for, as we have seen, the banks have considerable powers to create credit. Keynes was primarily interested in short-term policy, and changes in the supply of money in the short-term may be disregarded.

(b) Liquidity preference, the desire to hold money
If the supply of money is fixed then the rate of interest according to this theory will depend upon the desire of the public and business and financial organizations, to hold their assets as cash. Keynes identified three motives for holding cash.

(i) **The transactions motive.** Most people receive their incomes in 12 or 52 more or less equal instalments. Since they do not spend all their money the moment they receive it, but try to spread expenditure over the whole period between pay-days, they need to hold some income in money balances either in cash or in bank accounts. The average amount held in this way will depend partly on the frequency of pay-days (the more frequent they are the lower the average holding of cash) and partly on the level of income of the individual (the wealthy generally hold more than the poor). In addition all kinds of businesses will need to hold some cash assets to meet their day to day commitments. In the case of both the individual and the business, the balance that is held for transaction purposes is likely to be fairly stable, each of them knowing their probable commitments for the next few weeks and holding cash accordingly.

(ii) **The precautionary motive.** The careful individual and the prudent businessman will hold a further reserve of cash to meet irregular or unexpected expenditure. A sudden rise in the cost of raw materials might otherwise find a business short of cash, and an individual might be somewhat embarrassed by an expensive repair to his car. Once again, however, the amount held in the form of money will probably be fairly stable and depend mainly upon income. For most purposes indeed, there is no reason why we should not regard the precautionary motive as part of the transactions motive. Beyond holding cash for these reasons it might seem sensible for firms and individuals to invest their money to earn a return, and the majority of them do so. But wealthy individuals and most financial institutions have a further reason for holding assets in the form of cash.

(iii) **The speculative motive.** As we saw in Section 17.4 (*h*) the price of fixed interest securities varies inversely with the rate of interest. A £100 bond

R–I

(security) issued at 4 per cent will have a market value of only £50 if the rate of interest rises to 8 per cent, but will be worth £200 should the rate of interest fall to 2 per cent. It now becomes clear why cash may be held for speculative purposes. If the rate of interest is low the price of bonds will be high and speculators will want to hold cash in anticipation of a rise in interest rates which will bring down the price of bonds. When this happens the speculators will buy bonds, reducing their cash holdings, and receiving a greater yield on their bonds because they have bought at lower prices. At high rates most of their cash is tied up in the form of bonds which they hope to sell when interest rates fall and the price of bonds rises. It is for

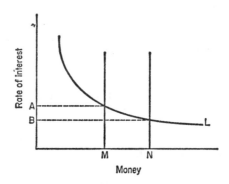

Fig. 21.1 Liquidity preference

this reason that the liquidity preference curve slopes downwards as in Fig. 21.1, levelling at a very low rate of interest since it is unlikely that speculators would allow the rate to fall to zero before demanding extra cash for speculative purposes, or that anyone would be willing to part with the liquidity unless some interest were payable. In Fig. 21.1 if the supply of money is fixed at OM, the equilibrium rate of interest would be OA. An increase in the money supply to ON reduces interest to OB. Changes also occur as a result of changes in tastes or income distribution, and these cause the liquidity preference schedule to move.

The chief significance of this theory of interest is that it shows quite clearly how the authorities can influence interest rates by altering the supply of money, a topic to which we return in Unit Twenty-eight.

21.5 Profits

We noticed above that the interest paid to some investors incorporates an element of profits rewarding the investor not for the loss of liquidity but for the risk he takes that he might lose his money altogether. The risk is really the uncertainty embodied in any industrial or commercial project that it will fail for any one of a number of reasons:

(*a*) The managers of the enterprise may have misjudged the market and may discover that by the time their product is on the market tastes and the pattern of demand has changed in favour of something else. Clearly such risks will be greater in the fashion industry than in many branches of the food industry, but the wise manufacturer, whatever his field, will make a careful survey of his likely market before committing himself to heavy investment to produce the goods for it.

(*b*) In many industries it will be some years before a particular investment pays for itself. There is always the possibility in such cases that changes in technology will render the original investment obsolete at considerable cost to the company concerned.

(*c*) Finally social and political changes may reduce the chances of profit or, if the business becomes nationalized, eliminate them altogether.

During the industrial revolution the risk of loss through such causes was borne by the entrepreneur himself. Today it is largely borne by the holders of ordinary shares in a limited company and as compensation for this risk or uncertainty they expect an element of profit to be included in their income. Profit, though, is not only the reward for risk-taking. It may also be regarded as the reward for enterprise, for spotting the right opportunity and taking advantage of it. It is this rather than the risks that explains the relatively high profits that accrue to the developers of fairly simple inventions.

21.6 The Nature of Profit

Although profit is a reward to one of the factors of production it differs in several respects from the income payable to the other factors.

(*a*) Profit is a residual reward payable to the entrepreneurs after all the other costs of the business have been met. Whereas labour can be certain of its weekly wage or monthly salary, and the debenture holder fairly certain of his interest, there is no guarantee that the ordinary shareholder will receive a dividend. There is no contractual obligation on the part of the company to pay a dividend.

(*b*) Indeed it may sometimes be impossible for the company to pay a dividend as profits are negative and a loss is incurred. While it is impossible not to pay wages to employees, the company may survive for some time without actually making a profit as we saw in Unit Fourteen.

(*c*) Although profit is, in an accounting sense, a residual payment to capital we saw in Unit Ten that the economist regards it as a cost that has to be met if the firm is to stay in business. While the firm may survive one year and perhaps two without making a profit it will in the long run go out of business if it cannot pay a dividend to its shareholders.

(*d*) Just as interest sometimes embodies an element of profit, the payment that is itself referred to as profit may, especially in the small business, have several components. Consider the case of the local grocer, who at the end of the year, having paid all his bills and the wages of his staff, has a surplus of £2500. To describe this all as profit is to exaggerate the payment he is receiving for his enterprise or risk-taking. The following modifications must be made:

(i) The entrepreneur could probably have earned £1500 per year in some other occupation. This opportunity cost should be set against the £2500 before a realistic assessment of profits can be made.

(ii) The money that the entrepreneur has invested in his business could have earned him an income in the form of pure interest if he had invested it in gilt-edged securities. There is thus an element of interest included in the surplus of £2500.

(iii) Part of the £2500 might also be 'abnormal profits' or monopoly rent.

When these aspects have all been taken into consideration we are left with a very much lower figure for profits than the original £2500.

21.7 The Role of Profits

The capitalist or free enterprise economies are said to be based on the profit motive. We will try to determine the meaning of this, and the actual role of profits in these economies.

(a) If there were no reward for enterprise or risk-taking no one would invest in anything but pure interest yielding stock. The risk inherent in investing in plant and machinery to meet future demand which might not materialize, must be balanced by the prospect of a return higher than that available on risk-free investment. It is the prospect of this extra return 'profit' that brings forth a continual succession of large and small businesses each to make its own contribution to the growth and development of the economy.

This view of profits should be tempered by remembering the distinction between normal profits and abnormal profits. It is unfortunately the case that the greatest profits frequently seem to go to the least obviously productive enterprises. Property companies and Holding Companies often show enormous returns on capital, largely because of the ability of their proprietors to manipulate the market. It is one of the weaknesses of the economic system that the rewards for this kind of initiative are so much greater than the rewards for more obviously productive investment.

(b) The fact that profit may be either positive or negative allows it to serve as a useful guide to the allocation of resources. Where an industry is expanding to meet an important demand its profits are likely to be buoyant and the funds necessary for the further development will be easily acquired. In many cases this will not necessitate the entry of new firms, but simply the expansion of existing ones. As more capital is employed, seeking a share in the profits available from the industry, so extra units of the other factors of production will be employed. On the other hand unprofitable firms will find it difficult to attract extra capital, or even to renew their existing capital equipment and will therefore release other resources as they decline. It needs no single decision-making body to decide where funds should be allocated or when they should be withdrawn; the votes of consumers as expressed by their purchases determine this. They may be regarded as the hidden hand of the capitalist system.

We should not regard the profit motive as infallible in the allocation of resources. There are many activities in which it is almost impossible for private capital to make a profit and operate on a scale large enough to meet the public's requirements. In these circumstances the industry may be left in private hands but receive a subsidy from the Government, as with shipbuilding. More frequently, as in the case of the public services, they are taken over by the state and run as Public Corporations. On the other hand the profits in many firms, and the resources employed in those firms, are excessive because the firms do not have to meet all the costs of production. The costs of environmental pollution are often incalculable, and even where an estimate can be made, the burden of meeting such costs falls on taxpayers and ratepayers rather than the consumers of the product. Apart from these special cases the profit mechanism operates in the vast majority of industries, both large and small, with surprising smoothness.

(c) In competitive industries profits may be regarded as a sign of efficiency, as profits will be larger if costs are kept to a minimum. But we must be wary of assuming that the firms making the largest profits (however measured) are necessarily the most efficient. They may be in fairly strong monopolistic positions, or they may be members of a cartel able to cover their inefficiency by charging exorbitant prices.

(d) A feature of the capitalist (profit-seeking) system, that has been heavily criticized is that it is the owners of capital or their agents who take all the decisions relating to the business, even though there could be no production without the co-operation of all the factors of production. Related to this is the criticism that incorrect judgements lead not only to the unemployment of capital but also to that of labour. We cannot pursue this debate here. It must be said in defence of the system that it can only react to the wishes and demands of the consumers. No matter how skilful the textile industry's entrepreneurs, there is little they can do to create jobs if no one wants their product. Nor would there be much difference under any other system except to the extent that the man in Whitehall (or Moscow) could dictate both the patterns of consumption and the allocation of resources. While acknowledging the deficiencies of the system we may conclude that the profit motive serves its purpose fairly well in the absence of any proven alternative.

21.8 The Rent of Land

The reward earned by the factor land like those earned by labour and capital depends on the conditions of supply and demand. We have already examined the concept of economic rent as a surplus payment to any factor of production. In order to distinguish the rent of land from economic rent we may call it *commercial rent*.

David Ricardo, who developed the early theory of rent, regarded it as a surplus in much the same way as we now regard economic rent. This was because in common with his contemporaries he believed that the value of a product depended upon the labour embodied in it and on nothing else. Moreover the supply of agricultural land was fixed and had virtually no other use. Accordingly anyone who received rent for land was receiving a payment which he had not earned, and he only received because land was in short supply.

It became apparent during the Napoleonic Wars that as more and more land was brought into cultivation to meet the demand for wheat, the owners of the most suitable land were able to increase the rent they charged to farmers. The reason for this was that the price of wheat was determined by the cost of producing it on the least suitable land. No one would farm the land if they could not at least cover their costs. The farmers on the most suitable land were then making very high profits, not because they were better farmers but because they happened to be farming the most fertile land. These high profits Ricardo regarded as being attributable to 'the original and indestructible properties of the soil' and it was these that the landlord was able to absorb in rent.

Under these conditions rents are clearly determined by the price of the product, for the higher the price of wheat the higher the rent chargeable. But the payment of rent also depends upon the shortage of land. Early settlers in America and Australia paid no rent because there was no shortage of land. This is indeed very different from the position on land for housing purposes today. Landowners deliberately withhold land from the market, waiting for the price to rise, and by contributing to the shortage ensure that rents rise in the face of heavy demand. When the demand for housing is highest the price of land is highest and thus rents in London are higher than those in places like Devon. Landlords in London are therefore able to command an economic rent as well as a commercial rent because of the shortage of land and accommodation there.

Although rents in general seem to be demand-determined, it is possible to regard them as price determining. The proprietor of a new business taking a site in London will have to include the rent he pays as part of his costs and it will be embodied in the prices he charges for his products. Similarly the landowner will be faced with a number of alternative outlets for his funds and will only invest them in property if he can earn at least as much from it as he could in the best alternative occupation. So the cost that the proprietor has to meet will be at least as high as the transfer earnings of the landowner.

So far we have considered the payment made to individual units of various factors. Before leaving this topic we should consider the way the National Product is divided between the broad groups labour, land and capital.

21.9 The Shares of Labour, Land and Capital in the Gross Domestic Product

Table 21.1 Division of the GDP 1949, 1959, 1969

	1949		1959		1969	
	£m	% of GDP	£m	% of GDP	£m	% of GDP
Income from employment and self-employment	8,609	77·1	15,980	76·0	30,347	76·8
Profits and surplus of public enterprise	2,101	18·7	3,868	18·4	6,527	16·6
Rent	475	4·2	1,170	5·6	2,692	6·6
GDP	11,185	100·0	21,018	100·0	39,566	100·0

Source: *National Income and Expenditure*, HMSO

Table 21.1 shows that the share of labour in the national product remained fairly constant between 1949 and 1969 despite strenuous efforts on the part of Trade Unions to increase it. Although profits have declined most of that decline occurred during the period 1964–9 when the economy was deeply depressed for balance of payments reasons. Reference to the *Blue Book on National Income* shows that profits have now regained some ground, and we may conclude that despite interminable debates about the relative rewards of the factors very little change in the distribution of these rewards has occurred, although we may note that in 1938 income from employment and self-employment amounted to only 70 per cent of the Gross National Product.

21.10 Questions and Exercises

1. Distinguish between interest and profit.
2. Why is it sometimes dangerous to speak of the rate of interest?
3. What are the main determinants of the rate of interest?
4. Outline the loanable funds theory of interest.
5. Outline the liquidity preference theory of interest.
6. What are profits? What determines their level and what is their role in the economy?
7. 'Profits are a residue.' (Accountant.) 'Profits are a cost.' (Economist.) Comment.
8. Distinguish between economic rent and commercial rent showing the main determinants of the former.
9. Why are rents higher in the centre of London than in Middlesbrough?

10. From the financial pages of your newspaper obtain these current rates of interest: (*a*) the minimum lending rate; (*b*) the base rates of the Clearing Banks; (*c*) the rate offered by Local Authorities on 12-month loans; (*d*) the rate of Treasury Bills; (*e*) the rate offered for Building Society Deposits. How do you account for the differences you find?

Unit Twenty-Two
The Level of Employment

22.1 An Introduction to Macro-Economics

In the previous Units we have examined the constituent parts of the economy, such as banks, firms, industries, and factors of production, and observed various aspects of their behaviour. Such studies belong to the field of micro-economics. But each element in an economy contributes to the overall performance of that economy in respect of such matters as the level of employment, the general level of prices, the rate of economic growth and the balance of payments. Macro-economics is the study of whole economies, and here we enter the field of public economic policy. In the remainder of the book we will consider macro-economic problems, paying particular attention to the policies which the Government may invoke in its attempts to achieve its economic objectives.

We may summarize these main objectives as follows:

(*a*) Full employment.

(*b*) An acceptable rate of economic growth, leading to a high standard of living.

(*c*) Stable prices.

(*d*) Balance of Payments equilibrium.

(*e*) A more equal distribution of income and wealth through the system of taxation and social security payments.

In addition any particular Government will have a whole variety of subsidiary aims each of which may be expected to have an influence on the general allocation of resources and economic performance.

One of the difficulties facing the Government is that the main policy objectives are not necessarily complementary. Full employment may be achieved only at the expense of rapidly rising prices; or a balance of payments surplus may result in an unacceptable level of unemployment; and a more equal distribution of income may reduce savings and have an adverse effect on the level of investment and economic growth. These and other conflicts make the formulation of policy a hazardous task, but before we consider the policies that the Government may use we must in this and the next few Units look more closely at some of the objectives. First we shall consider the domestic aims of full employment, stable prices, and a reasonable rate of growth. Later on we shall look at the complications introduced by international economic relationships and finally we shall make a full appraisal of Government policies in the concluding Units.

22.2 Why the Level of Employment is Measured

The level of employment has long been accepted as one of the principal indicators of the general level of economic activity. If the factories are all fully occupied with full order books, then few people will be out of work. On the other hand when the general level of activity is low and when both consumers and industrial organizations are delaying purchases for some reason then unemployment will tend to be high. It is by reason of this broad relationship between output and employment that the level of employment has long been felt to give an indication of the state of the economy. Later we shall see that the concept of the 'number of unemployed' has come in for some severe criticism. First we must consider what determines the level of employment.

22.3 The Determination of the Level of Employment

We saw in Section 20.3 that a single firm's demand curve for labour slopes down from left to right, indicating that when wages are low the number employed will be high. Not surprisingly it was held for a long while that the way to cure unemployment was to reduce wages to induce employers to take on more labour. At times of heavy unemployment such a solution was widely advocated. Unfortunately it overlooks the fact that the demand for labour is a derived demand, and that people have a dual role in the economy: they are both workers and consumers. If no one wants to buy the products of an industry, then that industry must rid itself of some of its factors of production, thus causing a rise in the number of unemployed. If sufficient people are forced to accept wage reductions the level of demand for some products must also fall, producing a rise in unemployment. This brings us back to the dual role of people as factors of production. On the one hand they have the task of producing the goods and services required by society; but, as members of that society they spend their income on the goods they have helped to produce. If they decide to save some of their income rather than spend it goods will be left in warehouses and as stocks accumulate employers will be forced to lay off labour. This applies not only to a single factory or industry but also to the economy as a whole. If no one wants to buy the goods a factory produces then it cannot afford to continue employing the factors of production. In the case of the economy it is the total demand for the goods and services produced by its factors of production that is important. This total demand, which is the same as the Gross National Product, is known in this context as *aggregate monetary demand*. It may be broken down into four components: consumption, investment, Government expenditure on goods and services and exports.

22.4 Aggregate Monetary Demand

The Gross National Product is generated by the production of goods for consumption, investment, purchase by the Government, or for export. For all practical purposes exports may be ignored as, although they amount to approximately 5 per cent of the Gross National Product, they are almost exactly balanced by imports so the net effect on aggregate monetary demand is rather small. In the initial stages of our examination the Government's demand for goods and services will also be excluded, partly because it is determined by social and political as well as economic factors, and partly because the Government may regulate its demand according to the level of consumption and investment.

We are thus left with an economy in which aggregate monetary demand depends on the demand for consumer goods and the demand for investment goods. We may infer that in the relatively short run the level of employment also depends on consumer and investment demand. (We have seen in earlier units that short-run increases in output result from an increased employment of the variable factors. In the long-run increases may come about through the use of extra machinery which may result in *less* labour being employed. It is the short-run we are concerned with here and it is an acceptable approximation to say that an increase in the Gross National Product is accompanied by an increase in the level of employment.) It follows from this that if we know what determines the level of consumption and the level of investment then we shall also know what determines the level of employment.

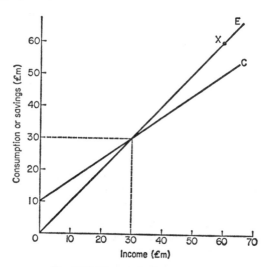

Fig. 22.1 The propensity to consume

22.5 The Propensity to Consume

(a) The level of income

In Fig. 22.1 the line OE is a convenient guideline linking all the points where the whole of national income is spent on consumption. Thus we can see that if national income is £50 million and it is all spent on consumption the situation is plotted on OE. When consumption is £30 million the horizontal axis shows that the national income will be £30 million. In fact it is unlikely that in most households expenditure on consumption will be exactly equal to income: at very low levels of income households may be expected to spend more than their current income either by drawing on their previous savings or by borrowing other people's savings via the banks or hire purchase companies. This would be the case for the community as a whole if income were below £30 million in Fig. 22.1. On the other hand at higher levels of income households will be able to save and their consumption expenditure will be lower than their income. This is the case for the community as a whole when income exceeds £30 million. For example when national income is £50 million, £43·5 million is spent on consumption and £6·5 million is saved. The gap between the line OE and the consumption function C therefore gives a measure of the amount of savings.

It is clear from the diagram that as income rises consumption rises but the proportion of income spent on consumption falls. Thus when income is £30 million it is all spent on consumption, but when income rises to £60 million the proportion spent on consumption is 83·3 per cent, and with income at £65 million, 81·5 per cent is spent on consumption. The relationship between the level of consumption and the level of national income is known as the *average propensity to consume* (APC). It is always expressed as a fraction of 1: $APC = \dfrac{C}{Y}$ where C = consumption; Y = national income. Thus when national income is £40 million:

$$APC = \frac{37}{40}$$

The proportion of income not spent on consumption is known as the *average propensity to save* and is likewise expressed as a fraction. The sum of the average propensity to save and the average propensity to consume is always one.

The *marginal propensity to consume* (MPC) and the *marginal propensity to save* must not be confused with average propensities. The marginal propensity to consume is that proportion of an increase in income spent on consumption; thus if income rises by £1 million and consumption by £500,000 the marginal propensity to consume is ½. The marginal propensity to save, which is the proportion of an increase in income not spent on consumption, is also ½. If the consumption function is a straight line as it is in Fig. 22.1 the marginal propensity to consume is constant and whenever income increases by a specific amount consumption rises by a smaller constant amount. This is an approximation of actual behaviour which need not worry us too much. The propensity to consume curve is more likely to resemble that in Fig. 22.2 where it rises steeply at first and then flattens out. This is because at very low levels of income a community will spend the majority of any increase in income, while at very high levels the bulk of a similar increase will be saved (thus giving a low MPC). For example if

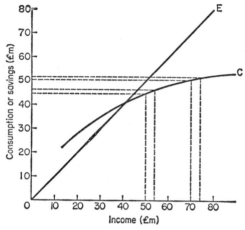

Fig. 22.2 Diminishing marginal propensity to consume

income rises by £4 million from £50 million, consumption rises by £2 million and MPC = $\frac{1}{2}$. A similar rise in income from £70 million to £74 million causes consumption to rise by rather less than £2 million and MPC is less than $\frac{1}{2}$. Similar considerations apply to sections of a given community: consider, for example, the likely reactions of Old Age Pensioners and millionaires to an increase in income of £1·00 per week.

(b) The distribution of income

While the level of national income is the most important determinant of the level of consumption expenditure the distribution of that income between rich and poor will also exert an influence. If income is concentrated largely in the hands of the relatively wealthy then we would expect the average propensity to consume to be lower than if income were more evenly distributed. The Government's taxation policy will also have some influence on the level of consumption. Let us suppose that the very rich who have a marginal propensity to consume of $\frac{1}{2}$ are forced to pay extra taxes of £1 million. As they would have saved £500,000 of this anyway, they cut their consumption by only £500,000. If the £1 million spent in taxes is used to increase social security benefits for those with a marginal propensity to consume of 1, they will spend the whole of the £1 million and consumption will rise by £500,000. By such a policy the Government can move the consumption function vertically upwards or downwards.

(c) Institutional factors

These are of significance. We noticed in Unit Twenty-one that the supply of savings will depend partly on the encouragement given to them by financial institutions. Accordingly we can say that the position of the consumption function depends partly upon the encouragement given to savings in the community.

(d) Expectations of rising prices

These will persuade consumers to buy today rather than tomorrow, causing an upward movement of the consumption function. In the same way if prices are expected to fall, consumers will delay their purchases thus causing the consumption function to fall. To simplify our investigation we shall assume that consumption varies directly with income and that the other factors discussed above are constant.

Returning to Fig. 22.1, if consumption is the only kind of expenditure the equilibrium level of national income must be £30 million. Once again we mean by equilibrium a situation which is unchanging, and if national income is £30 million the consumption function shows the £30 million will be spent, and this will generate income in the next period of £30 million and so on until some disturbance is introduced from outside. Should some other level of national income be arbitrarily established forces will be set up to return the economy to equilibrium. Thus if income is £50 million, consumption will be £43·5 million, which generates income of £43·5 million and consumption of £39 million and so on until income and consumption are both £30 million again. The level of employment within the economy will be that necessary to produce a national income of £30 million but there is no reason why this should constitute anything approaching full employment. Let us assume that there is a direct relationship between national income and the level of employment and that a national income of £30 million leads to an unemployment rate of 50 per cent, so that £60 million would constitute the full employment level. Thus point X is the full employment equilibrium in Fig. 22.1 and for this to be reached the consumption function (or Aggregate Monetary Demand curve) must pass through this point. One possibility is that the Government will intervene and cause the consumption function to move upwards by redistributing income. But before we consider that line of approach we must consider the other part of aggregate monetary demand, investment.

22.6 The Level of Investment

In order to produce consumer goods effectively businessmen need to purchase machinery. This demand for investment goods, like the demand for consumer goods, creates employment and the greater the level of investment the greater the level of employment in the investment goods industry. (Note that we are concerned only with investment in new, real productive assets since it is only this kind of investment that influences employment; the purchase of second-hand machinery or of securities has no direct impact on the level of employment.) What factors determine the level of investment?

(a) Replacement investment. Some investment occurs almost automatically as existing equipment wears out and has to be replaced. In flourishing firms this causes few problems but in marginal firms it will be subject to the same considerations as new investment.

(b) The most important factor in the determination of investment is the expectation of future profits. We learnt in our study of the theory of the firm that any firm will go out of business if it does not make a profit. We might amend this and say that if entrepreneurs see no prospect of profit then there will be no investment. Thus expectations play a crucial part in determining the level of investment and these expectations depend largely on underlying economic conditions, as we shall see in our subsequent examination of economic fluctuations.

(c) The rate of interest was once felt to be the main determinant of the level of investment, but although it is still of significance it is not of crucial importance. This is partly because the prospective yield on investment, technically called the *marginal efficiency of capital*, has long been far above the prevailing rate of interest. Thus entrepreneurs have been able to reduce the weight attached to the rate of interest when planning investment, as marginal changes in the cost of borrowing capital are not likely to turn profitable projects into unprofitable ones.

(d) It is also frequently true that this year's investment is part of a five year schedule of, and is therefore unlikely to respond to, marginal changes in the rate of interest.

(e) A large proportion of investment is now undertaken by the Government or by Government agencies, and is therefore unlikely to respond to ordinary economic factors. Council house building for example continues for social and political reasons irrespective to a large extent of changes in economic conditions.

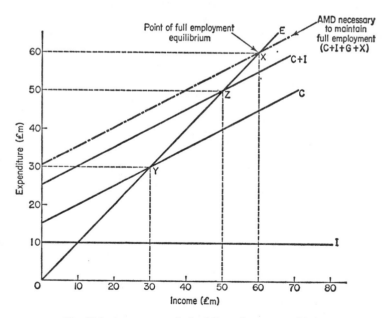

Fig. 22.3 A move towards the full employment equilibrium

It is thus impossible to be precise about the determinants of investment. We must be contented with the assumption that in the real world investment generally rises as national income rises. For the purposes of our discussion of the level of aggregate monetary demand and the determination of the level of employment, we shall consider investment fixed at £10 million. We will now consider the effect of the introduction of such an investment to the economy illustrated in Fig. 22.1 and reproduced in Fig. 22.3.

22.7 Towards Full Employment

In this diagram the horizontal line I represents the constant level of investment, and the line C+I shows the level of expenditure of both kinds at various levels of national income. It is parallel to C and £10 million above it. Notice that, compared with the previous equilibrium level of expenditure total expenditure initially rises by £10 million, but that the equilibrium level of national income rises from £30 million to £50 million. This is a most important feature of investment expenditure, and it is mainly attributable to the *multiplier*. A separate example follows to illustrate the process.

We will assume an economy which has been in equilibrium for some time with a national income of £20,000 million and a constant average and marginal propensity to consume of $\frac{4}{5}$. Thus the national income consists of £16,000 million of consumption and £4000 million of investment. (Savings are also equal to £4000 million and we will return to this equality later.) What will happen if investment rises by £1000 million? The owners of factors of production employed in the investment goods industry receive an increase in incomes of £1000 million, as all expenditure creates a corres-

Table 22.1 The multiplier in action: effect of extra investment of £1000 million with MPC = $\frac{4}{5}$

Rise in income (£m)	Rise in consumption	Rise in savings (£m)
1000·0	800·0	200·0
800·0	640·0	160·0
640·0	512·0	128·0
512·0	409·6	102·4
.	.	.
.	.	.
.	.	.
.	.	.
.	.	.
Total 5000·0	4000·0	1000·0

ponding income. (In practice there may be long time lags before some of these incomes are received but we may ignore this difficulty.) We know that the marginal propensity to consume is $\frac{4}{5}$, and therefore the recipients of the extra £1000 million of income will spend £800 million and save £200 million. This expenditure will be largely on food and clothes, television sets and motor cars, and the owners of factors of production working in these industries will then receive extra incomes of £800 million. The following changes in income are summarized in Table 22.1.

Each increase in consumption leads to successively smaller increases in income and eventually national incomes rise by £5000 million as a direct result of the initial increase in investment of £1000 million. The ultimate rise in the level of national income depends upon the size of the marginal propensity to consume. If the community is so rich that its marginal propensity to consume is zero, then the rise in national income will be the same as the rise in investment as the recipients of the original £1000 million will save it all and it will not generate a secondary increase in income. But if the marginal propensity to consume is 1 then there will be no end to the rise in national income, at least in money terms (see Unit Twenty-three).

The general formula for the multiplier is that:

$$\text{The rise in national income} = \text{The rise in investment} \times \frac{1}{\text{MPS}}$$

Thus in our example:

$$\text{The rise in income} = £1000 \text{ million} \times \frac{1}{\frac{1}{5}} = £5000 \text{ million}$$

Table 22.2 gives some examples assuming a rise in investment of £1000.

Table 22.2 The effect of the size of MPS on the rise in national income

Rise in investment	MPS	Multiplier	Rise in national income
1000	$\frac{1}{5}$	5	5000
1000	$\frac{1}{4}$	4	4000
1000	$\frac{1}{3}$	3	3000
1000	$\frac{1}{2}$	2	2000

The multiplier works both ways and a fall in the level of investment will lead to a larger fall in the equilibrium level of national income. The extent of this fall will again depend upon the size of the marginal propensity to consume. Before returning to Fig. 22.3 we may note that other *injections* of expenditure into the economy such as a rise in exports or a rise in Government expenditure will also give rise to a multiplier effect. Savings are also not the only way in which money leaks from the economic system. As far as the employment of factors of production is concerned money taken by taxation and not spent on goods and services, and money spent on imports is all equivalent to savings, for it cannot create employment in the domestic economy.

The model in Fig. 22.3 has been constructed so that the multiplier has a value of 2, and the initial investment of £10 million leads to an ultimate rise in income of £20 million. At the new equilibrium level of income, consumption is £40 million and savings are £10 million. Another way of looking at this is to say that the consumer goods industries are producing £40 million worth of goods and there is investment to the extent of £10 million. As the nation's savings and investment are undertaken by two different sets of people we are entitled to ask the question 'why are they equal?' In fact this question has almost been answered in the analysis of the multiplier effect. In order to clarify the process by which equality comes about a distinction should be made between planned savings and investment, and actual savings and investment.

There is absolutely no reason why planned savings on the part of households should be equal to planned investment by firms. The only real equality is between actual savings and actual investment. If we look back at the economy in the past we can see that the savings and investment that have occurred are equal in amount. There is no great mystery about this, as the terms are just different names for that part of the national product which is not spent on consumption. If we suppose that an economy is at point Y, in Fig. 22.3, with an equilibrium of £30 million and no savings or investment, planned savings are zero. If entrepreneurs decide to spend £10 million on investment, planned savings and planned investment at the beginning of the period are unequal, but they become equal by the end of the year with the action of the multiplier. This generates sufficient increases in household income to enable savings of £10 million to be made out of the new national product of £50 million. On the other hand, if the economy is at Z with planned savings of £10 million, and entrepreneurs decide to invest nothing then the negative multiplier will reduce income to £30 million so that actual savings will fall to zero.

22.8 Injections and Leakages

The economy we have been discussing has been unrealistic in that it has engaged in no foreign trade and there has been no Government interference. The introduction of either of these, however, does not materially alter the theory. Any injection of expenditure into the economy leads to a higher level of aggregate monetary demand (AMD): any leakage from the system leads to a lower level of AMD, and each of the movements is associated with the appropriate multiplier effects. Fig. 22.4 illustrates the possibilities. We have added to the basic circular flow diagram of Unit Two. Leakages from the economy now consist of savings, taxation and expenditure on imports and these may be lumped together since each of them indicates that money is taken out of circulation and no longer contributes to AMD in the domestic economy.

On the other hand investment by industry, financed through the banks (which embraces all kinds of financial institutions discussed in Unit

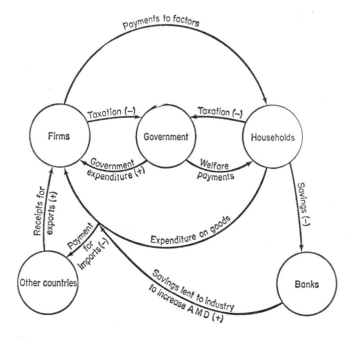

(−) Leakages from the economy
(+) Injections into the economy

Fig. 22.4 Injections and leakages

Nineteen), Government expenditure, and the demand for exports are all injections into the economy. They all increase the pressure of AMD and thus generate more employment.

If the economy is to be in equilibrium at X in Fig. 22.3 certain conditions must be satisfied. Savings do not have to equal investment, but savings (S) + taxation (T) + Imports (M) must equal investment (I) + government expenditure (G) + exports (X).

$$S + T + M = I + G + X$$

If S + T + M is greater than I + G + X the economy will be moving to a lower equilibrium. If S + T + M is smaller than I + G + X so that there are net injections into the economy then it will be expanding and moving towards a higher equilibrium income.

22.9 The Achievement of Full Employment

As a result of the introduction of investment into our economy, national income has increased to £50 million and employment is much nearer the target of full employment. The responsibility of the Government is to ensure that the aggregate monetary demand curve (AMD) passes through point

X (arbitrarily assumed to be the full employment equilibrium). There are two broad courses open to the Government.

(a) It may add its own demand for goods and services to that of the households and entrepreneurs, thus causing the AMD curve to move upwards. There are, of course, likely to be multiplier effects from any such injection of expenditure.

(b) It may exercise its monetary and fiscal policy to encourage higher consumption and investment expenditure, thus causing the C + I curve to move upwards. The discussion of the way in which these policies may be used is postponed until Units Twenty-eight and Twenty-nine.

To complete the picture we should mention the demand for exports. A further way of increasing the level of demand is to persuade countries overseas to buy more of our goods. The equilibrium of the economy can now be achieved at an income of £60 million generated by consumption, investment, Government expenditure and exports so that Aggregate Monetary Demand goes through point X in Fig. 22.3. Should demand rise above £60 million which represents the physical limit of the output of the economy then inflation will occur. This is discussed in Unit Twenty-three.

22.10 The Nature of Full Employment

Implicit in our discussion of employment has been the idea that the term full employment should be taken to mean a situation in which all units of labour are employed and the economy is up against the physical limit of its output. In practice no Government is likely to aim for such a high level of employment, for a certain amount of unemployment is necessary to allow flexibility in the labour market. Governments frequently accept that a certain percentage of unemployment, perhaps 2 per cent of those normally working, constitutes full employment and are satisfied if they can keep unemployment down to this level. Some find such arbitrary figures unacceptable and prefer to relate the number of unemployed to the number of vacancies; equality between these two being taken as an indication of full employment. An excess of unemployed over the number of vacancies is a sign of unemployment and an excess of vacancies over the number of unemployed is taken as a sign of over-full employment.

But such measures depend upon the accuracy of the numbers of registered unemployed. The figures, however, have recently been the subject of much criticism. The main purpose of a person registering as unemployed is the receipt of those social security benefits to which he is entitled. He is not trying to assist the Government in its assessment of the number of people actively seeking work. This being so the figures obtained may be criticized for several reasons.

(a) Many part-time workers, particularly women, do not bother to register as unemployed. They seek jobs by replying to advertisements and visiting private employment agencies rather than through the appropriate Government departments. Similarly, many people over the official retirement age probably do not register even though they are actively looking for work.

(b) At times of particularly high unemployment school-leavers, who are potential employees, may be forced to continue their full-time education rather than join the ranks of the unemployed.

(c) Both (a) and (b) tend to reduce the registered unemployed. On the other hand a large number of companies find it expedient to retire managerial employees on pensions at the age of sixty. Many of these people then register as unemployed because they are entitled to benefits even though they have little intention of taking another job.

(d) There are also a few whose potential earnings are so low in relation to the social security benefits they receive when unemployed that they opt for unemployment.

(e) There is also some evidence to show that some of those registered as unemployed have part-time or even full-time casual employment. This suggests that the number of registered unemployed may be an overestimate and certainly that a careful appraisal of the published figures is necessary.

The unemployment figures are published on a national basis but there may be a number of causes of unemployment, many of them local in origin. To conclude this unit we shall consider the various types of unemployment.

22.11 Types of Unemployment

(a) General unemployment

If the Government fails to generate an adequate level of aggregate monetary demand (£60 million in our imaginary economy in section 22.7) there will be general unemployment in the economy. (This is sometimes called *cyclical unemployment* as it is associated with cycles or fluctuations in economic activity.) During the inter-war period unemployment was persistently above 10 per cent and at times rose above 20 per cent; but in the post-war period the level has rarely risen above 3 per cent. One reason for the much lower post-war figure has been the implementation of Keynesian policies for the management of aggregate monetary demand. It has been possible for Governments to influence (but not really control) the level of AMD by the manipulation of tax rates and public sector expenditure so as to maintain full or nearly full employment. This is more difficult than is at first apparent for any action that the Government takes to stimulate demand in one sector will have repercussions elsewhere. As we shall see later the headlong pursuit of full employment almost certainly leads to inflation and an adverse balance of payments. The elimination of this inflation may then restore the unemployment.

(b) Structural unemployment

This will occur as some industries contract and force men out of work. If labour were perfectly mobile it would quickly be re-absorbed by other expanding industries. But a man who has spent 25 years as a coalminer does not easily adjust to the intricate work of wiring electronic equipment and cannot automatically find new work even when it is within easy travelling distance of his home. To minimize this kind of unemployment the Government makes extensive funds available for the retraining of labour to enable men to move into expanding industries. Any Government success helps to remove friction in the labour market, though there will always be some *frictional unemployment* where men moving between jobs are temporarily out of work.

(c) Regional unemployment

As was shown in Unit Nine, this type of unemployment occurs when the basic industries of an area go into decline without being replaced by others. One way of reducing regional unemployment is to increase the geographical mobility of labour so that the work force migrates towards areas of high economic activity, but as we have seen, the general policy is to move industry and jobs to the regions of high unemployment.

(d) Technological unemployment

Whereas structural unemployment results from a change in the pattern of demand, technological unemployment is a result of a change in the methods of production. In the dock industry the introduction of containers has enabled a given volume of cargo to be handled by a much smaller work force. Dockers who leave the industry in consequence may be considered to be unemployed because of changes in technology. One of the dilemmas of economic policy is that the Government when encouraging economic efficiency, which normally involves substituting capital for labour, actually generated technological unemployment. If this occurs on a wide front it has serious consequences. Let us suppose that as a result of the introduction of labour-saving techniques 200,000 men find themselves out of work. Before the improvement in techniques they could have been re-absorbed by a given increase in AMD, but with changing conditions a much larger increase in total demand is necessary to create employment for the same number of men.

(e) Seasonal unemployment

Seasonal factors may cause unemployment in some industries. Many building workers are temporarily unemployed in January and February when inclement weather prevents outside working. The tourist industry employs most of its labour during the summer holidays and, for a much shorter period, at Christmas. Much of the labour force is not required for the rest of the year and may be regarded as seasonally unemployed.

For most of the post-war period, however, the Government has not had much trouble in maintaining full employment. Its main domestic preoccupations have been the prevention of over-full employment, and the related excess of aggregate monetary demand and inflation. These inter-related problems will now be considered.

22.12 Questions and Exercises

1. Outline the main factors determining the level of employment.
2. Examine carefully the main determinants of the level of consumption.
3. What is the multiplier? Assess its importance in modern economic policy.
4. What are the main determinants of the level of investment?
5. In what ways can the Government move the economy towards a full employment situation?
6. What is meant by full employment?
7. Examine the various types of unemployment and the remedies for them.
8. On what assumptions does the multiplier rest? How far does the removal of these assumptions invalidate the theory?
9. The Government authorizes a programme of investment in nuclear power stations costing £3,000,000. If the economy is closed and there is no taxation what will be the rise in GNP if the marginal propensity to save is $\frac{1}{5}$? What will be the rise in GNP if $\frac{1}{4}$ of all consumption expenditure is on imported goods?
10. There are one million unemployed in Agraria, which is a closed economy with a GNP of £30,000 million. It has been established that the MPC of the economy is constant at $\frac{1}{2}$ and that an increase in the GNP of £1,000,000 reduces unemployment by 200,000. What increase in investment is necessary to reduce the level of unemployment to 200,000?
11. By referring to the current edition of National Income and Expenditure, establish the level of personal incomes, the level of personal expenditure, and from this the Average Propensity to Consume.
12. Utopia has long been a closed economy. It suddenly finds that without importing anything it can export £100,000 of goods. If the marginal propensity to save is $\frac{1}{4}$ and the Government takes $\frac{1}{4}$ of income in tax what will be the final effect of the exports on GNP?

Unit Twenty-Three

The General Level of Prices: Inflation

23.1 Introduction

The economic spectre of the inter-war period was the demoralizing level of unemployment. The implementation of Keynesian policies following the war allowed over 20 years of full employment, when the level of unemployment only reached 3 per cent on two isolated occasions. The worry of unemployment has given way to a concern over inflation; the condition of generally rising prices, and the bulk of post-war economic policy may be seen as a continuing fight to restrain price increases and the distortions created by them. Before examining these distortions we must consider some of the causes of inflation.

23.2 Monetary Demand and Inflation

In Unit Twenty-two the model economy was pushed into a position of full employment equilibrium through the introduction of various kinds of demands for goods and services, and we assumed that in this position it was physically incapable of producing any more goods and services per production period. (Of course it might be possible to overcome this by the introduction of more efficient machinery, but this is a long-run process and we are here concerned mainly with the short-run.)

The model economy is reproduced in Fig. 23.1, and is again in a position

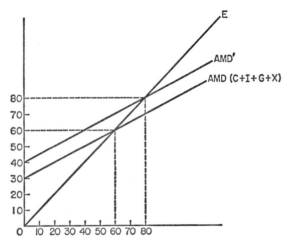

Fig. 23.1 An inflationary increase in aggregate monetary demand

of full employment equilibrium at £60 million. (The individual curves comprising AMD have been omitted to simplify the picture.) What would now happen if there were an increase of £10 million in the demand for machinery?

Extra orders are placed with the manufacturers of machinery, but we know that they are physically incapable of producing more goods. Reference to Unit Thirteen shows that in conditions of excess demand, firms and industries choke off the extra demand by increasing their prices. In practice the process will be accelerated as buyers try to jump the queue by offering higher prices than those being asked. As a result factors of production in the machine goods industry receive higher money incomes with which they can bid up the prices of consumer goods. There is in effect a change in the conditions of demand in individual consumer goods markets and so the demand curves move to the right. The multiplier works itself out in the same way as before except for the fact that the rise in national income is a monetary rise only. At the new equilibrium national income of £80 million the economy is producing the same number of goods as it was when equilibrium was £60 million and the *real* national income remains unchanged.

Although no economy is likely to reach a position of absolutely full employment it is not uncommon for them to reach a position where any increase in AMD leads to a relatively small increase in output and a relatively large rise in prices. This leads many people to regard inflation as *a condition of excess AMD over aggregate supply in conditions of full employment*. We know from painful, recent, experience that inflation can also occur in conditions of considerable unemployment. The importance of this definition of inflation (for it is not the only one) lies in the fact that it draws attention to aggregate *monetary* demand and consequently to the supply of money. It is fairly obvious that if there were no such thing as money then inflation would cease to be a problem. A certain number of goods would be produced and consumed but they would have no prices. As soon as one good is used as money the problem of inflation arises. The relationship between the supply of money and the level of prices has long been of interest to economists. The best known expression of the relationship is known as the *quantity theory of money*. This has many variations, but we shall examine it in its simplest form.

23.3 The Quantity Theory of Money

The great inflation that swept through Europe in the sixteenth century was attributed to the influx of gold and silver from South America and the consequent increases in the money supply. During the Second World War the Germans produced remarkable counterfeits of British Treasury notes, and hoped to ruin the United Kingdom economy by getting them into circulation. The Japanese had the same motive when they flooded some Pacific Islands with cowrie shells, which were still widely used as currency. Underlying each of these situations was the belief that the general level of prices depends upon the supply of money.

We will imagine a simple economy which produces 1000 units of output a year which are all purchased by the residents. The money supply in the economy consists of 200 one-pound notes and the goods are priced at £1 each. As the goods are all bought it follows that each one-pound note is involved in five transactions on average. The value of all the transactions is equal to the supply of money multiplied by the number of times each unit of money changes hands. Another way of looking at the total value of the transactions is to consider the average price of the goods sold (the general level of prices in our simple economy) and the number of transactions that occur (here, £1 × 1000).

These two aspects of the same events constitute the basis of the quantity theory. If we call the money supply M, the average number of times that each unit of money changes hands V, the general level of prices P, and the number of transactions T we may summarize the previous paragraph by saying that

$$PT \equiv MV$$

The sign $\equiv$ indicates that these two items are exactly the same. We are not proving anything; we are just looking at the same transactions from two different angles. If we divide each side by T, however, we obtain

$$P = \frac{MV}{T}$$

This implies that the general level of prices P is related to the supply of money, the rate at which that money changes hands and the number of transactions that occur. Now it is clear that an increase in prices *might* follow from an increase in the money supply, or its velocity of circulation, or a reduction in the number of transactions. There is, however, nothing in the theory to say that an increase in the supply of money *will* lead to a rise in the general price level, since a rise in M may be compensated by a fall in V, or prices may rise merely because the velocity of circulation increases. Moreover the money supply may be increased in conditions of heavy unemployment and will lead to an expansion of output without rising prices. Thus the quantity theory is important in drawing attention to one aspect of inflation, but it does not fully explain how inflation occurs. We must approach the problem in a different way.

23.4 Cost Inflation and Demand Inflation

While acknowledging the importance of the money supply to the inflationary process it is useful to consider other powerful forces which make their contribution to rising prices. The standard distinction is between 'cost-push' inflation and 'demand-pull' inflation. The names indicate the main causes of the particular inflation although it is usual for one kind of inflation to lead to the other kind in a particularly unpleasant circle.

Cost-push inflation occurs when prices rise as a result of the costs of production increasing more rapidly than output. An increase in the cost of imported raw materials, a rise in wages unmatched by a rise in output, or an increase in profits to meet the demands of shareholders will all tend to push prices up. The process may sometimes be encouraged by taxation. The Selective Employment Tax increased the cost of employing labour and led to rising prices. An increase in company taxation must either be absorbed by extra efficiency in the firm, or lead to reduced profits or be met by a price increase in the hope of maintaining profits. If prices rise for any of these reasons over a broad front of the economy, then workers and shareholders whose incomes have not increased will find their real incomes fall (they will be able to buy fewer goods from their current income) and they will demand extra payments as compensation. If this second round of pay rises is not accompanied by at least proportional rises in productivity then more prices will rise and so the spiral will be maintained.

When cost inflation of this kind is widespread, it necessarily leads to demand inflation as the recipients of extra income want to increase their purchases. In conditions of anything like full employment a further increase in price occurs. The initial impetus may, of course, come from the demand side. If the banks freely create credit for consumers these people will be able to spend in excess of their current income. The aggregate monetary demand is likely to rise ahead of output, prices will increase, and the Trade Unions will demand wage increases to compensate. These wage increases will lead to further price increases and again the spiral will continue.

Once an inflationary atmosphere is established the process is in danger of becoming not only self-perpetuating but self-accelerating. It is not uncommon for wage claims to have three distinct elements: a 'cost of living' rise to compensate for increases in prices which have occurred since the last round of negotiations; a further increase to give a rise in the workers' real income; and last and most dangerous, a further rise to compensate for anticipated price rises before the next round of wage negotiations. The dangers of such a process lie mainly in the precedent it sets and in what we previously termed the 'league table effect'. If one group of workers secures a large increase in wages and becomes the fifth best paid group in the country rather than the tenth, those who have been overtaken will exert pressure to regain their traditional place in the league. This can happen even if the original large increase was entirely justified by an increase in output. When one group of workers moves out of line there are spill-over effects on other workers and industries.

Inflation in recent years is thought to be partly due to the fact that the Trade Unions are beginning to exercise their considerable power in particular areas of the economy. As industry becomes more capital intensive it becomes more expensive for employers to resist strikes and industrial action as they have very heavy fixed costs to meet whether there is any production or not. Trade Unions have realized this and taken action. Once a strongly placed group of workers has secured a large rise in income, others try to follow their example and there are many recent examples of traditionally acquiescent unions adopting more forthright attitudes. A second view, and one cogently put by the Trade Unions, is that the inflationary process is best countered by rapid economic growth and that the main cause of rising prices has been the inability of the Government to generate sufficient increases in output.

Finally we must return to the supply of money. For many years discussions about inflation centred on cost-push and demand-pull ideas and the supply of money was regarded as being adjustable to the demands of the economy. Many economists have put forward the suggestion that if the supply of money were properly controlled then inflation could be killed at the outset. If a company concedes an excessive rise in wages to its employees it will often finance the increase initially from a bank overdraft. It will have to borrow until it begins to obtain an extra flow of revenue from rising prices; a process which may take some weeks. If the banks were not prepared to allow an overdraft the company would either have to find another way of financing the pay rise or be forced out of business. When taken to extremes such a policy could no doubt drastically reduce the pace of inflation, but only at the expense of very heavy unemployment. Clearly the Government must try to steer a course between heavy unemployment and rapid inflation. While unemployment represents a loss of production and consequently a reduction in the standard of living, inflation has many insidious effects.

23.5 The Effects of Inflation

While there is a minority view that a degree of inflation is a necessary stimulant to the economy (slightly rising prices encouraging investment) the strenuous efforts of Governments to restrain its pace suggest that it produces many undesirable side effects.

(a) Lurking in the background are fears that moderate inflation will give way to very rapid inflation and a complete collapse of confidence in the monetary unit. If prices rise at 2 per cent per year it is 35 years before they double; but if they rise at 10 per cent per year they double in only seven years. And as we have seen once prices start rising rapidly wage rises begin to anticipate price rises.

(b) The effect on the distribution of real income is, however, of greater immediate importance. In general the wages of workers in strong Trade Unions tend to rise more rapidly than prices, so the real wages of these people increase while those of people in loosely organized unions fall. Those who suffer most in this respect are people on fixed incomes, notably Old Age Pensioners. Similar distortions occur with other forms of income: those receiving fixed-interest payments on gilt-edged securities lose while holders of a similar investment in equity will possibly find their dividends keeping pace with inflation and will have the compensation of rising capital values. It is frequently said that no one gains from inflation since increased incomes are quickly absorbed by increased prices. But suppose that the coalminers obtain a handsome pay increase unmatched by any simultaneous increase in productivity. The result is an increase in the price of coal which is a much greater burden to pensioners than to miners, who in any case do not buy all their coal. This constitutes an important reason for trying to control price rises. The effects of inflation are not confined to those who have generated the inflation.

(c) The real burden of borrowing money is reduced by inflation, since repayment is made in money of reduced purchasing power. Inflation thus encourages people to borrow and is a strong disincentive to thrift. It has other effects on the capital market as well. One hundred pounds invested in gilt-edged securities might yield £7 or £8 per year, but if prices are rising at 10 per cent a year the holder of such securities is in a worse position at the end of the year than at the beginning even if he is still able to sell the securities for £100. Consequently during periods of inflation savings are diverted away from the orthodox channels provided by the financial institutions into non-productive channels: vintage cars, postage stamps, plots of land, anything that can be expected to yield a capital gain. While this is fine for the individual it does nothing to improve industrial investment and growth and in the long term may be one of the more harmful aspects of inflation.

(d) It is not only the domestic effects of inflation that demand Government action, for a rapid rate of inflation has an adverse effect on the Balance of Trade. Imagine two countries whose trade with each other is in equilibrium when the level of prices in one country rises rapidly. Our knowledge of the principles of demand indicates that it will now be more difficult for that country to sell its goods abroad. Further, imports will now be more attractively priced in relation to home goods and the demand for imports will increase. In this way inflation may lead to a Balance of Trade deficit and, as we shall see in Unit Twenty-six, considerable difficulty for the Government.

23.6 The Control of Inflation

The policies available to the Government to moderate the rate of inflation are the reverse of those used to encourage full employment. The banks' freedom to create credit is restricted and interest rates are increased in an attempt to reduce the pressure of Aggregate Monetary Demand. Increases in taxation have the same intention and may be accompanied by a reduction in Government expenditure (or more likely a reduction in the rate of growth of Government expenditure). Both policies are aimed fundamentally at demand inflation and seek to reduce the pressure of demand. Such policies are generally known as deflationary since they seek to deflate demand.

These controlling methods contribute little or nothing to the reduction of cost inflation. Indeed rising interest rates add directly to the costs of companies, while increases in taxation are often a spur to inflated wage demands, and may thus be counter-productive so far as the price level is concerned.

The control of cost inflation rests with policies of income restraint or with the encouragement of greater productivity. 'Productivity deals' have fallen from the high favour they enjoyed in the 1960s but since cost inflation arises when costs rise more rapidly than output such deals which lead, or should lead, to greater output per man-hour (the usual measure of productivity) contribute to cost control.

A productivity bargain usually involves the labour force relinquishing traditional methods and adopting a more flexible approach to work. They may surrender established demarcation lines or accept technical innovations, in return for considerably higher incomes, and normally a reduction in the work force. A deal of this kind—which will be enormously complicated and the result of many months' negotiations—can produce significant results for particular firms or plants and enable them to reduce their costs and gain ground in the struggle against inflation. But even here there is a danger that the higher earnings received via higher productivity will have spill-over effects in other parts of the economy. The workers involved will almost certainly have moved up the wages league and other groups will seek what they regard as their rightful position, irrespective of productivity.

It is this kind of development that is partly responsible for demands from all sides for a prices and incomes policy. Such policies may take many forms but in the United Kingdom they have always been regarded as part of the general armoury against inflation; with other considerations such as protection for lower-paid workers reduced to a secondary role. The prices and incomes policy of 1965-9, which is the best example of this kind of policy, was embodied in the National Board for Prices and Incomes which adjudicated upon proposals for price or income increases referred to it by the Government. The Board developed a considerable expertise in the assessment of such claims but had probably only a marginal effect on the rate of inflation.

The principal difficulty surrounding a formal incomes policy of this kind is the establishment of appropriate criteria for increases in income. It is usual for the Government to set a basic 'norm'—the average increase that should occur. This immediately becomes the accepted minimum increase. Increases above the norm are allowed to those showing increases in productivity, or to those demonstrably underpaid. Difficulties arise in respect of certain groups whose productivity is not measurable in meaningful terms, but who live in the same inflationary world as everyone else and have the same need to maintain their real wages. Policemen, teachers and nurses are examples of such groups. It is even more difficult to cope with the problem of the low-paid with an incomes policy designed to fight inflation. If it is conceded that the low-paid should receive increases sufficient to bring them up to an established national minimum wage of perhaps £20 per week, and there are powerful social arguments for such a policy, cost-push and demand-pull inflation are each likely to receive a considerable impetus, particularly when the 'league table effect' begins to exert itself. A further difficulty is the timing of the introduction of an incomes policy: it is never the right time because there is always a group which has to be the first to be subjected to restraint.

However, the alternatives are hardly any more attractive. On the one hand all prices and incomes could be left to market forces, which means the monopolistic bargaining power of employers and Trade Unions with the poorly organized workers as the losers. On the other hand the Government may follow a policy of exhortation and persuasion. This frequently results in a policy of restraint in the public sector, where the Government is the monopsonistic employer, and a free-for-all in the private sector, a situation obviously unfair to the public employees whose wages rapidly fall behind.

Of the three possibilities the first seems the more desirable if only because a Prices and Incomes Board has an important educational role to play as an opinion-forming body. In any case prolonged industrial disputes are usually resolved by some kind of arbitration and this can best be expedited by a properly constituted and established body rather than an ad hoc committee. In the last resort it may be argued that an incomes policy is necessary to control inflation largely because of the failure of monetary and fiscal policy in this field. We shall return to these problems in Units Twenty-eight and Twenty-nine, and in Unit Thirty we examine the forces which led to the re-establishment of a formal policy in 1972.

23.7 Questions and Exercises

1. Distinguish between the various kinds of inflation, relating your answer to the United Kingdom economy.
2. To what extent is it possible to regard inflation as a purely monetary phenomenon?
3. 'Trade unions cause inflation.' Comment.
4. Examine the quantity theory of money. Does it offer an adequate explanation of inflation?
5. Consider carefully the economic effects of inflation.
6. 'Those who cause inflation are rarely those who suffer its effect.' Comment.
7. To what extent is a formal Prices and Incomes policy likely to control the rate of inflation?
8. For what reasons do Governments seek to control inflation?
9. Comment on the following indicators which relate to the United Kingdom:
   Index of Retail Prices.
   Index of Average Earnings.
   Index of Wage Rates.
   Money Supply.
   Bank Lending.
   Government Borrowing.
10. Explain the distortions likely to arise from the implementation of a prices and incomes freeze.

Unit Twenty-Four

The Basis of International Trade;
Tariffs; Trading Blocs

24.1 Introduction

So far we have examined domestic economic problems and noted that the Government has monetary and fiscal policies at its disposal for dealing with them. The technicalities of these policies will be considered in Units Twenty-eight and Twenty-nine, but the extent to which they are effective is frequently limited by the repercussions they may have on the external trading position of the country. Indeed it is frequently the balance of payments, which may be regarded as the annual summary of the trading position, rather than the level of unemployment or the rate of inflation, that dictates the direction of economic policy. Thus before considering the effectiveness of monetary and fiscal policies we shall examine the basis of international trade in this Unit, and the financial and economic problems to which this gives rise in the following two Units.

24.2 The Reasons for International Trade

The pattern of world trade is so complicated that no single explanation will satisfy this question. The most obvious reason for trading with other countries is to obtain goods which cannot be produced in our country, or can only be produced at great expense. Climatic and geological differences account for a proportion of international trade and for a large proportion of the exports of the poorer countries of the world. Less obviously, perhaps, differences in the skills of labour and in the accumulation of capital account for some of the exports of the wealthy countries. It was differences in factor endowments that underlay the traditional pattern of world trade, with manufactured goods from the United Kingdom being exchanged for food and raw materials from Commonwealth countries. This pattern of trade is gradually being superseded today by exchanges of manufactured goods. Countries frequently import goods that they can quite easily produce for themselves, for example 25 per cent of new cars purchased in Britain are manufactured abroad, and an enormous amount of industrial machinery is imported, while overseas countries have a considerable demand for British machine tools.

The explanation underlying such trade is to be found in the 'law' of comparative costs. This shows that trade will be beneficial to a country if it concentrates (but not necessarily specializes entirely) on the production of those goods in which it has the greatest relative advantage over its trading partners.

24.3 The Law of Comparative Costs

The Law of Comparative Costs shows that even where one country is more efficient than other countries at producing everything—that is to say it has an *absolute* advantage in all industries—gains will be available to it from international trade so long as it transfers resources towards the industry in which its absolute advantage is greatest. It will then sell the surplus to other countries who in their turn will channel resources towards those industries in which their deficiency is least.

But first let us consider a more simple example. We shall imagine a two-country world, with Industria and Agraria which each produce two goods, food and machines. Industria is an efficient producer of machines but finds food production expensive as there are land shortages. Agraria is an efficient farming nation but is backward in producing machines. In order to avoid the major complication of international trade, that of exchange rates of domestic currencies, we shall deal in terms of 'days' of labour when measuring the costs of production and prices of goods. We will suppose that the following cost pattern is established in each country in the absence of international trade:

	No. of days' labour in Industria	No. of days' labour in Agraria
One unit of food	2	1
One unit of machinery	3	4

Each country has an absolute advantage in the production of one good but the reasons for trade are best explained in terms of the comparative costs within each country. If the domestic comparative cost ratios ($\frac{2}{3}$ and $\frac{1}{4}$ in this case) are different then trade is beneficial to each country. Agraria can produce 4 units of food by giving up the production of 1 unit of machinery and transferring resources to agriculture. Industria produces 1 unit of machinery by giving up the production of $1\frac{1}{2}$ units of food and using the resources for manufacture. If Agraria can acquire more than 1 unit of machinery in exchange for the 4 units of food that she now produces instead, she will be better off. If Industria can acquire more than $1\frac{1}{2}$ units of food in exchange for one machine, she too will be better off. They can obviously meet each other's requirements: Agraria produces an extra 4 units of food, keeps, perhaps, 2 units for domestic consumption and exchanges 2 units for one machine produced in Industria. Industria then has $\frac{1}{2}$ unit of food more than she did in isolation. The total gains from trade will depend on the size of the two economies, the extent to which specialization occurs, and on the relative strength of demand in the two countries for the two traded commodities. If we assume that each country has 1000 days of labour at its disposal and in the absence of trade each divides its resources equally between the production of food and machines, total production could be summarized as follows:

	Output of food	Output of machines
Agraria	500	125
Industria	250	166⅔
Total	750	291⅔

(Each output figure is obtained by dividing the total days allocated to the production of a commodity by the number of days required to produce a single unit.)

If international trade begins and complete specialization occurs output is increased:

	Output of food	Output of machines
Agraria	1000	—
Industria	—	333⅓
Total	1000	333⅓

Thus the maximum gains, given the distribution of factors before trade began are 250 units of food and 41⅔ machines.

While one would intuitively expect trade to give general benefits where each country had an absolute advantage in one commodity, it is perhaps a little more surprising to find that each country can benefit when one of them has an absolute advantage in both products.

Suppose that the cost pattern in the two countries is as follows:

	No. of days' labour in Industria	No. of days' labour in Agraria
One unit of food	2	3
One unit of machinery	1	4

Industria, therefore, has an absolute advantage in each product, but enjoys a much greater advantage in the production of machinery than in the production of food. If trade occurs, then Industria would export machinery in exchange for the food exports of Agraria. But how can this help each country? Industria can produce two extra units of machinery by giving up

1 unit of food. If she transfers resources in this way and can exchange 2 units of machinery for more than 1 unit of food she will have gained from trade. In Agraria the resources necessary to produce 1 machine can be transferred to food production and could produce 1⅓ units of food. Clearly Agraria will be happy to channel resources into food production and exchange, say, 1⅙ units of food for 1 machine from Industria, and Industria will have gained by having slightly more food and the same amount of machinery. In this case however complete specialization would be unlikely owing to the relative inefficiency of Agraria. If she concentrated entirely on food production she would produce only 333⅓ units, and, unless this were sufficient to meet world demand, Industria would continue to produce both food and machines. The gains likely to be made in these circumstances depend on the distribution of resources within each country before trade begins and the level of demand for each product. If each country divides its resources (1000 days' labour) evenly to begin with 'world' production will be:

	Food	Machinery
Agraria	166⅔	125
Industria	250	500
Total	416⅔	625

The output figures are again derived by dividing the labour available (500 units per industry) by the number of days necessary to produce 1 unit of the commodity. Let us suppose that the demand for food is fully met at 416⅔ units. In this case the pattern of production emerging from trade would be as follows:

	Food	Machinery
Agraria	333⅓	—
Industria	83⅓	833⅓
Total	416⅔	833⅓

Agraria specializes completely in the production of food but cannot meet world demand. Therefore Industria devotes sufficient resources (166⅔ days labour) to food production to make up the deficiency, but channels most of her factors of production into the machine industry and produces 833⅓ machines. The gain from trade in this case is 208⅓ machines. The reader should establish for himself the gains from trade in terms of food if the world demand for machines is satisfied at the pre-trade level of 625.

The rate at which goods are exchanged between the two countries cannot be fixed precisely from the information we have available but we can establish the range in which it must lie if trade is to take place. In Industria a unit of food is worth 2 units of machinery, and Industria would be no worse off from trade if she could receive 1 unit of food for every 2 units of machinery she exports. Agraria on the other hand would be willing to offer 1 unit of food in exchange for ¾ of a unit of machinery (or fractionally more if she is to gain from trade). The rate at which the goods are exchanged must therefore be between the rates of 1 unit of food for 2 units of machinery (most beneficial to Agraria) and 1 unit of food to ¾ units of machinery (most beneficial to Industria). The rate that will prevail depends upon the strength of demand for the products in each country. Notice that if trade is conducted outside these rates, one of the countries would be better off if it were not trading. Suppose the rate is 1 unit of food for 4 units of machinery: Industria exports 4 units of machinery and receives 1 unit of food in exchange; but if she had not produced those machines in the first place she could have used her resources to produce 2 units of food herself. Similarly if the rate of exchange were 1 unit of food to ½ a unit of machinery Agraria would best serve her own interests by transferring resources back to the machine industry and producing ¾ of a machine for each unit of food foregone.

The rate at which goods are actually exchanged for each other is known as the terms of trade. This is a concept to which we shall return briefly later in this Unit.

24.4 The Limitations of the Comparative Cost Theory

The theory outlined above depends on the existence of certain rarefied conditions, and complications arise if these conditions are relaxed.

(a) There are of course more than two countries and more than two commodities. Although the introduction of more countries and commodities complicates the exposition of the theory, it does nothing to deny the basic tenet that a country should concentrate on the production of the good or goods for which its comparative advantage is greatest (or its disadvantage least).

(b) We have implicitly assumed that factors can transfer freely from one industry to another. This process is in practice likely to take many years and necessitate considerable retraining as we discussed in Unit Nine. In these circumstances the benefits of trade may be reduced, and will almost certainly be delayed.

(c) We have also assumed constant costs during the transfer of resources from one industry to another. This means that the transfer of a given amount of resources from one industry to another always results in the same rise of production. In fact if the expanding industry in each country is subject to decreasing costs (increasing returns to scale) the gains from trade will be greater than implied in our earlier discussion. In conditions of increasing costs in each country the benefits available will be reduced.

(d) Transport costs, which we have ignored, will also limit the benefits available, as will the introduction of tariffs and other obstacles to trade. These are discussed in Section 24.6.

(e) Some of the greatest problems of international trade are connected with currencies. By omitting this aspect from our picture we have exaggerated the smoothness with which trade can occur.

(f) Finally there are frequently political, social or strategic reasons why Governments will not permit free trade to develop into over-specialization. There are, for example, strong strategic reasons for the United Kingdom to maintain a viable home agriculture. On the grounds of the most economic allocation of world resources, however, the case is less compelling since many other countries have a comparative advantage in agriculture.

24.5 The Terms of Trade

In Section 24.3 we spoke of the terms of trade in relation to the numbers of goods that could be acquired for a given volume of exports, and it is these *real* terms of trade that are important. Since Britain imports and exports thousands of different goods each year it is impossible to obtain a meaningful measure of the real terms of trade, so we have to resort to a monetary measure. The terms of trade are found by the following formula:

$$\frac{\text{Index of export prices}}{\text{Index of import prices}} \times 100$$

In the base year the index of both import prices and export prices will be 100 (see Section 16.5), and the terms of trade will be 100. If in the following year the import price index rises to 102 while the export price index remains at 100 the terms of trade will become 98·0. Thus the terms of trade will have deteriorated for the exporting country since they now have to export a slightly greater volume of exports to pay for a given volume of imports.

Table 24.1 United Kingdom terms of trade

Year	Index of export prices	Index of import prices	Terms of trade
1961	100	100	100
1962	101	99	102
1963	104	103	101
1964	106	107	99
1965	109	107	102
1966	113	109	104
1967	114	109	105
1968	123	121	102
1969	127	126	101
1970	136	132	104
1971	148	134	110
1972	160	144	111

Source: *Monthly Digest of Statistics*, HMSO

In general the terms of trade of the industrial countries have improved in the post-war period at the expense of the primary producers. This shows that inflation has occurred in the industrial countries and also that there is an excess supply of many primary products and that synthetic substitutes have been adopted for many industrial raw materials.

24.6 Obstacles to Trade

Economists have frequently praised the virtues of free trade—trade unhampered by any artificial barriers such as tariffs—and they have seen it as a means of inducing the most economic allocation of resources. We saw in Section 24.3 how the opening up of trade between two countries leads to greater world production of traded goods and by implication to an increase in economic welfare. Conversely any measures which inhibit trade between nations may be expected to reduce the volume of output and economic welfare.

Despite this all Governments take steps to reduce the volume of imports entering their country; they do this for a variety of reasons and in the certain knowledge that they invite retaliation from their trading partners. There are a number of ways of protecting the home economy from overseas competition. Those most frequently used include the following.

(a) Tariffs

These may be imposed on goods entering the country so that their price is increased and if demand is not perfectly inelastic the quantity purchased will fall. Tariffs may be either specific—50p per unit imported—or *ad valorem* —10 per cent of the price of the good. The former type is less significant as prices rise, while the latter increases in proportion to price rises. While the imposition of a tariff enables home producers to compete more effectively with overseas producers, not all tariffs are imposed for protective purposes: many are essentially revenue tariffs which may be imposed on goods for which there is no effective domestic substitute.

(b) Subsidies

When these are applied to home producers the effects are similar to taxes on overseas goods. If, for example, American wheat costs £25 delivered to the United Kingdom and British farmers need £35 to cover their costs and make a living then they clearly cannot compete. The Government can assist British agriculture either by imposing a tariff of £10 on American wheat or by giving the British farmer a subsidy of £10 to enable him to sell at £25 and thus be able to compete with the imported wheat. While the form of protection that is chosen will make little difference to the ultimate pattern of trade, it does have an important effect on the direction of support for home agriculture. If tariffs are imposed on imports the consumer supports home agriculture directly by paying a higher price than he would in the absence of the tariff; but if home producers receive a subsidy to enable them to compete, the support for agriculture comes indirectly from the taxpayer. As most of the tax is paid by the relatively wealthy (see Unit Twenty-nine) support through the taxpayer may be considered preferable to direct support from the consumers, many of whom have very low incomes.

(c) Quantitative restrictions

Tariffs seek direct influence on the price at which goods are sold, but an alternative form of protection, quantitative restrictions, seeks to control the level of imports and allows them to find their own price. The main difference between this kind of protection and tariffs is that whereas with a tariff the Government gains extra revenue, with quota restrictions the benefit of the higher price from limited supply goes to the seller of the goods unless the Government takes prior steps to prevent this. The easiest way is to auction import licences so that the highest bidders pay the Government for the licence to sell the volume of goods required. Alternatively licences may simply be sold (or merely issued) to overseas suppliers on the basis of the pattern of supplies in earlier years.

(d) Non-tariff barriers

Until recently the commercial barriers to trade were assessed purely in terms of tariffs and quantitative restrictions. With the complete removal of such barriers in the European Economic Community the significance of non-tariff barriers to trade has increased. These may take the form of discriminatory administrative practices, such as deliberately channelling Government contracts to home companies even where their tenders are not competitive or insisting on different technical standards. In a Common Market of nine members this could force producers to manufacture to nine different sets of technical specifications if they wished to sell in each member country. They might have yet more specifications if they wish to sell outside the area. Steps are already well underway to eliminate such situations within the EEC, but their existence will probably remain a hindrance to world trade in general for a long while.

(e) Exchange controls

As an alternative to imposing various limitations on traded goods Governments may reduce the flow of those goods by restricting the amount of foreign exchange (currency) available for their purchase. If the Government wishes to reduce the flow of imports from America it may set limits to the amount of US dollars that citizens can acquire to buy American goods. In 1967 the Government, anxious to conserve foreign currency, imposed a limit of £50 per year on the value of foreign currency that could be acquired for holiday purposes. Such restrictions require a mass of controls and invite evasion. For this reason they are discouraged by such bodies as the International Monetary Fund.

Whichever restraint is employed the intention (except in the case of purely revenue raising tariffs) is to increase the demand for home goods at the expense of foreign suppliers. The implication is normally that foreign producers can market goods in this country more cheaply than home producers are able to do. By imposing tariffs or other controls the Government enables home producers to maintain or increase their sales, maintain employment in the industry and perhaps increase profits. While those connected with the protected sector benefit partly at the expense of overseas producers, they also benefit at the expense of home consumers who are forced to pay a higher price than would be necessary in conditions of free trade. Since one (perhaps small) sector benefits at the expense of two other groups it is worth asking ourselves what are the arguments advanced in defence of tariffs and other obstacles to international trade.

24.7 Why are Tariff Barriers Erected?

(a) The Balance of Payments
Tariffs and other controls are often imposed as a short-term expedient method of correcting a deficit in the Balance of Payments. While this may be effective it does little to deal with the underlying cause of the deficit (see Unit Twenty-five) and will almost certainly invite retaliation, more often through non-tariff obstacles than through a retaliatory tariff. The import surcharges imposed by Britain in 1964 and the USA in 1971 were each imposed for Balance of Payments reasons and each induced a retaliating response.

(b) Anti-dumping
Closely connected with Balance of Payments is the imposition of a tariff to counteract the practice of dumping. This is the case of monopolistic discrimination where price is raised in the protected home market and part of the proceeds of the higher prices is used to subsidize exports so that they can undercut competitors in the world market. Such competition is clearly unfair and other Governments are perfectly justified in imposing a tariff to protect their own producers.

(c) Infant industries
In developing countries tariffs are frequently imposed to protect new or infant industries until they are strong enough to meet foreign competition, with its great advantage of mass production. The case for such tariffs seems reasonable enough; the difficulty is that once the tariff is established the protected industry can always find very good reasons for its continuation and many infant industry tariffs still protect full grown, but relatively inefficient industries.

(d) Employment
An important reason for the imposition of tariffs in the 1930s was to reduce the level of unemployment. Tariffs could be so high that they virtually excluded imports altogether, and thus maintained a high level of employ-

ment in the home industry. Although there are often important social justifications for such a policy it results in a lower level of imports and exports and postpones the readjustments that must occur in the economy in response to structural changes in the pattern of demand.

(e) Strategic tariffs

These are imposed to protect industries whose products may be essential in times of war or international crisis. British agriculture is protected partly for this reason though it would accord with the Comparative Costs argument if the United Kingdom were to import more food from abroad. Governments do not like to be entirely dependent upon foreign suppliers for essential materials. The coal industry too has enjoyed protection, not only for employment reasons, but also because it was until recently the only indigenous source of energy.

We may conclude that there are frequently important social or strategic reasons for the protection of home industries but that on a world basis the economic losses are likely to be greater than the economic gains. For this reason it is the object of one organization—the General Agreement on Tariffs and Trade (GATT) to secure the reduction of tariffs and quantitive restrictions to trade.

24.8 GATT and Tariffs

The General Agreement on Tariffs and Trade grew to importance through the failure of signatories of the Havana Charter of 1948 to ratify the establishment of the *International Trade Organization* whose object was to be the reduction of tariffs and the elimination of quantitative restrictions and subsidies. GATT took over the supervision of this programme, working mainly through periodical 'conferences' at which concessions are traded between members. The most notable of these was the 'Kennedy Round' of negotiations, instigated by President Kennedy, which successfully reduced industrial tariffs but did little to reduce obstacles to trade in primary products. This is indeed a major criticism of GATT—that it has benefited the interests of its wealthier members rather than the poorer. (The industrial countries have always found very good social or strategic reasons for maintaining tariffs on primary products that they can produce for themselves.)

This bias within GATT has led the bulk of the underdeveloped countries to establish their own organization, *The United National Conference on Trade and Development* (UNCTAD), which specifically demands the removal of tariffs on their exports, even if it is at the expense of aid which they regard as largely political in intent and as having only a marginal effect on economic development.

One important direction in which free trade has developed has been the establishment of trading blocs such as the European Economic Community and European Free Trade Association. GATT has tried to encourage them to be outward-looking in their policies, but as we shall see in the next section their success has been somewhat limited.

24.9 The European Economic Community and European Free Trade Association

One of the most notable post-war economic developments has been the division of Western Europe into these two distinct economic groupings. It is important to be clear about the difference between them. The EEC is a *Customs Union* which aims not only at the elimination of all internal barriers to trade, but also at the establishment of a Common External Tariff so that the exports of non-members receive the same treatment wherever they enter the Customs Union or common market. There is also normally a commitment within a Customs Union to pursue common economic policies in a number of fields. The Free Trade Area is a much looser organization which aims solely at eliminating internal trade barriers (and even then not for all categories of goods) while leaving each member free to pursue its own policy towards non-members, or *third countries*, as they are called in this context.

While EEC and EFTA are not the only examples of large trading blocs we shall confine ourselves to them in our analysis of the economic effects of such organizations.

(a) The European Free Trade Area

This was founded in 1959 (Convention of Stockholm) by those countries, Denmark, Norway, Sweden, Austria, Portugal, Switzerland and the United Kingdom, who felt at that time unable to commit themselves to the full economic integration envisaged by the EEC. Subsequently Finland became an associate member in 1961 and Iceland a member in 1970. The United Kingdom, Ireland and Denmark withdrew from EFTA when they joined the EEC in 1973. The principal target of free trade in manufactured goods between members was achieved ahead of schedule in 1967 except insofar as non-tariff barriers (Section 24.6) remained. The social and strategic arguments in favour of maintaining a viable agriculture proved far too strong for those hoping for free trade in primary products.

Members of EFTA adopt their own individual policies towards third countries so the central problem facing the authorities is to determine which goods qualify for 'area treatment'—i.e. are traded between members free of tariffs—and which goods must bear the tariff. Britain, for example, allowed certain Commonwealth manufactures into the country free of tariffs, but the Scandinavian countries imposed a tariff on them. Since there was free trade in manufactures between Britain and Scandinavia it would be sensible for Commonwealth producers to export to Scandinavia via Britain, thereby avoiding the tariff and increasing their sales. To prevent this, fairly strict *rules of origin* were drawn up to ensure that only goods originating substantially within EFTA received area treatment. Basically 50 per cent of the value of the goods must have been added within the area for goods to qualify, although the value of most raw materials may be deducted before making the calculation. Thus goods costing £100, but embodying £30 worth of imported raw materials and £10 of imported manufactured components are valued at £70 when being assessed to determine whether or not they pay a tariff. As $\frac{6}{7}$ of that £70 has been added within the area the goods would qualify for area treatment. The whole character of a Free Trade Area is determined by its rules of origin. If the rules are very rigid so that goods only qualify for area treatment if entirely produced within the area, very few will qualify, and there would be little point in establishing the organization. But, equally, if one went to the other extreme and allowed all goods to receive area treatment then the operation would be pointless.

The object of establishing a Free Trade Area is to increase trade between members with accompanying increases in welfare, in accordance with the theory of comparative costs. The impact of EFTA in generating extra trade between members may be assessed from Table 24.2.

Table 24.2 The growth of trade 1960–70 (dollars 000m)

	1960	1970	% rise
EFTA exports	19·8	43·1	117·6
EFTA exports to EFTA members	4·1	12·0	192·6
EEC exports	29·7	88·5	197·9
EEC exports to EEC members	10·2	43·3	324·5
UK exports	3·7	8·0	116·1
World exports	128·3	312·5	145·3

Source: General Agreement on Tariffs and Trade

It will be seen that whereas the total exports of EFTA countries rose by 117 per cent during the period 1960–70, their exports to other EFTA countries rose by 192 per cent. Impressive though this may be the benefits were limited by the modest size of the EFTA market—a population of 100 million dominated by Britain's 55 million. We might expect greater benefits from specialization within the EEC which now has a population approaching 300 million.

(b) The European Economic Community

This was established by the Treaty of Rome in 1957, and should be seen as fundamental to the cause of European unity. Although it has great economic significance it is primarily a mile-post on the road to political unity. Our job, though, is to examine the economic motives that underlie the Community, the extent to which they have been achieved and the effects that the Community has on non-members.

The immediate economic aim of the EEC was the elimination of internal barriers to trade so that both trade and output of members could increase in accordance with the principle of comparative costs. At the same time the common external tariff was established at the arithmetic average of tariffs existing at 1 January 1957. Both targets were achieved by 1968 and as both tariffs and quantitative restrictions existed until that date free trade has had a very short life within the market.

Nevertheless the expansion of intra-market trade has shown an impressive growth and as we can see from Table 24.2 there has been a growth of 324 per cent in exports of EEC countries to other members, whereas world trade has grown by 145·3 per cent and Britain's exports by 116·1 per cent. Of course this may be attributable to factors other than the existence of EEC but there are other indicators which suggest that the EEC has been beneficial to members.

One of the greatest aspirations of the founders of EEC was that free trade would lead to high output and thereby to higher living standards for the peoples of the community. We have already noted (Unit Two) the difficulties of measuring changes in the standard of living but if we use the general criterion of changes in the Gross National Product we can see significant differences between the EEC and the United Kingdom. These are indicated in Table 24.3.

Table 24.3 The growth of gross national product: EEC and
United Kingdom (dollars 000m)

	1958	1969
EEC	163·4	427·4
UK	64·8	109·8

Source: European Economic Community

The more rapid growth shown by members of the EEC may be attributable to a number of factors such as a higher level of investment, and a more flexible labour supply, which we cannot analyse here. We should also expect the elimination of trade barriers to act as a spur to output and to result in a further growth of the national product. Those who seek an economic justification for the Common Market must base their case partly on this comparative costs argument. Closely related to this is the economies of scale argument. This states that modern capital-intensive industry needs large markets to ensure the long runs of production that result in low unit costs. Furthermore, competition will lead to rationalization and concentration, which, it is claimed will allow yet more economies. The main basis for such a case lies in the example of the USA where a large market and large firms are associated with a high level of economic efficiency. There is no proof that the same recipe will yield the same results in Europe and, indeed, if the argument depends upon having tariff-free access to a large market the same results might be achieved by the bilateral abolition of tariffs between Britain and other countries.

The EEC, however, seeks to achieve more than the establishment of a large market. Its aim is the complete integration of the constituent national economies which it hopes to achieve by the gradual harmonization of various aspects of economic policy. While moves are already underway for the formulation of common policies on taxation, monetary policy, transport, monopoly and many other fields the most significant advance has been the establishment of the Common Agricultural Policy. In order to understand the nature of the policy it is necessary to realize that the rural population has always occupied a position of central importance in European (and especially French) politics. In 1958 as the EEC began its life 20 per cent of the working population derived its living from the land.

The Agricultural policy is consequently very protective. For each product covered, the authorities establish a *target price* designed to allow even marginal producers to make a living. Set at 5 per cent or 7 per cent below the target price is the *Guaranteed Price*, the minimum that all producers receive. If the market price falls below the guaranteed price the difference is made good by the Agricultural Guarantee and Guidance Fund. The main contributions to this fund are derived from a levy imposed on imported food to prevent it undercutting European produce. The purpose of the Fund is not simply the support of prices; some of its resources are used for the promotion of technical improvements and rationalization of the farming industry, while some is used to subsidize the export of surplus produce.

It is worth noting the main economic effects of such a policy:

(i) The price of many products is considerably higher within EEC than outside owing to the need to set high prices to satisfy the least efficient producers. This represents a gain in welfare for the agricultural sector but a loss to the consumer whose real income is reduced. However the alternative policy of supporting the farmer via the taxpayer would be prohibitively expensive because of the very large numbers of farmers involved.

(ii) As prices are made high for the benefit of marginal farmers, the intra-marginal farmers make very good profits and expand their production. This results in a considerable excess supply of some commodities. Necessarily, resources are retained in agriculture beyond an economic level.

(iii) Much of the surplus produce is exported at heavily subsidized prices with a detrimental effect on non-EEC producers, who already suffer the loss of the lucrative European market for their products.

(iv) The policy highlights the use of *trade creation* and *trade diversion* when a Customs Union is established. As a high levy is imposed on imported food consumers are forced to buy EEC produced food in substitution for more economical purchases from abroad. A great deal of trade is thus created within the EEC, but most of this increase is diverted from the traditional channels. When assessing the achievements of the Common Market the economist should take account of the losses that are incurred elsewhere as a result of the EEC policies.

(v) In so far as Britain is concerned, in 1971 over 75 per cent of her trade was with non-EEC countries. To the extent that membership of the EEC forces the United Kingdom to buy from EEC countries rather than from outside she may expect some loss of export orders from outside countries.

(vi) As a final point the agricultural policy involves a considerable balance of payments cost for those members who import a high proportion of their food, for the proceeds of the levy have to be handed over to the Agricultural Guarantee and Guidance Fund in foreign currency. It has been estimated that this alone could cost Britain a few hundred million pounds.

The problems confronting the EEC authorities in their attempts to harmonize policies in other areas are as great if not greater than those present in agriculture. (The problems associated with the concept of a monetary union are outlined in Unit Twenty-six.) Members acceding to the Treaty of Rome now all have a voice in the formulation of such policies through the institutions of the Community. Those who believe that this is insufficient safeguard for the future must answer the question 'What would have happened to the United Kingdom if she had remained outside the Community?'

24.10 Questions and Exercises

1. Explain carefully the circumstances in which nations find it beneficial to trade with each other.
2. 'Governments can always justify the establishment of trade barriers.' Examine critically the economic arguments in favour of tariffs.
3. Describe and comment on the significance of various forms of barriers to trade.
4. Examine the role of the General Agreement on Tariffs and Trade in the post war economy.
5. Distinguish between a Customs Union and a Free Trade Area.
6. Explain the likely economic effects of any country joining a Customs Union.
7. Examine critically the working of the European Agricultural policy.
8. Consider the problems of trade diversion and trade creation in relation to the European Economic Community.
9. In what ways do the figures in Table 24.2 suggest that Britain might benefit from membership of the EEC?
10. Refer to various copies of the *Annual Abstract of Statistics* to trace changes in the United Kingdom pattern of trade since 1950.

The Balance of Payments: Its Structure and Meaning

25.1 Introduction

In the analysis of comparative costs in Unit Twenty-four we confined ourselves to trade by barter, deliberately excluding any idea of money or
currency. In practice it is the use of different (token) currencies that causes
most of the problems associated with international trade. This Unit shows
the need for careful recording of international transactions, the nature of
these transactions, recent changes in their structure so far as the United
Kingdom is concerned, and the methods available for dealing with short-
term Balance of Payments difficulties.

First of all we must consider the differences between a purely domestic
transaction and an international purchase. In day to day transactions we
simply hand over cash in exchange for goods and services. The cash has no
intrinsic value but only exchange value; it is acceptable to vendors because
they know that they will subsequently be able to exchange it for other goods
and services.

The British importer of a £1000 German car will find that his German
supplier will not have much use for 1000 pound notes and will insist on
payment in Deutschemarks. It is this fact that distinguishes international
transactions from domestic trade. The importer will instruct his bank to
buy £1000 of Deutschemarks from the foreign exchange market. The
currency is provided essentially by the Bank of England in its role as
custodian of the foreign exchange reserves. The Bank's holding of foreign
currency falls and its holding of sterling rises by £1000; the importer's
sterling balance falls by £1000 but he can now use the Deutschemarks
to pay for his car. If many people are buying foreign goods the Bank of
England may have difficulty in meeting demands for foreign currency. In
general, however, the demand for foreign currencies is approximately
balanced by the supply of foreign currency to the market from those overseas residents who need sterling to buy goods from British exporters. It is,
as we shall see in Unit Twenty-six, only in cases of considerable imbalance
that the Government is forced to intervene.

All these transactions between one country and the rest of the world
involving the exchange of currency are brought together annually under
the heading of the Balance of Payments. This effectually summarizes the
country's economic relationships with the rest of the world during the
preceding 12 months.

25.2 The Reasons for Measuring the Balance of Payments

(a) To measure performance

A country's Balance of Payments may be likened to the annual income and expenditure of a household, although the comparison must not be carried too far. The household receives income by supplying the services of factors of production and spends that income on the purchase of goods and services it requires. If the household spends all its income, no more and no less, it is in the same position as a country whose Balance of Payments just balances; if it spends more than its income either by borrowing or drawing on past savings the household has a balance of payments deficit for the year; if its expenditure falls short of income it may regard itself as having a balance of payments surplus.

This illustrates one reason for assessing the Balance of Payments: it shows whether or not the country as a whole is paying its way in the world. If Agraria produces £10,000 million of goods, exports £2000 million of them and imports £3000 million its inhabitants have consumed more goods than they have produced. On the other hand the inhabitants of other countries must have consumed fewer goods than they have produced, a situation which as we shall see cannot continue indefinitely. (Again the comparison with the household is appropriate.) The Government needs an assessment of the Balance of Payments in order to check that the community is living within its means.

(b) To protect the foreign currency reserves

Goods imported from abroad have to be paid for in currency acceptable to the supplier. Individual importers do not keep stocks of foreign currencies needed to buy goods overseas but they can acquire them from the Bank of England. A second important reason for keeping track of the Balance of Payments is that a deficit leads to reductions in these reserves and prolonged deficits force the Government to take restrictive action in order to preserve the currency for essential purposes.

Although it is this currency aspect of the Balance of Payments that arouses most concern, yet even where countries use the same currency it is possible for Balance of Payments problems to exist. England, for example, has a balance of payments surplus with Wales, but this does not create a currency problem for Wales. The Balance of Payments deficit appears as a higher level of unemployment in Wales, since the demand for Welsh goods is relatively low.

25.3 The Structure of the Balance of Payments

The Balance of Payments is a record of all the financial and economic transactions between one country and the rest of the world. It is normally divided into two accounts, the Current Account and the Capital Account. Table 25.1 summarizes the British Balance of Payments for 1971.

Table 25.1 The United Kingdom balance of payments 1971

	£m	
Current account		
Exports	+8790	
Imports	−8491	
	———	
		+ 299
Invisible services		
Government	− 527	
Private	+ 798	
Interest profits and dividends (private)	+ 669	
(Government)	− 199	
	———	
		+ 741
Investment and other capital flows		
Official long-term capital	− 273	
Overseas investment in the UK public sector	+ 181	
Private long-term capital invested overseas	− 725	
Overseas investment in the UK private sector	+ 921	
Other capital flows	+1735	
	———	
Total investment and other capital		+1839
Balancing item	+ 349	
Total currency flow	+3228	
Allocation of special drawing rights	+ 125	
Gold subscription to IMF	—	
	———	
	+3353	
	———	
*Official financing**		
Repayment to IMF	− 554	
Repayment to other institutions	−1263	
Addition to gold reserves	−1536	
	———	
		−3353

Source: *United Kingdom Balance of Payments*, HMSO

* Minus indicates reduction in liabilities or increase in assets.

(a) The Balance of Payments on current account

This is the trading account and it summarizes all purchases and sales of goods and services. It is divided into two parts, one showing visible trade, the other invisible trade or services. 1971 was one of those rare years when the United Kingdom showed a surplus on the visible trading account. In fact goods worth £8491 million were imported, but exports amounted to £8790 million. (The surplus is indicated in the accounts by the + sign.)

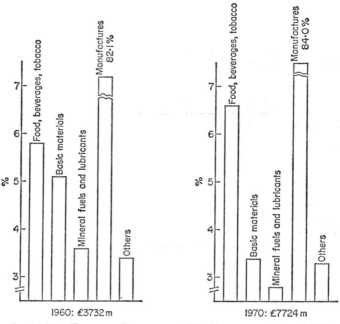

Fig. 25.1 (a) The commodity structure of British exports, 1960 and 1970

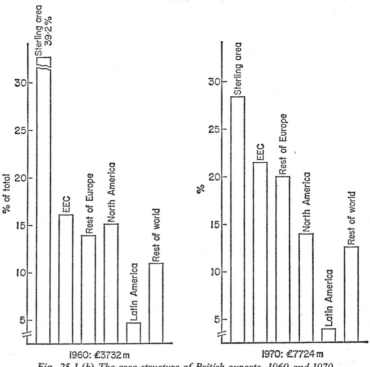

Fig. 25.1 (b) The area structure of British exports, 1960 and 1970

The structure of visible exports and imports and recent changes which have occurred may be judged from Figs 25.1 and 25.2. The commodity structure of exports has not changed very much and manufacturers still constitute the major share. The destination of those exports, however, has gradually changed in emphasis so that a larger proportion of United Kingdom exports are now destined for Europe and a smaller proportion for the Commonwealth. In this sense British membership of the EEC may be regarded as swimming with the tide; whether she joined or not an increasing proportion of her trade would be with the industrialized European countries.

One aspect of Britain's foreign trade that has given cause for concern is the fall in her share of world trade from 11 per cent of exports in 1949 to 6 per cent in 1970. Many factors account for the decline. Among the least important are possibly those which were once advanced as the principal causes—the heavy dependence on the Commonwealth market and on traditional exports such as textiles and heavy engineering goods. Research suggests that these matters had little particular influence and that the decline was due to a general lack of competitiveness, and to the remarkable post-war resurgence of Germany and Japan.

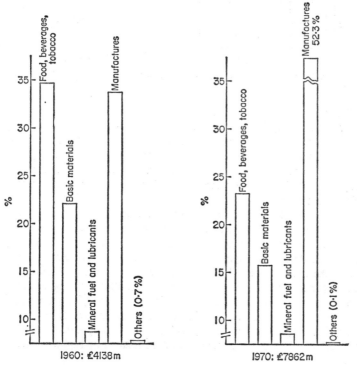

Fig. 25.2 (a) The commodity structure of British imports, 1960 and 1970

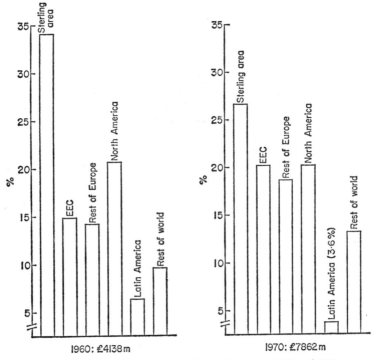

Fig. 25.2 (b) *The area structure of British imports, 1960 and 1970*

This general lack of competitiveness is reflected in the changing pattern of British imports, for whereas food and raw materials used to predomin- ate, manufactured goods are now more important. We thus import a large volume of goods which we could produce for ourselves, although there may of course be good comparative cost reasons for doing this. An increasing proportion of UK imports are from industrial countries as opposed to primary producers.

While visible trade is the trade in tangible goods, the invisible account assesses the value of services provided for people in other countries or purchased from them. Historically this has always been the strongest part of Britain's Balance of Payments and although its performance has declined during the post-war period it still makes a mammoth contribution to any surplus (or to reducing the extent of a deficit). Whenever services are pro- vided by people or organizations in the United Kingdom for people resident in other countries it helps the United Kingdom Balance of Payments as money is brought into the country. On the other hand, when United King- dom residents buy services from abroad the Balance of Payments deterior- ates.

A glance at Table 25.1 shows that the invisible surplus would be even stronger if the Government did not have such a huge deficit. The main items of expenditure are on administrative and diplomatic services overseas, defence and economic grants to underdeveloped countries. The grants appear here and not in the capital account. Compensating for this public sector deficit are interest, profits and dividends, and other services. These other services are largely provided by the City of London. If overseas companies run bank accounts in London, as many of them do, they have to pay for the service they receive and thus contribute towards the surplus here. In the same way overseas residents using the services of Lloyds or other service professions make monetary contributions. Interest or dividends received on overseas investments is regarded as payment for an invisible export—the service provided by the capital investment. It is clear that overseas investments by British industry in previous years results in a substantial and important inflow of funds today.

The net effect of other invisible transactions is not so important as those mentioned above but considerable sums may be involved. In the case of travel and tourism the sums involved may be expected to rise fairly rapidly in future.

Taken together, the visible balance and the invisible balance give the Balance of Payments on current account. This is sometimes called the Balance of Trade, although care is needed when using this term as it is more commonly used to describe the balance of visible trade. While it is not a disaster for the current account to be in deficit (since the deficit may be made good elsewhere in the account) deficits recurring year after year do indicate a general lack of competitiveness and an unhealthy economic state. Governments tend to regard it as a sign of economic weakness if the current account is too frequently in deficit.

(b) Investment and other capital flows

Not all international payments are in exchange for goods and services; a great deal of money flows to and fro in relation to international investment of various kinds, and a summary of the transactions appears in this part of the Balance of Payments. Investment by a UK resident or organization in an overseas project results in money leaving the economy and the initial effect is harmful to the Balance of Payments. In subsequent years, however, interest and profits earned, and the repayment of the investment, will be helpful to the Balance of Payments accounts. The following are the most important types of capital flow that occur.

(i) Official long-term capital includes long-term loans between Governments, especially international aid, and the repayment of these (but not the interest on them which belongs in the current account); and the Government's contributions to organizations such as the International Bank for Reconstruction and Development.

(ii) United Kingdom private investment overseas may be divided into 'direct investment' undertaken by companies in their overseas subsidiaries and plants, and 'portfolio investment' undertaken by individuals and institutions in this country buying paper securities issued by overseas Governments and companies. Historically the United Kingdom has been a net exporter of capital (i.e. has had a deficit on this account) but the surpluses which may occur elsewhere in the account often derive from the outflow of capital. A firm establishing a subsidiary overseas will probably buy much of its equipment in the parent country thus boosting the current account; and in later years profits remitted to the parent company contribute to the invisible account surplus.

(iii) Other Capital Items includes both long-term and short-term movements, almost entirely in the private sector. Here are recorded the import credits (in effect a loan to Britain and therefore temporarily helpful to the balance of payments) and export credits which reduce the surplus or increase the deficit. Such credits are often Government-financed and are given where an importer needs time to pay for his goods and the exporter needs cash to continue production. A variety of transactions undertaken by the Banking Sector are also included in this account. If overseas traders deposit money in a London Bank (perhaps because of their trading commitments in the United Kingdom) the Balance of Payments is affected in the same way as when the Government or someone else negotiates a long-term loan from abroad. These movements of money into and out of the banking sector are essentially short-term movements (they can be withdrawn at any time) and they used to appear in a separate account from the long-term items discussed earlier in this section.

After summation of the investment and capital flows a balancing item is added. This is nothing to do with the size of the Balance of Payments deficit or surplus, but merely indicates the errors and omissions which have occurred. That part of the accounts which we have so far overlooked, is called 'Official Financing'. This records the changes that have occurred in Government holdings of foreign currency (or liquid claims to currency) over the year. If there is no deficit elsewhere in the accounts there will be no change in the foreign exchange reserves as the amount of money coming into the country will equal the amount required by importers and overseas investors. A deficit, however, will result in a net outflow of funds while a surplus will increase the amount of foreign exchange held by the Government. Thus the total of official financing should be of the same size (but of the opposite sign) as the sum of the current account and investment and other capital flows. A year's international transactions are so complex that the two items rarely coincide, but the accuracy of the official financing figure is more trustworthy than the others. The balancing item is, therefore, placed with the capital flow. If it seems rather large in relation to the overall Balance of Payments surplus one must realize that it is very small in relation to the total value of transactions that occur in each year.

(c) Official financing
Before we examine this we must investigate the two items, Allocation of Special Drawing Rights and Gold Subscription to the International

Monetary Fund. These are both Government transactions: Special Drawing Rights are a new form of international currency which is periodically issued to members of the International Monetary Fund (IMF); and the gold subscription to the IMF is a payment made every few years to increase IMF reserve funds. Each of these items is discussed more fully in Unit Twenty-six. The official financing itself shows how the Government has dealt with a deficit or surplus arising in the other sections of the Balance of Payments. For this reason the items appearing here are often called *accommodating payments* (they accommodate the imbalance arising elsewhere) while other transactions are known as *autonomous* in that they occur irrespective of their likely effect on the Balance of Payments. This division is somewhat arbitrary in that many of the 'other capital flows' occur for 'accommodating reasons'. Import and export credits are examples of this.

The United Kingdom had an overall surplus in 1970 so the Government was able to pay back £134 million to the International Monetary Fund; and repay loans to other monetary authorities (overseas Central Banks) of £1161 million and add £125 million to the foreign exchange reserves. (Do not be confused by the minus signs which indicate an *increase* in assets or a *reduction* in liabilities.) In a deficit year the official financing would show the extent to which the deficit had been met by drawing on the cash balance (a 'plus' movement in the foreign exchange reserves) or by borrowing from the IMF or other organizations.

It should now be clear that the Balance of Payments always balances because it is constructed to balance. The current account and the Capital flows conceal a vast number of transactions whose net effect is either for money to leave or for money to come into the country. The official financing shows how the deficit is raised or the surplus distributed.

We shall now consider the following example: at the beginning of 1970 Agraria has foreign exchange reserves of £1000. During the year the following transactions occur:

(a) Residents import £3000 of food from America;

(b) Residents export £2500 of machinery to Africa;

(c) Overseas residents pay back charges of £350 to Agrarian Banks;

(d) The Government receives £750 from India in repayment of a loan;

(e) An African firm acquires an overseas subsidiary for £1000.

The total transactions of £7600 have largely cancelled themselves out, but the demand for foreign currency from Agrarian residents has been £400 greater than the demand for Agrarian currency by overseas residents wishing to buy Agrarian goods. It is the Government's duty to ensure stability in the foreign exchange market, and to ensure an adequate supply of currency there. In this case they have borrowed £100 from both the IMF and the USA and reduced the foreign exchange reserves by £200 to £800. They could, of course, have met the deficit entirely by reducing reserves but might have felt that this would deplete the currency reserves too severely.

The transactions could be recast as follows:

		Balance of Payments of Agraria 1971
Current Account		
Visible trade	−500	
Invisible trade	+350	
		−150
Investment		−250
Total currency flow		−400
Official financing		
Loan from IMF	+100	
Loan from USA	+100	
Drawing on reserves	+200	
		+400

If is of the greatest importance to understand that financing a deficit does nothing to cure the underlying causes of it. It is essentially a holding operation which allows the Government sufficient time to introduce policies necessary to prevent the deficit recurring in the following year. The 'cures' available for a Balance of Payments deficit depend upon the overall international monetary system. In the last 100 years three different systems have been tried. We shall now consider these and the problems associated with them.

25.4 Questions and Exercises

1. Why is it necessary to make an estimate of the Balance of Payments?
2. Examine the structure of the Balance of Payments showing why it must always balance.
3. Consider the changes in the structure of Britain's external trade that have occurred in recent years.
4. 'In joining the EEC Britain is merely recognizing the changing nature of her external trade.' Comment.
5. If the Balance of Payments always balances why does the Government worry about the Balance of Trade?
6. Consider the contribution of invisible trade to Britain's Balance of Payments.
7. What are the limitations of the Balance of Payments as an indicator of economic performance?
8. Find the Balance of Payments figures for last year, and comment on the areas of strength and weakness.

Unit Twenty-Six

The Finance of International Trade

26.1 Introduction

The official financing figures shown in Table 25.1 represent the immediate reaction of the authorities to a Balance of Payments surplus or deficit. If the underlying domestic policies are not changed it is likely that the surplus or deficit will recur in the following year. If a country shows a persistent deficit then drastic action may become necessary to generate a surplus. The pressure on surplus countries to restore equilibrium is not so great but adjustment is also necessary.

The policies by which deficit and surplus countries achieve equilibrium may be divided into two broad groups, *expenditure switching* policies which try to channel the demand for goods and services towards the deficit country, and *expenditure reducing* policies which aim at reducing expenditure in the deficit country in the hope that this will result in a fall in imports. The weight attached to each type of policy will depend upon the system of international adjustment that is in use at the time. Whichever system is considered international prices are made up of two elements, the domestic price of the goods and the rate of exchange against foreign currencies. For example if £1 = DM8·30, goods exported from Britain with a domestic price of £100 will cost DM830 in Germany. There are two ways of altering the price of such goods in Germany. If their cost of production in the United Kingdom were reduced to £90 they would sell for DM747 in Germany, or if the exchange rate fell to £1 = DM8 the German price of the goods would fall to DM800.

During the past 100 years three systems of adjustment have been used to achieve Balance of Payments equilibrium:

(*a*) the gold standard where exchange rates are rigidly fixed and adjustment comes about through expenditure reducing policies;

(*b*) flexible exchange rates where the exchange rates may vary from day to day and adjustment is achieved via the expenditure switching that such changes induce;

(*c*) the present system of managed flexibility which is a compromise between the two previous systems and which allows for adjustment by means of both expenditure reducing and expenditure switching policies.

We shall examine each of these in turn.

26.2 The Gold Standard

Although it is now largely of historical interest the Gold Standard is important in illustrating the effects of sacrificing domestic economic balance to external balance. Under the Gold Standard the Balance of Payments had to be given absolute priority over the domestic aims of full employment and economic growth. Although there are many possible variations of the Gold Standard each would have had to comply with the following rules:

(*a*) The monetary authorities of each country must fix the value of their domestic currency in terms of gold, and be prepared to exchange gold for notes at the fixed rate.

(*b*) No restrictions must be placed upon the import or export of gold.

(*c*) The domestic money supply must be closely linked to the supply of gold so that an increase in the supply of gold (because of a Balance of Payments surplus) results in a rise in the domestic money supply, while an outflow of gold leads to a reduction in the domestic money supply.

The implication of each monetary authority fixing the value of its own currency in terms of gold is that exchange rates between currencies became fixed within very narrow limits which are determined by the cost of shipping gold between the two countries. If the pound sterling is fixed at £1 $= \frac{4}{35}$ ounce of gold and the American dollar at 1 dollar $= \frac{1}{35}$ ounce of gold then the exchange rate between the two currencies will be £1 $= 4$ dollars. If it moved far from this it would be worthwhile buying gold in one country and shipping it to another country. Thus the rate of exchange under the gold standard can be taken as being fixed.

Since this is so, adjustment to the Balance of Payments had to be achieved by altering the relative domestic price levels in the countries concerned. In the days when the Gold Standard operated Governments also always sought to balance their budgets so this adjustment had to be achieved through monetary deflation in the case of deficit countries or inflation in the case of surplus countries.

We shall now consider the process in the deficit country:

(*a*) It suffers a loss of gold as it is shipped out to settle the deficit.

(*b*) The loss of gold forces the authorities to cut back the supply of money in the economy. The extent to which the domestic money supply is reduced depends upon the ratio that the authorities are required to maintain between the gold reserves and the note issue. If the ratio of gold to total note issue is thought adequate at 10 per cent then a reduction in the gold reserves of £1000 leads to a fall in the note issue of £10,000. (This is achieved by means of open market operations, see Section 28.2.) Bank credit becomes more difficult to obtain since the base for creating credit is reduced. As it is more difficult to acquire loans the level of demand falls and with it the demand for imports causing some improvement in the Balance of Payments.

(*c*) This movement will be reinforced by the rise in interest rates which will have occurred owing to the reduction in the supply of money. Some consumers or entrepreneurs will postpone expenditure, and there will also be a beneficial Balance of Payments effect of attracting money from abroad in search of higher interest rates.

(*d*) The fall in the level of domestic demand will lead to a rise in unemployment (with the accompanying negative multiplier effects discussed in Unit Twenty-two). This will result in cuts in wages, lower costs of production and lower prices in the deficit country.

(e) Whenever the elasticity of demand for exports and imports is greater than one equilibrium will be restored, and indeed the deficit country will move into surplus, and the surplus country into deficit.

Ideally the surplus country will have been implementing policies diametrically opposite to those above: an increase in the money supply and an expansion of credit leading to lower interest rates, higher prices and a deteriorating balance of payments. If both countries (or all countries) stuck to the rules of the system, the gold standard did provide an automatic means of adjustment to the Balance of Payments and did not allow deficits to become too large before corrective action was taken. This was its principal advantage although many commentators feel that the fact that the exchange rate of one currency against another was fixed was an important consideration since importers and exporters alike knew how much they would have to pay or would receive from a given transaction. The significance of this may be more fully appreciated from the discussion of flexible exchange rates which follows (Section 26.3).

Despite the apparent smoothness with which the gold standard could operate it was finally abandoned in 1931 having shown itself inadequate to meet the requirements of the international economy. The main weaknesses which had been revealed were as follows:

(a) There was no pressure on surplus countries to obey the rules. An inflow of gold could be frozen, and not allowed to swell the domestic money supply. After all why should the economy be inflated specifically in order to create Balance of Payments troubles? But if surplus countries refused to inflate, the pressure on deficit countries became much greater: the whole burden of adjustment fell upon them and necessitated massive domestic deflation, with consequent cuts in living standards and political and social difficulties, as millions of workers found themselves unemployed.

(b) Here we can appreciate the second weakness: millions of units of output were lost merely to maintain an arbitrarily established exchange rate. Public opinion became gradually less tolerant of such an order of priorities.

(c) In any case the smooth working of the system depended on wage costs being flexible in a downwards direction. In the 1920s the Trade Unions began to flex their muscles and were less prepared to accept the cuts in wages necessary to secure a Balance of Payments surplus. Their resistance weakened the Gold Standard. Moreover it was more difficult to cut prices because industry was becoming gradually more capital intensive and consequently fixed costs formed a higher proportion of total costs. In many cases what was required to cut unit costs was an increase in output not reductions.

(d) Ultimately it was the absolute rigidity of the exchange rate, linked to gold, a metal almost irrelevant to economic welfare, and the changing order of political and social priorities that led to the abandonment of the system in 1931.

It was temporarily replaced by a system of flexible exchange rates, unhappily and perhaps unjustly associated with high levels of unemployment.

26.3 Flexible Exchange Rates

While adjustment under the Gold Standard was achieved by policies designed to reduce expenditure in deficit countries and increase expenditure in surplus countries a system of fully flexible exchange rates would achieve adjustment by alterations in the exchange rate brought about by market forces and resulting in expenditure being switched towards the products of the deficit country.

Under such a system the price of one currency in terms of another would be determined by the forces of supply of and demand for the two currencies which in turn depend on the level of demand for each country's exports. Suppose that Britain and America have been trading happily in equilibrium at an exchange rate of 4 dollars = £1 for some time. The implication is that British importers supply £1000 to the foreign exchange market to buy necessary dollars; and USA importers supply 4000 dollars to acquire sterling. If the currencies are supplied to the market in these proportions then the rate of 4 dollars = £1 will be maintained. What will happen if the rate changes? Let us imagine that 4000 dollars and £1000 are the absolute amounts of each currency which have been regularly supplied to the market, but that owing to a reduction in the demand for British goods abroad at prevailing prices American importers now supply only 3000 dollars, implying that they wish to buy £750 worth of British exports at the rate of 4 dollars = £1. If British importers persist in supplying £1000 to maintain the level of imports the foreign exchange market will receive 3000 dollars and £1000. Naturally the exchange rate then becomes 3 dollars = £1. Goods produced in America for 4 dollars and previously costing £1 when sold in the United Kingdom, now cost £1·33 to the British importer, and unless demand is perfectly inelastic fewer American goods will be imported. British exports to America, costing £1 to produce, will now sell at 3 dollars instead of 4 dollars and we should expect an increase in demand for them. This reduction in British imports and increase in British sales will restore equilibrium.

There are many possible causes of a disturbance to the equilibrium position, but a new exchange rate will always be established to equate the demand and supply of the two currencies. The implication of this is that the Government will be able to pursue its domestic economic policies without worrying about the Balance of Payments. A Balance of Payments crisis of the familiar kind represented by a shortage of foreign currency cannot occur as the price of the currency is such as to bring demand and supply into equality.

While the Gold Standard provided automatic adjustment via the domestic economy a system of flexible exchange rates allows automatic adjustment without impairing the domestic economy. This is a seemingly desirable achievement, yet the system was abandoned. In actual fact a system of completely flexible exchange rates has never really been allowed to operate. Throughout the 1930s Governments were deeply concerned about the level of unemployment and one way in which an individual Government might alleviate the problem was to force a depreciation of its currency relative to others. It hoped that by doing this foreign importers would buy more of its now cheaper exports and that its own importers would switch from relatively expensive imports to relatively inexpensive domestically produced goods. No one country gained by these moves, as a manipulated depreciation in one country sparked off identical compensating moves in other countries.

Furthermore even when exchange rates were allowed a reasonable unhampered degree of flexibility Governments were quick to employ other expenditure-switching policies in the form of tariffs on imports and subsidies to exports intended to stimulate demand for home products. Indeed it was the paraphernalia of such controls that characterized the pattern of international trade in the 1930s and necessitated the establishment of GATT (see Unit Twenty-five) and the International Monetary Fund.

Although a system of completely free rates has never actually operated, there are many objections to even an experimental introduction of the system. The most important of these objections are as follows:

(a) Free exchange rates introduce an extra element of uncertainty into the already hazardous business of international trade. If we suppose that a United Kingdom manufacturer invoices an American customer for £1000 when the exchange rate is 4 dollars = £1, and that the customer pays with a draft for 4000 dollars. By the time the exporter receives the draft the exchange rate has altered to 4·50 dollars = £1 so the exporter will receive only £888·88 $\left(\dfrac{4000}{4·5} \right)$ when he converts it into sterling. If the exporter had insisted on payment of £1000 in sterling then the risks associated with an alteration to the exchange rate would be transferred to the American importer for if on the day he wished to settle the account the rate has become 4·50 dollars = £1 he would have to hand over 4500 dollars rather than 4000 dollars to acquire £1000. If the rate moves in the opposite direction then exporters or importers can make windfall gains.

It would be possible for traders to insure against such fluctuations in the foreign exchange market, but it is likely that a number of potential exporters would be deterred by the risk of loss or would seek higher profits, through higher prices, to compensate for the extra risk. Higher prices would lead to lower demand and a reduction in world trade. It is easy to exaggerate the likely effects on world trade especially as the 1930s, the period in which the international payments system most closely approximated to complete flexibility, was a period of intense trading depression. That trade depression, however, resulted from factors far more complex than the adoption of free exchange rates.

(*b*) Free exchange rates tend to be inflationary. As we saw in Unit Twenty-three one of the main reasons for controlling inflation is to prevent a heavy Balance of Payments deficit. Under free exchange rates a deficit cannot occur because the exchange rate itself rations the available currency, so a powerful incentive for inflation control is removed. It is, however, erroneous to believe that Governments control inflation only for Balance of Payments reasons. During the period 1970–2, when there was one of the strongest Balance of Payments situations, strong steps were also taken to control inflation.

(*c*) It is feared that free exchange rates would lead to a destabilizing of currency movements. Speculators might be able to profit from selling sterling at the rate of 4 dollars = £1, waiting for the rate to fall to 3 dollars = £1 and then buying back more sterling than they had originally sold.

(*d*) An important, related point is that the exchange rate does not respond only to current account transactions. Large movements of capital (perhaps in connexion with an international merger) would affect the rate as would all kinds of Government payments. Either of these might be expected to move the exchange rate away from a well-established equilibrium position, thus causing trading difficulties.

(*e*) A further difficulty is that if fluctuations in the exchange rate are to be kept within reasonable limits the elasticity of demand for imports and exports needs to be quite high. The adjustments in demand can then occur quite rapidly. If the elasticity of demand is low, however, the exchange rate may have to alter considerably before it has a marked effect on the pattern of trade.

(*f*) Finally, and in some ways most important, the depreciation of the exchange rate of a deficit country causes the terms of trade to move against it as the price of its exports falls in international markets and the price of its imports rises. If it is heavily dependent on imported raw materials this will almost certainly lead to a new round of price increases which will again lead to a reduction in the demand for its exports. This will then lead to a further depreciation of the exchange rate and the cycle will be repeated.

Difficulties and fears such as these lead to a disenchantment with free exchange rates. Since few authorities wanted a return to the Gold Standard a compromise was sought and embodied in the rules of the International Monetary Fund.

26.4 Managed Flexibility

A long war-time series of meetings between the Allies laid the foundations of the post-war international monetary system, which was finalized at the Bretton Woods conference in 1944 and is frequently referred to as the 'Bretton Woods' system. The need for a compromise arose, as we have seen, from the dissatisfaction with the rigidity of the Gold Standard under which all adjustments had to be made via the domestic economy, and the unhappy association of free exchange rates with the economic difficulties of the 1930s.

At this point we must recognize that there are in fact two problems here rather than one.

(a) First, there is the Balance of Payments or exchange rate problem itself. How was it possible for countries to get away from the absolute rigidity imposed by the Gold Standard without falling into the trap of what might be called the manipulated flexibility of the 1930s.

(b) Second, there is the liquidity problem. If exchange rates are completely free this is no difficulty as the demand and supply of currency are equated. If rates are fixed, however, a deficit country must have, or have access to, a stock of currency, with which to finance its deficit while corrective action is taken.

The Bretton Woods conference addressed itself to both these problems.

(a) Exchange rates

It was agreed that while the value of the American dollar, far and away the strongest currency at the end of the war, should be fixed at $\frac{1}{35}$ of an ounce of gold, the par value of other currencies should be declared in terms of the dollar. (A strong currency is one that is generally in demand because there is no danger of its losing value.) A small degree of flexibility was built in, as currencies were allowed to fluctuate within 1 per cent above or below their declared par value. In the United Kingdom the job of maintaining the rate within the permitted limits is entrusted to the Exchange Equalization Account (EEA), a department of the Bank of England established in the 1930s to damp down fluctuations in the exchange rate. The EEA consists of a stock of foreign currencies and sterling. When the price of sterling in the foreign exchange market approaches its ceiling, because it is in heavy demand as overseas importers want to buy British goods, the Bank will announce its willingness to sell sterling at a price fractionally below the official ceiling. It thereby ensures that the price will not rise higher for if the Bank of England will sell at 2 dollars no one will pay a higher price to someone else. When the price of sterling falls towards its lower limit the Account will undertake to buy it at a price fractionally above the official limit, thereby fixing the floor below which the price will not fall.

Changes in the par value itself were to be allowed only in cases of fundamental disequilibrium which may be indicated by persistently large deficits or surpluses, but of which there is no absolutely reliable indicator. Changes of up to 10 per cent from the original par value could go ahead so long as the IMF, the guardian of the system, was informed but larger changes needed the prior agreement of the fund.

The original post-war declaration of par values was a fairly arbitrary matter governed more by nostalgia for the pre-war rates than an assessment of economic realities. In consequence the parity of the pound was declared at 4·03 dollars. This was a rate which made it virtually impossible for

British exporters to compete with American producers, and it led to devaluation in 1949 to 2·80 dollars.

There are really two ways to establish the 'correct' exchange rate if one wants a fixed rate between two currencies. The first, now largely discredited, is known as the 'Purchasing Power Parity Theory'. This says that the exchange rate should be such that if a given representative bundle of goods costs £1000 in the United Kingdom and 3000 dollars in the USA the exchange should be 3 dollars = £1. If the prices subsequently rise by 10 per cent in the United Kingdom but are stable in the USA the rate should alter to 2·72 dollars = £1 $\left(\frac{3000}{1100}\right)$. This is all very well on paper but in practice it is beset by all the difficulties associated with the compilation of index numbers, and the very great problem of finding a representative bundle of goods.

The second way of establishing the rate is much simpler and increasingly popular: it is to allow market forces to operate freely for a few weeks or months in order to establish the equilibrium on the basis of which the new par value can be declared.

(b) International liquidity: the role of the International Monetary Fund (IMF)

We have already established that there will be no need for liquidity in the form of official reserves if a system of free exchange rates prevails as there will never be a shortage of currency. But if rates are fixed within very narrow limits the Government of a deficit country will need to provide foreign currency to importers in that country to the extent of the deficit. It may be able to do this from currency acquired in the past, but if the deficit is large this may not be possible. The IMF acts as a second line of defence in this connexion.

All members are allotted a quota based on their economic importance and determining the size of their subscription to the fund. The quotas are increased from time to time and in 1971 the United Kingdom's quota was 2800 million dollars. Twenty-five per cent of the quota must be submitted to the IMF as gold or American dollars and the balance in a member's domestic currency. If a member should fall into Balance of Payments difficulties it would automatically be entitled to draw foreign currency from the Fund to the extent of 25 per cent of its quota (that is the amount previously contributed in gold or dollars). The currency withdrawn is purchased rather than borrowed, since a corresponding amount of domestic currency must be deposited with the Fund at the time of withdrawal. Money drawn from the Fund will normally be in a number of currencies, those in greatest supply in the Fund and those of the area to which the deficit country must make payment. Two British drawings in 1964 and 1965 were spread over 12 different currencies. Only member *Governments* are able to obtain foreign currency from the Fund. They then put the money at the disposal of importers who wish to settle their overseas debts. The money can also be used to fight off speculation about the value of the country's own currency (see Section 26.5). Once the first 25 per cent of a country's quota has been withdrawn the country may if Balance of Payments troubles persist apply for further instalments. Each time it does this it must hand over corresponding amounts of domestic currency, until the Fund holds its domestic currency to the extent of 200 per cent of its quota. Each successive instalment is likely to be more difficult to acquire than the previous one and the Fund is likely to attach conditions concerned with domestic economic policy before making later advances. Once the IMF holds 200 per cent of a country's quota in its own currency no further advances are allowed until some repurchases have been made. Normally a time limit of five years is set for repayment and a small service charge is made.

It should be apparent that this part of the Fund's activities does not increase the amount of foreign exchange or liquidity available for the finance of international trade. It does, however, gather the money together so that it is more easily available, and thereby obviates the need for deficit countries to try to borrow from the surplus countries.

Before we consider the effectiveness of the Bretton Woods system there is one other aspect of the Fund's work that merits attention.

(c) The IMF and convertibility
One of the methods used in the period 1931–45 to safeguard the Balance of Payments was to impose limitations on the convertibility of a country's currency. Thus a Government could impose a variety of exchange controls to prevent its residents buying goods abroad. Just as GATT was established to secure the removal of tariffs and other commercial barriers to trade, one of the tasks of the IMF was to secure the full convertibility of currencies as restrictions on convertibility are just as effective in distorting patterns of trade in one country's favour as are tariffs. The task was a difficult one owing to the overwhelming trading superiority of the USA, and it was not until 1958 that sterling finally became fully convertible to foreign holders. Sterling was then followed by most other currencies.

26.5 The Weaknesses of the Bretton Woods System
As the post-war period has witnessed an unprecedented growth of world trade it may be felt that there was little wrong with the payments system which supported such an expansion. In reality, the system contained the seeds of its own collapse particularly with regard to liquidity. We may consider the weaknesses under the headings of exchange rates and liquidity.

(a) Exchange rates
The very name of the system, 'managed flexibility', indicates that its founders intended exchange rates to be flexible, but flexible in a managed or orderly way. In practice the exchange rates of the major trading nations, and especially of USA and the United Kingdom, proved

extraordinarily inflexible. The pound was devalued in 1949 and then again in 1967; but the dollar maintained its pre-war gold value until the end of 1971. Such rigidity was, of course, the fault of the users of the system rather than of its designers. With economies growing at different rates, and especially prices inflating at different rates in different countries, alterations to exchange rates must occur if distortions to individual economies are to be avoided. We shall imagine that in 1949 the rate of 2·80 dollars = £1 was exactly right, in that goods produced in the United Kingdom for £100 could compete on level terms with similar American goods produced for 280 dollars. If over the next ten years United Kingdom prices doubled while those in the United States rose by only 50 per cent, the goods that could be traded on level terms in 1949 would now cost £200 to produce in the United Kingdom and 420 dollars in USA. If British goods were now exported at that rate they would cost 560 dollars in USA while American goods would cost £150 in the United Kingdom. It is clearly very difficult for United Kingdom producers to compete if the exchange rate is fixed permanently at 2·80 dollars = £1. If the pound were devalued to 2·10 dollars = £1, however, the goods could again be traded on level terms.

The rigidity of the exchange rate meant that Balance of Payments adjustments had to be achieved through the domestic economy and by attempts to reduce prices in the deficit country. Thus adjustment involved monetary and fiscal policies, which also have the job of inducing full employment and controlling inflation. We must once again defer our examination of these policies until Units Twenty-eight and Twenty-nine, but it is sufficient to say here that the longer the exchange rate remains fixed the more difficult it is for domestic policies to secure Balance of Payments equilibrium.

Why were the monetary authorities so determined to maintain stable exchange rates? Part of the answer to this lies in Section 26.3 where we dealt with the disadvantages of flexible rates. But in the case of the United States and the United Kingdom there was the added complication that their currencies were used as Reserve Currencies by many other countries. Thus instead of holding their reserves in the form of gold, these other countries held them in the form of paper dollars and sterling, attached to which were promises from the respective Governments to pay cash (gold or other acceptable currency) on demand. Overseas holders of sterling prior to 1967 expected to be able to convert their sterling into dollars at a rate of 2·80 dollars to the pound. One of the main reasons for the British Government's resistance to devaluation was that the real sufferers would be these overseas holders. When devaluation did take place and the pound dropped to 2·40 dollars they found that the dollar value of the sterling part of their reserves had fallen in proportion.

A further reason for avoiding devaluation is the inflationary effect that it has on the domestic economy. Devaluation increases the price that must be paid for imports, thereby putting up the cost of essential raw materials and causing costs to rise. It also persuades domestic consumers to switch the demand from imports to home produced goods. There are therefore elements of both cost and demand inflation inherent in devaluation; and the demand inflation may be magnified with the increasing overseas demand for the now cheaper exports of the devaluing country.

There might also have been some fear that the conditions necessary for successful devaluation were absent. There are two of these, as follows:

(i) The elasticity of demand for exports and imports should be greater than one, so that foreign exchange earnings will rise and foreign exchange expenditure will fall. A net improvement will result if the sum of these elasticities is greater than one, but we must consider them separately in the light of the discussion of elasticity in Unit Three.

If the elasticity of demand for exports is greater than one earnings from exports will increase, and the higher the elasticity the higher the increase in earnings. If the elasticity of demand for imports is greater than one, total expenditure on them will fall when their price rises. As a substantial proportion of Britain's imports are food and raw materials there was some fear that the elasticity of demand for them might be too low for devaluation to be beneficial.

(ii) The elasticity of supply of exports should also be greater than one. If it is not, price rises will go on eradicating the benefits of devaluation. In the extreme case where the elasticity of supply is zero there is no point in reducing the unit price of one's exports by devaluation, as the induced extra demand could not be met.

Thus for a number of reasons devaluation was resisted by the authorities for a long while. This had the unfortunate effect of leaving Governments to achieve Balance of Payments equilibrium by means of domestic expenditure reducing policies which became progressively irrelevant. It also threw into sharp relief the problem of international liquidity.

(b) International liquidity

This now consists of gold, reserve currencies, IMF drawing facilities, and Special Drawing Rights. In the post-war period trade has expanded more rapidly than liquidity and has given rise to fears that an adequate expansion of trade might be inhibited by lack of finance. The debate about the appropriate level of reserves is unresolved. No clear-cut answer has been found as to whether they should be related to the level of trade or to the size of the deficits; or how they should be related to either concept. All that is certain is that trade has grown much more rapidly than liquidity has expanded and that the expansion of liquidity has been derived largely from the increasing use of the Reserve Currencies, the Dollar and the Pound. It is this that constituted the great weakness of the Bretton Woods scheme: there was no provision for the planned expansion of reserves closely related to the growth of trade. An inspection of the four forms of liquidity will reveal the difficulties.

(i) **Gold** is historically the most important means of settling international debts. One easy way to increase the amount of liquidity would be to increase the price of gold, or the productivity of gold mines. Both courses still have their advocates but majority opinion is that either would be reactionary and would only postpone the introduction of a more rational means of supporting the world monetary system.

(ii) **Reserve Currencies.** The shortage of gold for monetary use and the justifiable reluctance to increase its price led to a greater dependence on reserve currencies as internationally liquid assets. For a number of historical reasons countries became willing to accept payment in paper dollars or sterling on the understanding, but not with the guarantee, that they could be converted on demand into gold at a fixed rate of exchange. The success and continuation of such a system depends upon the confidence of overseas holders of dollars and pounds that the issuing Governments have sufficient liquid assets to pay gold (or whatever currency is acceptable instead) on demand. If that confidence is lost the system comes under great pressure. We will consider the following hypothetical example.

At the beginning of the year overseas holders of sterling have claims of £5000 million against the United Kingdom Government. (These are normally called Sterling Liabilities, thus indicating the obligation of the Government to honour the debts they represent.) Against this £5000 million the United Kingdom Government holds £1000 million in gold and convertible foreign currencies. The ratio of 20 per cent gold to liabilities may be perfectly acceptable to all overseas holders of sterling, in the same way as a bank's customers accept that the bank will not keep enough cash to cover all deposits. By the end of the year Britain incurs a Balance of Payments deficit of £500 million which she finances by issuing £250 million of sterling thus increasing her liabilities to £5250 million, and by drawing £250 million from the reserves, thus reducing her liquid assets to £750 million. The ratio of gold to liabilities has now fallen to 14·3 per cent $\left(\dfrac{750}{5250} \times 100\% \right)$.

There are two aspects to this. First the supply of liquidity has risen by £250 million, which is the amount of extra liabilities that has been created. This increase in liquidity may be desirable and indeed necessary. At the same time, however, the confidence of at least some overseas holders in Britain's ability to pay cash on demand will have been lost, and they will demand immediate conversion of their sterling into gold or dollars. This can and frequently does lead to speculation that Britain will economize on her reserves by devaluing the pound. There will be a rush to convert sterling into gold or dollars or almost any other currency before it becomes reduced in value.

Though much simplified, this example illustrates the central dilemma of the Bretton Woods system. There was no provision for the planned expansion of reserves, and they were increased largely by the deficits of the United Kingdom and the United States; but this expansion itself eroded the confidence on which the holding of reserve currencies depended. As the exchange rates were virtually fixed the United Kingdom and the US deficits tended to become progressively worse, thus feeding the speculation which ultimately forced the devaluation of sterling in 1967 and the dollar in 1971.

(iii) **IMF Drawing Rights** were discussed earlier (Section 26.4 (*b*)) and we need only note here that while the Fund has played an important part in channelling liquidity in the right directions, it is only in recent years that it has engineered the introduction of a new form of liquidity.

(iv) **Special Drawing Rights.** The increasing reluctance of all countries to hold reserve currencies accelerated the introduction of Special Drawing Rights (SDR) as a new method of adding to the liquid reserves of IMF members. SDRs are merely book entries credited to members of the Fund in proportion to their quotas up to a total in 1972 of 9500 million dollars. These credits are regarded by members as part of their reserves and they are transferable between Governments. If Britain has a Balance of Payments deficit with Canada she may, with the approval of the Canadian Government, transfer SDRs from her account at the IMF to the Canadian account in exchange for Canadian dollars, which would then be placed in the foreign exchange reserves of the United Kingdom. The dollars could then be placed at the disposal of people wishing to pay for their imports from Canada. The success of the scheme depends upon the willingness of surplus countries to accept SDRs in exchange for currency. They have no obligation to accept further SDRs once their holding of them is equal to 300 per cent of their quota.

As American deficits have continued on a large scale a rapid expansion of SDRs has not been required, but they represent a most important milestone on the road towards an internationally acceptable paper currency, whose supply can be geared to the needs of world trade.

26.6 Dealing with a Deficit under Managed Flexibility

Under the Gold Standard a deficit was eliminated by deflationary policies designed to reduce expenditure; while under free exchange rates a deficit is eliminated by a depreciation of the exchange rate resulting in expenditure being switched from surplus countries to deficit countries.

Under the system of managed flexibility both expenditure-reducing and expenditure-switching policies are permissible. The sequence of events in the deficit country may be summarized as follows:

(*a*) The deficit becomes obvious and is financed by means of loans or by reducing the level of the foreign exchange reserves. This merely gives the authorities time to introduce the policies necessary to eliminate the deficit; it does nothing to cure its underlying causes.

(*b*) On the assumption that inflation has led to the deficit, as so often happens, deflationary monetary and fiscal policies are implemented to reduce the demand for imports and reduce home prices relative to those of overseas competitors. The severity of the deflationary policy will be directly related to the size of the deficit.

(c) If necessary, direct controls in the form of tariffs, quota restrictions, and exchange controls will be imposed.

(d) If none of these actions are successful the currency will as a last resort be devalued in the hope of generating a surplus.

26.7 Recent Changes in the System

Once sterling had been devalued in 1967 there was recurring speculation that the dollar would be devalued in terms of gold. This happened in 1971. In May speculation was such that about 5 billion dollars were withdrawn from America in a gigantic speculation against the currency. By selling 1050 dollars a speculator could receive 30 ounces of gold (since 1 dollar = $\frac{1}{35}$ ounce of gold) and would hope to sell the gold back for more than 1050 dollars when the dollar was devalued in terms of gold.

In August America was forced to suspend the convertibility of the dollar and she also imposed a surcharge on imports. The ensuing negotiations led to a devaluation of the dollar by 7·9 per cent from $\frac{1}{35}$ ounce of gold to $\frac{1}{38}$ of an ounce. At the same time there were adjustments to most other major currencies and an extra degree of flexibility was added to the system by the extension of the permitted range of fluctuation of an exchange rate to $2\frac{1}{4}$ per cent above or below its declared value.

In fact the events of 1971–3 have effectively transformed the international monetary system into one of floating exchange rates, for the rates established in December 1971 had collapsed before the end of 1972. The main difficulty was that few people believed that the dollar had been sufficiently devalued in terms of gold or other currencies. Similarly they felt that the Deutschemark was undervalued. Accordingly institutions holding dollars were tempted to sell them for Deutschemarks. Such was the scale of this selling that in the winter of 1972–3 yet another re-adjustment of rates occurred. A striking feature this time was that although rates in general were becoming more flexible, six of the members of EEC decided to hold their currencies fixed in terms of each other. Britain did not immediately join this group but it may be regarded as the beginning of a monetary union embracing all EEC members, and aiming eventually at the establishment of a common European currency. While this would have the benefit of eliminating traditional Balance of Payments deficits between members, it would lead to an intensification of regional problems within the Community: one reason for regional unemployment in the United Kingdom is that the North, for example, has a Balance of Payments deficit with the South East. Similarly within an EEC monetary union: a United Kingdom trading deficit against the other members would result not in a currency crisis, but in a rise in the unemployment rate.

26.8 Questions and Exercises

1. Compare the working of the Gold Standard and a system of flexible exchange rates, showing the disadvantages of each.
2. Show how the system of managed flexibility is a compromise between the Gold Standard and free exchange rates.
3. What policies are available for curing a deficit under the system of managed flexibility?
4. Examine critically the system of managed flexibility.
5. What is meant by 'the international liquidity problem'? How has it been overcome in the post-war period?
6. Examine carefully the role of the IMF in the post-war international economy.
7. What are the economic effects of running a reserve currency?
8. Examine the main weaknesses of the Bretton Woods system, and the steps taken to overcome them.

Fluctuations in the Level of Economic Activity

27.1 Introduction

Economists have seen that the capitalist system has an inherent tendency to fluctuate between levels of high activity and levels of relative inactivity. Many explanations of such fluctuations have been offered, but there is little doubt that the post-war movements in the United Kingdom have derived their impetus mainly from the Balance of Payments. In this Unit we shall look briefly at the pre-war type of cycle before taking a closer look at the post-war fluctuations.

27.2 The Trade Cycle

The economic expansion which characterized the nineteenth century was marked by a series of fluctuations or trade cycles which lasted approximately eight years from peak to peak. One possible explanation of trade cycles and economic fluctuation in general is to be found in terms of the relationship between consumption and investment.

In Unit Twenty-two we noticed the importance of expectations in determining the level of investment. An alternative approach, the *accelerator theory* states more specifically that the level of investment depends upon the levels of consumption. The case becomes clear if we consider just one firm which uses 20 machines to produce 100,000 units of output per year, with each machine working to capacity. If each machine has a useful life of ten years the firm will on average invest in two new machines per year to meet the demand for the 100,000 units of output. But if demand for the product rises one year to 120,000 units and the firm maintains a constant ratio between investment and output it will now require not only its usual two replacement machines but also four new ones to meet the extra demand. Thus a 20 per cent rise in consumption leads to a 300 per cent rise in investment. If demand remains at 120,000 the following year, the demand for machines will fall back to two, the replacements for those bought ten years earlier. (If investment is to go on expanding, then consumption must expand at a faster rate.)

The accelerator theory offers an explanation of the occurrence of fluctuations in a given firm, and it is not difficult to imagine this working for the whole economy. The simple theory outlined above is certainly deficient in some respects, notably in assuming a fixed relationship between a change in consumption and a change in investment, and in ignoring the time lags which would occur between a change in consumption and the implementation of any investment decision. Between them, however, the accelerator and the multiplier can provide an elementary explanation of the trade cycle which occurs in the absence of Government interference. Let us break into the cycle at the trough or the depth of the depression.

A large proportion of the working force will be unemployed—perhaps 10 or 15 per cent, many factories will have closed down, some temporarily, others permanently, large parts of industry will be on short-time working, and factor incomes of all kinds will be at a low level.

After a period of months in this depressed state the economy begins to pick up. The cause of the change could be anything: the outbreak of war sparking off a massive rearmament programme; the rapid development of a new form of transport creating jobs for construction workers (both canals and railways had this effect at different times); a crash housing programme; or simply the resurgence of business confidence after the slump on the premiss that things must improve. If there is sufficient investment in one of these directions the expenditure will begin to lift the economy away from depression. The initial impetus might come from the side of consumption but we shall confine ourselves here to the case where investment takes the lead.

The initial spurt in activity will be followed by a time lag showing increases in consumption attributable to the multiplier. Trades other than those involved in the initial acceleration will feel the benefit of this and begin to re-employ labour previously laid off, or return to full-time working instead of part-time. Machinery previously lying idle will be brought back into use, and with a further lag the accelerator effect will be felt: as the demand for consumer goods rises the demand for machinery to produce them will rise. A wholly different atmosphere will pervade the economy; there will be optimism instead of pessimism, full capacity instead of idle resources, and employers searching for labour instead of long queues for the dole. The multiplier leads to rises in consumption which cause a rise in investment via the accelerator which has positive multiplier effects. Why should this seemingly desirable process ever be stopped?

The most obvious answer is that eventually the economy is fully employed with all employees working to the limit they set for themselves. When this full-employment ceiling is reached consumption stops growing; but if consumption stops growing the accelerator process shows that the level of demand for investment goods falls. This gives rise to negative multiplier effects. Seeds of doubt are sown, entrepreneurs postpone investment causing the depression to deepen until the bottom of the trough is reached and the process begins again.

Such cycles were a regular feature of the capitalist economy until 1940, although they were far from being identical and may have been generated by factors other than those suggested. The adoption of Keynesian policies of economic management resulted in the trade cycle and the waste associated with it in its traditional form largely disappearing, or rather being damped down. Its place was taken in the United Kingdom by fluctuations clearly related to the Balance of Payments and the operation of sterling as a reserve currency. These fluctuations became known as the stop-go cycle.

27.3 The Stop-Go Cycle

Figure 27.1 shows the fluctuations in the level of unemployment since the war, superimposed on the Balance of Payments current account surplus or deficit for each year. The immediate post-war years, characterized by the combination of war-time controls and the unsustainable exchange rate of 4·03 dollars = £1 do not fit into the pattern of the subsequent period. For most of the period from 1949 the Government's main priority was to maintain the exchange rate at its existing parity of 2·80 dollars = £1, primarily out of a feeling of obligation to overseas countries holding their reserves in the form of sterling, and ignoring the fact that exchange rates, like all other prices, need to be changed from time to time to meet changing economic conditions. The refusal to alter the exchange rates frequently led to Draconian measures to protect it and was responsible for the new style fluctuations in the level of economic activity. While the Government now had the means of dealing with the traditional trade cycle, another less severe but still wasteful cycle was developing.

The period from 1951 to 1964 falls into three clear cycles: if we use the Balance of Payments as our indicator they are 1951–4, 1955–9 and 1960–4. Each cycle begins and ends with a substantial Balance of Payments deficit, and unemployment at a low level which is indicative of a high level of economic activity. Different factors seemed to be operating after 1964 and we shall deal with these later.

When a large deficit in the Balance of Payments occurs the Government must take corrective action of the type outlined in Section 26.6. The usual expenditure-reducing policies will be introduced: increases in taxation, and cuts in Government expenditure (or at least in its rate of increase) will reduce the level of domestic expenditure; and increases in interest rates and reductions in bank lending will reinforce the reduction in expenditure as bank loans and even hire-purchase finance become more difficult to arrange. As the level of demand falls, the demand for imports will fall with it, though the extent of the reduction in imports will depend upon the marginal propensity to import. This is the proportion of an increment of income that would be spent on imports; the higher it is the greater the reduction in imports following a cut in expenditure.

As the policy measures took effect the cuts in Government expenditure would induce a negative multiplier effect, cuts in consumption would lead via the accelerator to a fall in investment so that a rise in the level of unemployment would result. Lower expenditure would slow down the increase in imports while the restricted home market would it was hoped persuade producers to seek markets overseas. Orthodox deflationary measures were

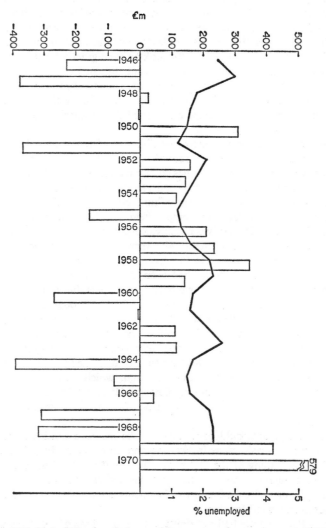

Fig. 27.1 United Kingdom Balance of Payments and unemployment, 1946–70

not the only ones employed: from time to time direct controls on trade and the use of foreign exchange became necessary in the effort to secure a Balance of Payments surplus.

During the 1951–5 cycle the turn round in the Balance of Payments was achieved very quickly, although the 1951 deficit was largely attributable to the Korean war and may be considered as a special case. The smaller deficit of 1955 was also removed quickly, but that of 1960 took longer to turn into a surplus and that of 1964 took an even greater length of time.

One of the main reasons for the increasing difficulty of removing trading deficits is the persistence of an exchange rate which increasingly overvalued the pound and consequently magnified the task of domestic policy. In fact the permanence of the exchange rate turned the Balance of Payments into a domestic economic problem.

A second reason for the increasing difficulty of removing deficits lay in the expansion of social security services and payments. If a person earning £30 per week loses his job and has to survive on perhaps £8 per week his level of consumption must drop considerably. If this cut in demand is multiplied by perhaps two or three hundred thousand the effect can be enormous. But if the level of unemployment pay is raised so that the man previously earning £30 receives £20 in various benefits the cut in his family's living standards is not so great. The very success of the Government's welfare policy is an obstacle to success for its Balance of Payments policy.

Once the deflationary policy had been in force for a sufficient time the Balance of Payments would move back into surplus, but at the expense of increased unemployment, as shown by Fig. 27.1. Although unemployment did not reach pre-war levels it was at its highest by post-war standards when the Balance of Payments was in surplus. The Government, fortified by the stronger Balance of Payments, took action to reduce unemployment by expanding the economy. A Government normally has the further motive that it will shortly have to seek re-election, which is more likely to be achieved in a condition of full employment. It is not by coincidence that elections generally coincide with Balance of Payments deficits and full employment.

Full employment could be induced by the methods considered in Unit Twenty-two and reductions in taxation and an easier credit policy would signal the beginning of the 'go' phase. The multiplier and the accelerator would soon therefore be operating in a positive direction. But in this case as unemployment fell the Balance of Payments deteriorated. Several reasons have been suggested for this:

(a) As the economy had run down in the stop phase of the cycle, manufacturers lacked confidence in the future demand for their products and kept stocks of raw materials to a minimum. As demand expands, confidence grows and manufacturers begin to build up their stocks of raw materials, but these are largely imported so there is a deterioration in the trade balance.

(b) At the same time manufacturers will have less incentive to export as they can now sell more readily at home. This will cause a further deterioration in the visible trade balance.

(c) As the boom gets into its stride it will generate inflation, thus reducing the competitiveness of British goods and making them more difficult to sell abroad. The inflation will also induce an influx of competing imports.

(d) As full employment approaches, order books will be full and waiting lists long. Customers at home and abroad become impatient and seek suppliers elsewhere with further harmful effects on the Balance of Payments.

By now the Government's policies have produced full or almost full employment, but the Balance of Payments deficit is now threatening the stability of the pound in the way described in Section 26.5.

Two alternatives face the Government: to induce another 'stop' phase to save the pound; or to allow the 'go' phase to continue, and if necessary devalue the pound in order to expand the economy. The pound was not devalued until 1967 but then it was not a devaluation designed to accommodate expansion; it was more truly forced upon the Government by their inability to correct the deficit in any other way and by irresistible speculative pressure against the pound. The devaluation marks a break in the traditional stop-go cycle, and since 1967 the economy has not followed the established pattern.

27.4 The Recent Trend of the Economy

The 1967 devaluation took longer than had generally been anticipated to turn the deficit into a surplus for this did not occur until 1969. Before this, as we have seen, the Government's main dilemma was that of combining full employment with a Balance of Payments surplus. The aim was frustrated by inflation, generated by full employment, eliminating the Balance of Payments surplus. After 1969 the Balance of Payments seemed to be on a sound footing, and the main problem became that of dealing with what had become known as 'stagflation'. This is a condition of high unemployment accompanied by rapid inflation.

Unemployment and the index of retail prices have moved in the same direction over the last few years. Instead of the traditional pattern of prices rising faster as the level of full employment is reached, prices and unemployment have now been moving upwards together. During 1970 and 1971 the Government followed the normal reflationary policies, lower taxation, greater Government expenditure, lower interest rates and an easier credit policy in an effort to reduce unemployment. Massive tax cuts failed to prevent unemployment rising to over one million during the winter of 1971. On all reasonable expectations based on previous experience the tax cuts that occurred were sufficient to have kept unemployment lower than this. Not only was unemployment high, but inflation set in at an almost unprecedented rate.

What were the causes of this unfortunate combination? The unemployment cannot be explained on a normal cyclical basis, for although a high level of unemployment is what we have learned to expect when there is a large Balance of Payments surplus as there was from 1969 to 1971, it is unusual for the number out of work to go on rising long after the implementation of heavy cuts in taxation. One of the reasons for the persistent high level of unemployment was the drive for industrial efficiency undertaken by the Labour Government during its term of office. We have seen that one of the results of productivity deals is that a given output is produced by a smaller labour force. Unless the economy is expanding sufficiently rapidly to absorb those made redundant unemployment is bound to rise. Moreover industry was increasingly using capital-intensive methods of production so that any given expansion of output absorbed fewer workers than before. Thus a faster rate of expansion is necessary than had earlier been the case. The incentive to resort to more capital-intensive methods also came from heavy increases in taxation during and after the struggle to save the pound. These increases forced employers to examine their labour requirements more closely with the result that many workers who were previously in a state of disguised unemployment joined the ranks of the completely unemployed. However, reflationary measures to the expansion of the economy began to reduce the level of unemployment from mid-1972 onwards.

If some of the unemployment was derived from measures taken to protect the pound so some of the causes of inflation may be traced back to the devaluation itself. Devaluation inevitably increases the pressure on domestic resources (see Section 26.5). By increasing the price of essential imported raw materials the cost of living was necessarily raised. Together with increases in taxation this led to demands for heavy compensating wage rises with all the effects discussed in Unit Twenty-three. Moreover, even against the background of rising unemployment Trade Unions were prepared to take militant action to secure what they regarded as just wages for their members. When it became obvious that prolonged resistance led to an increase in wages higher than that which could have been negotiated, other workers followed the example, and as we have seen the owners of other factors of production joined in the demand for higher incomes, and thus added fuel to the flames.

In such circumstances the Government is bound to intervene, either directly or indirectly, but it must consider aspects of policy other than inflation. Before considering in detail the measures that the Government employs to influence the general performance of the economy we must briefly consider the main aims of macro-economic policy.

27.5 The General Aims of Economic Policy

(a) Economic growth

The purpose of economic activity is the production of wealth, and one of the aims of Government economic policy is to secure the fastest possible growth of national wealth, or more precisely of the Gross National Product. Normally an increase in the Gross National Product measured at constant prices is indicative of economic growth, but we considered in Unit Two the difficulties of interpreting a rise in the GNP as a rise in the standard of living. Similarly when using the National Income Statistics as an indicator of economic growth we should not forget the disutilities that result from the pursuit of growth. Environmental pollution of all kinds, poisoned rivers, desecrated countryside, and ever growing traffic jams are a few of the more obvious disadvantages of the pursuit of growth. On the other hand a reduction in the size of school classes, an improvement in the National Health Service, higher retirement pensions and other welfare benefits are but a few of the desirable objectives that could be more easily achieved as a result of faster economic growth.

(b) Full employment
This is another of the important objectives of Government policy and to a large extent it is complementary to the pursuit of economic growth. In the short run extra output (which is really what comprises the economic growth) can be squeezed out of the economy only by employing more units of the variable factors of production. In the long run, however, economic growth is stimulated largely by investment, the immediate result of which is likely to be a reduction in the level of employment as labour-saving machinery is usually employed. This is, of course, a process that has been going on for the past couple of centuries without the pool of unemployed people continually rising.

(c) Stable prices
While the objectives of full employment and an adequate rate of economic growth may be complementary a conflict may arise between the achievement of these and the desire to keep prices stable. Although we have seen that a modest inflation may have a beneficial effect on the economy, too many difficulties arise from rapidly increasing prices for a Government to stand by and do nothing. The normal way to fight inflation is to reduce the level of Aggregate Monetary Demand by increasing taxation and reducing the power of the banks to create credit. This frequently results in a rise in unemployment and in the repressive atmosphere thus generated investment is likely to fall with an adverse effect on the rate of economic growth.

(d) Balance of Payments
For a period in the 1960s the Balance of Payments was of prime importance to the Government. We have discussed at length the implications of curing a deficit and it should be clear that the use of expenditure-reducing policies is likely to lead to unemployment, while expenditure-switching policies may well be inflationary. The Government must tread its ground warily if it is not to create trouble in the domestic economy through its Balance of Payments policy.

These four are the main macro-economic aims and the Government must determine the order of priority to be attached to them. The achievement of any one aim is a relatively straightforward matter; combining them in an acceptable way is very much more difficult as post-war British experience shows.

For example, if the Government determines that the Balance of Payments must be moved into surplus at all costs, but rules out the possibility of devaluation, it must be prepared for its restrictive policies, operating in the domestic economy, to reduce the pressure of demand and increase the level of unemployment as a negative multiplier takes effect. The restrictive measures may also have the effect of inhibiting economic growth since industry will find it difficult to expand in the face of higher taxation and interest rates. Moreover the very act of increasing taxation reduces real incomes and may encourage inflationary wage demands. Thus concentration on one of the central issues of economic policy may have undesirable consequences in respect of other policy objectives.

A further example is provided by the experience of the Conservative Government in 1970–2. Their main objective was to reduce the level of unemployment, and to this end taxes were reduced and restrictions on borrowing lifted. The success they may have had in reducing unemployment was tempered by the subsequent problem of inflation which resulted from a great increase in the supply of money, and by the deterioration in the external balance.

There are, then, great problems for any Government to resolve in the field of macro-economic policy, and they require careful value judgements. This is equally true of micro-economic policy. Questions of monopoly, industrial location, housing, taxation of people and goods, subsidies, the distribution of income and wealth are but a few of the economic matters on which the Government must take a view. As we shall see in Section 29.4 there are many claims on the Government's limited resources and there is no unique answer to the allocation of those resources. The choice, once it is removed from the market, is left to the Government.

27.6 The Policies Available

For most of the post-war period the Government has relied on a mixture of monetary and fiscal policies to move the economy in the desired direction. The balance between the policies has changed from time to time, and they have occasionally been reinforced by the imposition of direct controls such as rationing in the immediate post-war period and the incomes policy of 1965–9 which was discussed in Unit Twenty-three. In the remaining Units of this book we shall take a detailed look at monetary and fiscal policy and finally consider the extent to which the British economy is a planned economy and the difficulties associated with economic forecasting.

27.7 Questions and Exercises

1. Examine briefly the nature and causes of the trade cycle.
2. Consider the main determinants of the level of investment in the light of the accelerator theory.
3. To what extent was the 'stop-go' cycle induced by the Balance of Payments situation?
4. Examine the nature of the stop-go cycle and its effects.
5. What are the main aims of economic policy? To what extent do economic fluctuations inhibit their achievement?
6. Consider the importance of: (a) the multiplier; (b) the accelerator, in causing economic fluctuations.
7. A company produces 100,000 boxes of chocolates per year using five machines working at capacity. Each machine has a life of five years. Examine the economic effects of a 25 per cent increase in the demand for the company's products.
8. What are the main limitations of the accelerator theory as an explanation of economic fluctuations?

Unit Twenty-Eight
Monetary Policy

28.1 Introduction

As we have seen, if the Government needs to influence the level of employment, the rate of inflation or economic growth, or the Balance of Payments, it will implement some kind of monetary policy. Such a policy is designed to influence both the supply of money and its price. If the volume of money circulating in the economy is increased the level of Aggregate Monetary Demand (AMD) is likely to rise. If the price of money, that is the rate of interest payable for its use, is reduced the level of AMD is again likely to be stimulated. After prolonged discussions with interested parties important changes in the form of monetary policy were introduced in September 1971, and some of these were examined in Unit Seventeen. This Unit describes the traditional weapons of monetary policy, the reasons why they have been modified, and the nature of the new policy.

28.2 The Main Weapons of Monetary Policy

We can divide our examination of monetary policy into two parts, those measures intended to control the growth of the money supply and those measures to control the price of credit.

(a) Controlling the supply of money

Unit Sixteen showed that the bulk of the money supply exists only in the form of book entries in the banking system, and that from 1947 to 1971 the Clearing Banks had to maintain a cash ratio of 8 per cent which enabled them to create deposits of £1150 on the strength of every £100 of cash they received. It follows from this that if the Bank of England, which administers monetary policy for the Government, wished to control the supply of money, it needs to be able to control the level of cash held by the banks so as to ease or restrict their ability to create credit and deposits. The basic method of doing this has long been *Open Market Operations*. The Bank of England orders the Government Broker to buy or sell Government (gilt-edged) securities in the Stock Exchange.

If the Broker is instructed to buy £1000 of securities, he will always be able to find a vendor by pitching the price sufficiently high. The seller of the securities receives a cheque for £1000 drawn on the Bank of England, which he pays into his account at a Clearing Bank. His deposit at the bank thus rises by £1000, the bank presents the cheque for payment to the Bank of England who credit the Clearing Bank's account (held at the Bank of England under the heading Bankers' Deposits) with £1000. As a result of this the Clearing Bank can create further credit (or deposits) of £11,500 on the basis of an 8 per cent cash ratio. Not all of the extra money will go directly into the hands of the public, for as we know many of the loans created will be highly liquid and made principally to other financial institutions. Some will be invested by the banks themselves in securities. The effect of open market operations working in this direction is to add to the money supply and to the level of aggregate monetary demand. The banks are happy to co-operate in the expansion since they can please the Government, oblige their customers who require loans, and also satisfy their shareholders who require profits. When the Government is selling securities in the Open Market, however, the cash base of the banks is reduced and their lending powers are also reduced. One way of frustrating this aspect of Government policy is for the banks to hold a disproportionately high level of liquid assets. When their cash base is reduced they can then quickly restore it by turning the appropriate amount of liquid assets into cash.

In order to frustrate this activity the system of Special Deposits was introduced, whereby the Bank of England can call for Special Deposits from the Clearing Banks in order to reduce their holdings of liquid assets and their power to create credit. Such deposits are frozen at the Bank of England and not in any circumstances available to the banks, who receive interest on them equivalent to that which they would have received had the money been invested in Treasury Bills.

During the 1960s, in order to reinforce Special Deposits and restrictive open market operations, an increasing use was made of requests and directives to the Clearing Banks concerning the level and direction of their lending. The directives normally indicated that loans should be available to finance investment or export but not to finance personal consumption. Thus the control of the money supply which once depended largely on open market operations came to depend on more direct interference by the authorities.

This change in emphasis was formally acknowledged in 1971 when it was announced that in future the Bank of England would place its main emphasis on Special Deposits in the implementation of monetary policy. Thus an increase in Special Deposits would restrict the powers of the Banks to create credit and the implication was that there would be less use of qualitative controls, and a greater emphasis on changes in the rate of interest.

(b) **Controlling interest**
While open market operations themselves may influence the rate of interest prevailing in the money market by increasing (when securities are bought by the Bank of England) or decreasing (if the Bank of England sells securities) the supply of money, the most direct way of altering interest rates has traditionally been to alter bank rate, the rate at which the Bank of England would re-discount first class bills of exchange. One of the results of the changes in monetary policy of 1971–2 was that the bank rate itself disappeared and was replaced by the Bank of England's Minimum

Lending Rate (MLR) which still indicates the rate at which money will be advanced to the Discount Houses against first-class security. A significant difference is that whereas other rates of interest used to be adjusted in accordance with changes in the bank rate, the Minimum Lending Rate is adjusted in response to changes in the rate of interest on Treasury Bills; it is always $\frac{1}{2}$ per cent above the Treasury Bill rate, rounded up to the nearest $\frac{1}{4}$ per cent. Thus if the rate on Treasury Bills is 6·71 per cent the Minimum Lending Rate will be $\frac{1}{2}$ per cent higher at 7·21 and rounded up to 7·25 per cent. The authorities can thus ensure that the MLR is above the market rate on Treasury Bills, and that it is always a penal rate. For example if the Discount Houses borrow £100,000 from the Clearing Banks at 4 per cent and invest it in Treasury Bills yielding $4\frac{1}{2}$ per cent, MLR will be 5 per cent. If the banks demand immediate repayment of the £100,000 and the Discount Houses have to borrow from the Bank of England at the MLR their receipts from the Treasury Bills are £1125 $\left(£100,000 \times \dfrac{4\frac{1}{2} \times \frac{1}{4}}{100}\right)$ which is the interest on £100,000 invested for three months at $4\frac{1}{2}$ per cent, while they must pay the Bank of England £1250.

Thus the new policy is one that is designed to allow the market to set the rate, rather than the authorities, whose main influence is to control the supply of money rather than its price.

28.3 Domestic Credit Expansion

In some ways the measures on money supply used in Unit Sixteen are inadequate since they relate only to the domestic money supply. If we were operating a closed economy the measures previously discussed would be sufficient. But the money supply depends partly on the external position of the economy. If there is a Balance of Payments deficit the domestic money supply is reduced as importers run down their bank balances to acquire necessary foreign currency. Thus when there is a Balance of Payments deficit the growth of the domestic money supply underestimates the real growth of the money supply. Domestic Credit Expansion has thus been derived to give an additional indication of changes in the economy. It shows the increase in the money supply plus the sterling acquired by the Exchange Equalization Account as it provides foreign currency to imports (or minus the sterling accruing from an external surplus). It is thus a rather broader indication than the money supply itself.

28.4 The Need for a New Monetary Policy

(a) Ineffectiveness of traditional measures
The growing use of directives is an indication of the increasing ineffectiveness of traditional monetary measures.

(i) For example the purpose of bank rate was ostensibly to influence the amount of credit created by altering its price. Changes in bank rate, however, were influenced by considerations other than the desirable price for credit. Rates of interest were deliberately held down to maintain a 'tidy gilt-edged market'. Unit Seventeen showed that the price of gilt-edged securities varies inversely with the prevailing rate of interest, so if the Government feels that the rate of interest should rise it must accept the fact that the price of gilt-edged securities will fall. The Stock Exchange is a very sensitive instrument and the depression in gilt-edge prices tends to spread outwards with the danger that business confidence will tend to decline.

There is another side to all this. The Bank of England can prevent the rise in the yield of gilt-edged securities by supporting the market; that is by buying securities to stabilize their price and yield, and also to stabilize interest rates in general. If it does this it necessarily increases the supply of money and the power of the banking system to create credit. The alternative is to allow the price of gilt-edged securities to fall, while the yield on them and interest rates in general rise if a restrictive policy is to be followed, and vice versa if the economy is to be expanded.

This is an important issue. The demand for loans by industrialists and consumers is not likely to be much affected by a change in the rate of interest from 8 to 9 per cent, but a rise from 8 to 12 or 14 per cent could have a quite marked effect.

(ii) Restrictive open market operations have not been able to work effectively in the past because of the ease with which the banks have been able to acquire liquid assets. They have been able to obtain bills of exchange at times when the authorities have been trying to follow a restrictive policy.

(iii) We have seen that open market operations were supported by Special Deposits and Directives. The latter in particular were very arbitrary in their effect especially in the matter of ceilings on advances. In this respect they penalized banks that were relatively under-lent, rather than those who had not co-operated fully in previous restrictions, and were over-lent.

(b) Development of a two-tier banking system

It was not only the weakness of traditional policy that led to the changes. Most of the policy measures outlined above applied to the Clearing Banks only, while a whole range of secondary financial institutions were free from controls. The lending ceilings were applied to these other institutions but only gradually. As a result of this, the growth of the Clearing Banks rela-

Table 28.1 The growth of sterling bank deposits (£m)

	December 1967	December 1971
London clearing banks	10,262	11,735
Overseas banks (British and Commonwealth)	1,071	2,156
Overseas banks (American)	477	1,519
Other overseas banks	278	186
Other banks in UK	542	2,487

tive to other banks was inhibited. This is shown in Table 28.1 which indicates the growth of the sterling deposits of various groups of banks in London between the end of 1967 and the end of 1971.

The growth of the deposits of the non-Clearing Banks has been much more rapid than that of the Clearing Banks partly because the former were free to go out and seek business in a way that the Clearing Banks could not. But some of the Clearing Banks' difficulties were of their own making.

(c) The banks' cartel

The Clearing Banks had long operated a cartel on interest rates, which had been criticized by both the Monopolies Commission and the National Board for Prices and Incomes. It was to be part of the package of reforms that they should give up their collaboration in this respect.

(d) The monetarists

In the second half of the 1960s the work of economists such as Milton Friedman had led to monetary policy returning to favour as perhaps the main way of influencing economic activity. It was the combination of these factors that led to the introduction of the new system of control in September 1971.

28.5 The New Monetary Policy

In Section 17.6 we examined the changes as they affect the Clearing Banks, the abolition of the cash and liquidity ratios, and their replacement by a new Minimum Reserve Ratio of $12\frac{1}{2}$ per cent. This now constitutes the credit creating base for banks. Other financial institutions are brought within the framework of the new policy by allowing them to operate on a Minimum Reserve Ratio of 10 per cent, thus giving them slightly more freedom than the Clearing Banks over the disposition of their assets. At the same time the Clearing Banks' agreement on interest rates was discontinued, as was the practice of the Discount Houses of making a syndicated or joint bid for the weekly allotment of Treasury Bills.

The Bank of England had already announced the ending of its policy of supporting the gilt-edged market by buying securities, except where they had less than 12 months to run, and this is an indication that in future a much more vigorous interest rate policy is likely to be followed. It is indeed the intention that control should come through changes in interest rates brought about by a much tighter control over the money supply, reinforced where appropriate by the use of Special Deposits, and if necessary, perhaps, by varying the required Minimum Reserve Ratio.

The effectiveness of the new policy will be assessed when the Government seeks to restrict the level of economic activity by allowing interest rates to rise and the price of securities to fall. We may, however, indulge in a little speculation as to the course of the policy, by imagining a situation in which a fairly rapid inflation, resulting from full employment, has led to a Balance of Payments deficit which the Government decides to attack by a restrictive monetary policy.

The first indications of the restrictive policy would be the announcement of an increase in minimum lending rate and a call for Special Deposits. As the Government would not be supporting the gilt-edged market prices, they would be falling (and yields rising) as institutions of all kinds sought to sell securities to realize liquid cash in the face of difficulties in obtaining bank loans, arising from the call for Special Deposits. As the restrictions would apply to all the financial institutions, there would be little point in the customers of Clearing Banks looking elsewhere for loans. The restriction of the money supply would have a double effect on inflation. It would reduce the pressure of demand-inflation arising from consumers living beyond their means; and, to the extent that employers would be unable to finance large wage increases even temporarily by bank overdrafts, it might have a salutary effect on cost inflation. If the policy were pursued interest rates would go on rising, the familiar negative multiplier effects would become apparent and, hopefully, the reduction in the level of demand would bring about the desired improvement in the Balance of Payments. Thus although the structure of the new policy is different it will secure its objectives in the same way as did the old formula.

The great deficiency of monetary policy is that it is non-discriminating. While high interest rates may be necessary to deter high levels of consumption, they would also have the less desirable effect of reducing investment unless appropriate subsidies are available to encourage it. As industrial investment is the key to the successful running of the economy this may be something of a disadvantage in the new policy.

Interest rates do not distinguish between investment and consumption and neither can they distinguish between different industries in need of different treatment from the authorities. The motor industry may be fully employed, operating at a high level of profits, and well able to pay the market rate of interest for borrowed funds, while depression in the textile industry precludes borrowing even for vital purposes at the market rates. In such circumstances the new policy of control through interest rates rather than the more direct intervention via requests and directives may magnify the difficulties of some industries. This would probably necessitate Government assistance from another direction, via its fiscal policy. It is not therefore likely that monetary policy will depart from its established role of creating the economic atmosphere required by the Government. The objective of fiscal policy will be to secure more particular aims by means of changing taxation and Government expenditure.

28.6 Questions and Exercises

1. Examine critically the traditional weapons of monetary policy.
2. Why did the authorities find it necessary to introduce changes in monetary policy in 1971?

3. What are the main difficulties in using monetary policy as the chief means of economic control?
4. 'If you control the supply of money you control the economy.' Comment.
5. Explain the monetary steps that should be taken to induce conditions of full employment.
6. What were the weaknesses of monetary policy in the 1960s?
7. Why is it impossible for the Government to control the supply of money and the rate of interest at the same time?
8. What effects does a rise in the rate of interest have on the price of gilt-edged securities?
9. UK Government bonds were issued at 8 per cent. Assuming they are undated what will be the price per £100 of stock if the long-term rate of interest is now (a) 6 per cent; (b) 16 per cent.
10. Trace the changes in monetary policy that have occurred since 1971. Consider their effectiveness.

Unit Twenty-Nine
Public Finance and Fiscal Policy

29.1 Introduction

For most of the post-war period the Government has sought to control the level of economic activity by alterations in fiscal policy, and it has used monetary policy mainly to create the general economic atmosphere. 'Public finance' is the term applied to the study of the methods employed by the Government to raise revenue, and the principles underlying Government expenditure. It is important to understand that Government expenditure is just as much a part of public finance as adjustments to taxation.

In order to emphasize this we shall examine the expenditure side of Table 29.1 before investigating the main sources of revenue, and examining the role of fiscal policy in the economy.

Table 29.1 United Kingdom public sector income and expenditure 1971
Current Account (£m)

Income	
Taxes on income	7,953
Taxes on expenditure	
Central Government	6,610
Local Government	2,087
National Insurance	2,828
Trading surplus	1,604
Rent	1,272
Interest and dividends	375
	22,739
Expenditure	
Goods and services	10,287
Subsidies	923
Grants	4,981
Interest and dividends	2,244
Surplus	4,313
	22,739

Source: *National Income and Expenditure*, HMSO

Notes: The table relates to the whole of the public sector, including the Public Corporations. A more detailed analysis of receipts from taxation will be found in Table 29.3.

29.2 The Need for Government Expenditure

The oldest of the State's functions is that of defence, and with this may be linked today the maintenance of internal law and order. Each of these may be defined as a *collective service* in that they are necessarily provided to all members of the community at the same time. As it is impossible for individuals to opt out of using or benefiting from defence it would not make sense to talk of a market price for the commodity. Accordingly it must be supplied and paid for by the community as a whole. Similarly, since we all benefit from the maintenance of law and order, the cost of the police service is met by tax- and rate-payers rather than by those who happen to be willing to pay for protection. A conspicuous exception arises here when a private organization specifically requests police protection or assistance. For example, a large number of police officers are on duty in football stadiums on Saturday afternoons, and since they are there at the request of the football clubs, the latter are charged for the service.

The 'private' employment of police officers points the way to another group of services where the market mechanism would not be suitable. In the case of many services provided by the State there is clearly an individual who is the main beneficiary; but children in schools, students at University, patients in National Health Service hospitals, are not the only ones who benefit from the services they are consuming. Society as a whole benefits from the thorough education of its members and from their good health. If these services were provided privately many individuals would not take advantage of them as they would rather spend their money on other things in accordance with their own scales of preference. Social and humanitarian considerations dictate that the services should be provided free, or should at least be heavily subsidized by the State, both for the benefit of the individual and so that society may make the best use of its resources. The more important the public benefits obtained from the private consumption of services, the greater the degree of subsidy likely to be forthcoming. Primary and secondary education are now provided free of charge, while some direct charges are made for courses in Technical Colleges and the Universities. In the Health Services charges are sometimes introduced as a deterrent to consumption, to slow down the increase in public expenditure and reduce the burden on the taxpayer. The patient must, for example, make a contribution to the cost of dental treatment, thus reducing the costs to the Exchequer.

A further group of services taken over and provided free by the Government are those in which the cost of collecting revenue directly from users would be prohibitive. It would, for example, be possible to finance road-building and maintenance entirely via a complex system of tolls covering all roads. Since this would be enormously expensive, road-building is financed out of general taxation, to which, it is true, the motorist makes a substantial contribution. Similarly, a proposal to charge for National

Health Service prescriptions according to the value of the drugs prescribed met with great opposition because of the high administrative costs involved.

The Government also spends considerable sums on industrial projects and research in its efforts to increase industrial efficiency. The most important area of expenditure, however, is that covered by the welfare state, as illustrated by the great sums spent on social security benefits and welfare services shown in the table.

All these are examples of the inability of the price mechanism and market forces to meet the needs of the community. It would impose an intolerable burden on the poorer members of the community if parents had to meet the full cost of educating their children. It is therefore necessary for the State to intervene to ensure the provision of educational facilities at the expense of taxpayers in general.

29.3 The Growth of Public Expenditure

One of the most noticeable features of post-war economic activity has been the sharp rise in the proportion of the national product spent by the State. Fig. 29.1 shows the rise in money terms since 1959 and public expenditure as a percentage of Gross National Product.

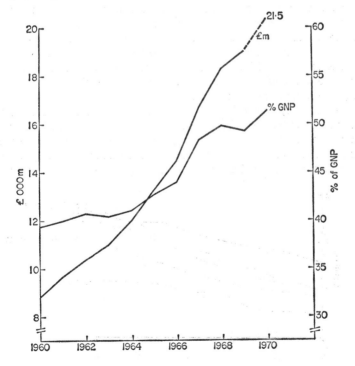

Fig. 29.1 United Kingdom public sector expenditure, 1960–70

It is clear that the actual expenditure and the proportion of GNP absorbed by the Government have both risen very sharply. There are several reasons for this.

(*a*) A rising population necessitates extra expenditure.

(*b*) As society becomes more complex and its institutions more sophisticated the tasks facing the Government increase. For example as the number of motor cars increases the cost of road-building and maintenance rises; consider for example the complexity of modern motorway interchanges compared with orthodox road junctions. There is also, unfortunately, a greater need for policemen and for accident wards in hospitals—all at public expense.

(*c*) Britain has become an increasingly humanitarian society since 1945 and has been prepared to allot increasing amounts of resources to the care of its weaker or more unfortunate members, via the social security system. Thus the introduction of the Redundancy Payments Scheme in 1966, while financed largely by employers, also involves substantial public expenditure.

(*d*) In more recent years the Labour Government of 1964–70 pursued a deliberate policy of greater State involvement in the economy, from which their successors have found it difficult to withdraw.

The increasing proportion of national income absorbed by the public sector is important because it reduces the economic power of the private sector. The balance between the two sectors is regarded by many economists as one of the most important determinants of the performance of the economy. At one extreme are those who believe that the profit motive, given a free rein, will lead to greatest efficiency, while at the other extreme are those who hold that the State should exercise the tightest control over as great a proportion of the economy as possible to prevent the exploitation of one group by another. Of course, in the arguments between these two factions, objectivity is often clouded by political views, and in trying to form a judgement of the 'correct' divisions of the national product between the public and private sectors the economist should consider the *opportunity costs* involved. If the Government increases its share of expenditure to

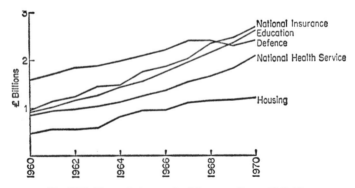

Fig. 29.2 The main items of public expenditure, 1960–70

provide services and thereby commands resources that could have produced far more in the private sector there is a strong economic argument that the resources should be retained by the private sector.

It is not only the balance between the sectors that has altered since the war; there have been changes in the pattern of public expenditure itself, as Fig. 29.2 illustrates. The same considerations apply: increasing the amount spent on one service limits the amount available for other services —opportunity costs are again important.

29.4 The Limitations of Public Expenditure

The 1960s saw a great acceleration in public expenditure as the Labour Government sought to implement many of its plans for economic and social reform. Such was the rate of growth of public expenditure that there were some years in which virtually the whole of the rise in the GNP was absorbed by the Government, as Table 29.2 shows.

Table 29.2 The rises in GNP and total public expenditure compared 1960–71

	Rise in GNP over previous year		Rise in public authorities' expenditure	
	£m	%	£m	%
1960	1395	6·5	493	5·8
1961	1594	6·9	820	9·1
1962	1145	4·6	650	6·6
1963	1594	6·2	590	5·6
1964	2169	7·9	1047	9·4
1965	2023	6·2	1326	11·0
1966	1649	5·2	1150	8·8
1967	1826	5·5	2230	15·3
1968	2002	5·7	1592	9·5
1969	1782	4·8	659	3·5
1970	3542	9·1	1795	9·1
1971	5116	10·6	1191	55·0

Source: *National Income and Expenditure*, HMSO

This inevitably gave rise to many demands for tighter control of Government expenditure. Successive Governments have tried to control the situation but they all meet the same obstacles: the electorate is in favour of falling taxation but supports rising public expenditure, and except in conditions of exceptional economic growth this is an impossible combination. To the majority of people an increase in Government expenditure is probably more beneficial than a fall in taxation, since they pay less in taxation than they receive in benefits. A reduction in taxation is more beneficial to the wealthy than to the poor.

All of the services provided by the State could be improved if the resources were available. Even the Chancellor does not have unlimited resources so he must make a choice between alternative policies:

(*a*) He can improve some services at the expense of others, keeping overall expenditure constant.

(*b*) It is possible to increase his expenditure by obtaining more revenue from taxation or by borrowing.

(*c*) He can withdraw the Government from the provision of certain services, and entrust them to the private sector.

(*d*) Part of the burden of financing certain services can be transferred from the taxpayer to the consumer or user.

It is the last of these that has attracted the attention of Chancellors in recent years; the Government has been more selective in giving assistance within the social security system. Many political disputes arise over the question of selectivity in awarding benefits. One view is that benefits should be available to everyone as a right. This is partly due to the administrative complexity and expense of establishing needy or deserving cases. The oppo-

Table 29.3 Central Government receipts (current account) 1971 (£m)

Taxes on income		
Income Tax	6,184	
Surtax	286	
Corporation Tax	1,491	
		7,961
Customs and excise duties		
Beer, wines and spirits	988	
Tobacco	1,102	
Hydrocarbon oils	1,434	
Purchase tax	1,394	
Other	431	
		5,349
Other taxes on expenditure		
Motor vehicles licence duties	468	
Less export rebates	− 5	
Selective employment tax	554	
Stamp duties	141	
Miscellaneous	103	
		1,261
Other receipts		
National Insurance contributions	2,540	
National Health contributions	236	
Redundancy Fund contributions	52	
Trading surplus and rent	161	
Interest and dividends	1,313	
		4,302
		18,873

Source: *National Income and Expenditure*, HMSO

site position is that as resources are limited, they should be concentrated where they will achieve the greatest benefit, and withheld from those whose marginal utility from the benefits is low. This is clearly the reasoning behind schemes for charging for National Health Service prescriptions, but exempting certain categories such as schoolchildren, the chronically sick, and the old age pensioners. In the same way the benefits of an increase in family allowances are now normally restricted to the poorer families in society, because the increase is 'clawed back' from the majority of income taxpayers.

However in one sense we are running ahead of ourselves, as before he can allocate expenditure between departments the Chancellor needs to know where his revenue is coming from. Before we consider some other aspects of public expenditure in connexion with the budget we must examine the income side of Table 29.1.

29.5 Taxation

Taxes may be classified in various ways. It was once the convention to divide them into *direct taxes* and *indirect taxes*. A direct tax is levied on a specific individual or institution and the burden of the tax falls on that individual; an indirect tax, however, is levied on the activities of an individual or institution and the burden of the tax may be shifted to the final consumer. (See Units Four and Twelve for a diagrammatic treatment of taxation.) Thus when you are assessed for income tax you must bear the burden; but when excise duty is levied on the distiller of whisky he is able to recoup the tax by increasing his prices so that the final consumer pays. But while such a distinction is clear in those cases it is not so clear for some other taxes. Corporation tax for example may be passed on in the form of higher prices even though it is a direct tax and the Selective Employment Tax was regarded by some authorities as a direct tax but by most people as indirect.

29.6 The Principles of Taxation

It has been the practice more recently to divide taxes into those levied on income or capital and those levied on expenditure (outlay taxes) as shown in Table 29.3.

A further way of classifying taxes, very useful to economists but not employed in Government statistics, is according to whether they are *progressive* or *proportional* or *regressive*. Fig. 29.3 illustrates the three possibilities.

Progressive taxes are those which are so arranged that the rich not only pay more tax than the poor, but also hand over a greater percentage of their income in tax.

Proportional taxes are those in which all taxpayers hand over the same proportion of their income in tax, with the rich paying a greater sum than the poor.

Regressive taxes are those which force the poor to part with a greater proportion of their income than the rich.

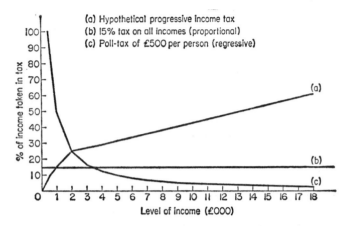

Fig. 29.3 *The impact of progressive, proportional and regressive taxes*

When Adam Smith first discussed the principles of taxation in *The Wealth of Nations* he said that taxes should be levied on people according to their ability to pay. He was clearly not thinking of regressive taxes, but there is room for discussion as to whether he wanted progressive or proportional taxes. The modern view is that equity (fairness) can only be brought to the fiscal system by incorporating a substantial proportion of progressive taxation.

The other principles laid down by Smith were that tax should be payable at a time and place convenient to the taxpayer; that the taxpayer should be certain of his liability in advance (no retrospective taxation) and that the tax should be economical to collect.

While these 'canons of taxation' provide a suitable framework for the tax system there are further desirable characteristics for modern systems.

(*a*) It should be easy to increase or decrease taxes according to the needs of policy, so that firmer control of the economy is possible.

(*b*) In general, taxes should be framed to interfere as little as possible with the allocation of resources: by taxing a good whose demand is highly elastic the Government could force it out of production despite the wishes of consumers. (See Unit Twelve for diagrams.)

(*c*) Taxes should not act as disincentives to effort. An economy has limited resources; and it is important that the Government does not deter their productive employment by excessive taxation.

It is unlikely that any system of taxation will meet all these requirements, but it is necessary that the taxes should be acceptable to the taxpayers.

There have been instances in some countries, even in recent times, of Governments provoking riots by the imposition of excessive taxes. Fortunately it has not happened in this country. We shall now examine the chief United Kingdom taxes in the light of the principles outlined above.

29.7 Taxes on Income

(a) Income tax

Income tax is the most important source of revenue to the Government. Until April 1973 high income recipients were also subject to surtax, but now the two taxes have been consolidated. The new tax is made progressive by the use of certain allowances which are deducted from total pay to arrive at taxable income.

For example in 1973–4 a married man is entitled to deduct £775 from his gross pay before calculating tax on the rest. Single persons may deduct £595. Further deductions may be made in respect of dependent children, and certain other commitments. Once these deductions have been made the first £5000 of taxable income is subject to 30 per cent tax, the next £1000 to 40 per cent and so on until any taxable income in excess of £20,000 is subject to a rate of 75 per cent.

There is a built-in discrimination against unearned income in that it is subject to the standard rate of tax up to an income of £2000, but has to meet an extra levy of 15 per cent thereafter.

Table 29.4 indicates the tax position of a married man on various levels of income under this system.

Table 29.4 Tax liability of a married man (no children) 1973–4. All income earned

Level of Taxable Income (£)	Income Tax (£)	Average Rate (%)	Marginal Rate (%)
500	150	30·0	30
1,000	300	30·0	30
3,000	900	30·0	30
5,000	1,500	30·0	30
10,000	3,950	39·5	55
15,000	7,100	47·3	65
50,000	32,500	65·0	75

The average rate of tax (the proportion of taxable income absorbed by taxation) increases as income rises and a smaller proportion of income is taxed at the standard rate. The rise in the marginal rate of tax (the proportion of an increment of income taken in tax) has the effect of increasing the average rate.

Under the present system of income tax this is likely to be more of a problem for those who are moving out of the standard rate band into the higher rate. For example Table 29.4 shows that the £5000 per year man pays tax at 30 per cent; if he received an increase in his taxable income of £1000 he would have to pay an extra £400 of tax. There may be circumstances in which he prefers to take extra leisure than to pay the higher rate of tax. This could only happen where the worker's income is well above subsistence level. At lower levels of income an increase in tax rates (say from 30 per cent to 40 per cent on the first £5000) could induce workers to work harder or longer in order to achieve the same net income – particularly where that income is fully committed to mortgage, hire purchase and other payments.

The tendency for the proportion of income taken in tax to rise as income rises is known as *vertical equity*, the achievement of which may be regarded as one of the objectives of the tax system. It must not be confused with *horizontal equity*, the principle that taxpayers with the same income and commitments should pay the same amount of tax. In the British system the discrimination against unearned income prevents the achievement of horizontal equity.

(b) Corporation tax

Just as the income of individuals is subject to taxation so is the income of Limited Companies. Since 1965 they have been subject to Corporation Tax, levied on profits in 1972 at a rate of 40 per cent. The overall rate of tax on profits distributed to shareholders, however, is much greater than this since any dividends must be paid net of income tax. Thus if a company had taxable profits of £1000 in 1972 it must pay Corporation Tax of £400; if it then distributed the remaining £600 to its shareholders in the form of dividends, it must hand over to the Inland Revenue a further £232·50 $\left(600 \times \dfrac{38·75}{100}\right)$ in respect of income tax payable by the shareholders. (Unearned income was taxed at a rate of 38·75 per cent in 1972.) In this case the overall rate of tax on the profits is 63·25 per cent, clearly a disincentive to pay large dividends. Once again the Chancellor of the Exchequer has foreshadowed reform of the system: in his budget speech of 1972 he announced his intention to remove the discrimination against distributed profits and proposals, to take effect in 1973.

(c) Taxes on capital

Taxes on capital may be regarded as taxes on accumulated income. There are two taxes which fall into this category in the United Kingdom.

(i) **Estate duty** is now payable at death only on estates exceeding £15,000. The tax is progressive being levied at 25 per cent on estates that just qualify but rising to 75 per cent on very large estates. The justification for such a tax is that the bulk of a large fortune will have been contributed by society as a whole and ought, accordingly, to be returned to society at an appropriate time. On the other hand it may be argued that the taxation of estates amounts to double taxation since the owner of the estate will have been paying income tax and surtax on his income for many years before. Whatever the merits of these arguments, a great deal of the opposition to the tax used to arise from the fact that it was almost an 'optional tax' in that it could be quite easily avoided by giving away the estate three years before death. The 'avoidance period' has now been increased to seven years which makes timing more difficult, but even so the duty produced only £400 million in 1971.

(ii) **Capital gains tax.** If you buy shares or a work of art today and subsequently sell them at a profit you are held to have made a capital gain. Until 1962 all such gains were exempt from tax. In the budget of that year short-term or speculative gains were made taxable, and in 1965 the majority of capital gains became liable to a tax which is now levied at 30 per cent on all capital gains irrespective of the period involved. Like all taxes the capital gains tax caused resentment on its introduction, but was justified on the grounds of equity, in that it was unreasonable for a man to work hard all week and find his income subject to tax, while he could read of others operating on the Stock Exchange and other markets making far more money and not contributing to the Exchequer, even though the gains made were clearly analogous to income.

The Capital Gains Tax, however, hits the small investor as well as the speculator and is frequently a tax on wealth itself. If shares are bought for £1000 today and sold for £1100 in a year's time when the cost of living has risen 10 per cent, and the £100 gain is taxed, the investor is worse off at the end of the year than the beginning.

29.8 The Economic Effects of Taxing Income and Capital

(a) The most important fact about these taxes is that they are progressive and play an important part in redistributing income from the relatively rich to the relatively poor, thereby having, as we saw in Unit Twenty-two, a significant influence on the level of aggregate monetary demand. By transferring money from those with a low marginal propensity to consume to those with a high marginal propensity to consume the Government can bring about an increase in the level of Aggregate Monetary Demand and employment. Furthermore as we noticed in Section 29.5 progressive taxes such as these are generally felt to be fairer than other taxes.

(b) Against this must be set the likely disincentive effects of heavy and steeply progressive income taxes. It has long been argued that the progressively higher rate of income tax persuades some people to take extra leisure rather than do more work. This would seem to depend on the relative strength of two conflicting forces: first on the average rate of tax rises, and second on the longer hours a worker must put in (or higher work-rate in the case of piece workers) to receive the same after-tax income per hour. If most of the worker's income is committed his response is likely to be to work harder or longer. On the other hand if tax rates rise the net income obtained from working an extra hour falls and where income is not fully committed the worker is likely to take more leisure. This is now a rather cheaper commodity (since he is forgoing less money to get an hour's leisure). One of the advantages of the proposed reform of income tax and surtax is that it will make the marginal rate of tax quite clear and thus reduce the disincentive effects.

(c) Similar considerations apply to the taxation of company profits. There comes a point when the disincentive effects of direct taxation lead to a fall in investment and business activity. The tax amounts to a tax on efficiency and enterprise.

(d) Progressive taxes are likely to inhibit personal saving. One cannot transfer money from the rich and still expect them to continue to save at a high rate. However, the Corporation Tax which is not progressive works in the opposite direction—it encourages company savings by discriminating against distributed profits.

(e) While progressive taxes do meet the basic requirements of a system of taxation their nature tends to encourage tax avoidance. In order to achieve the appropriate level of equity the rules have to be very tightly framed. The very tightness of the rules encourages some taxpayers to look for loopholes and, indeed, what might be called the tax-avoidance industry is a very lucrative one for accountants.

However, once all these aspects have been considered, our general conclusion must be that because of the equity argument and because of their heavy and economic yield taxes on income and capital must remain a major source of Government revenue.

29.9 Taxes on Expenditure

These fall conveniently into two categories, those collected by the Customs and Excise authorities, and those collected by other agencies. They are the so-called indirect taxes, which although levied on the manufacturer or wholesaler are passed on to the ultimate consumer by means of price rises. They may be explained more briefly than taxes on income:

(a) Customs duties
These are levied on imported goods on an *ad valorem* basis (i.e. according to the value of the goods). They may be divided into protective duties (designed to protect home industry from foreign competition) and revenue duties, designed to raise revenue for the Government.

(b) Excise duties
These are levied on a fairly narrow range of home-produced goods. They are normally *specific duties* (i.e. levied at so much per pound weight or gallon). For example, the duty on tobacco is levied at £5·04 per pound and on whisky at £18·85 per gallon.

(c) Purchase Tax
This tax is now of only historical interest since it is to be absorbed by the Value Added Tax. It was levied on a wide range of manufactured goods at

different rates. In the lowest range, for example, cutlery was taxed at 11¼ per cent; confectionery was in the next band taxed at 18 per cent; and motor cars and jewellery were liable to a tax of 25 per cent.

The most important of the other outlay taxes or taxes on expenditure are motor vehicle licence duties, collected by Local Authorities on behalf of the Exchequer, and the Selective Employment Tax, which disappeared in 1973. This latter tax is considered separately below in Section 29.11.

(d) The Value Added Tax

The Value Added Tax was introduced in 1973 as part of the comprehensive reform of the tax system undertaken by the Conservative Government. An example will best illustrate the operation of the tax.

Let us suppose that a farmer, obtaining his raw materials for nothing, produces and sells wheat to a miller for £200; the miller grinds the wheat and sells flour to a baker for £300; the baker (obtaining, for the sake of simplicity, other materials for nothing) sells bread to a retailer for £350; and the retailer sells to the consumers for £400. If there were a 10 per cent retail sales tax imposed it would yield £40 and be levied at the retail stage. However with a 10 per cent VAT the procedure is more complicated.

As the farmer has added £200 of value to the seed, the tax on the wheat is £20, and the farmer must remit this to the Government. He then adds £20 to the amount invoiced to the miller who is thus charged £220 (£200 + £20 VAT). The miller adds £100 of value to the original £200 of wheat that he bought and will therefore invoice the baker for £300 + £30 VAT since the value of the flour net of tax is £300. When remitting tax to the Government the miller is entitled to deduct the tax he has been charged by his suppliers (his input tax) from the tax he has collected from his purchasers (his output tax). He therefore pays £10 in tax: £30 received from the baker minus £20 he has had to pay to the miller. The baker having produced the bread sells it for £385 (£350 + £35 tax) and following the same principle remits £5 tax to the Government. On the same basis consumers pay £440 for their bread, and the retailer remits £5 to the Government £40−£35). Thus the Government receives £40 in tax (£20 + £10 + £5 + £5) but instead of receiving it from one producer or wholesaler, as with excise duties or purchase tax, it receives varying amounts from four different sources.

While arrangements for the levying of VAT vary in European countries the standard rate in the United Kingdom is 10 per cent though the rate is subject to a change of 10 per cent per year via the *regulator*, a device which gives the Chancellor power to adjust indirect taxes between budgets.

Some goods and services are *exempt* from the tax. They include small businesses whose taxable turnover is less than £5000 per year, transactions in land, insurance, postal services, betting, gaming and lotteries, financial services, educational services and burial and cremation services. Traders in these categories do not have to charge their customers any tax for the value they have added to their goods, though they will not be able to re-claim any input tax they have paid on goods and materials purchased.

Other goods are *zero-rated*, and again the trader will not charge his customers tax for any value he has added to his materials. The seller of zero-rated goods can reclaim tax paid on his purchases of materials, so the goods are in fact sold tax free. Some zero-rated goods are food, books, newspapers, fuel and power, construction of buildings, exports and services to overseas traders.

Apart from the distinction over reclaiming tax paid the main difference between the exempt and zero-rated categories is the necessity for the seller of zero-rated goods to keep the appropriate records to enable him to re-claim tax paid.

29.10 Economic Effects of Taxing Expenditure

(*a*) The most important aspect of these outlay taxes is that they are normally very regressive. The tax on a packet of 20 cigarettes is a much greater burden to the £1000 per year man than the £10,000 per year man, in that it absorbs a larger proportion of the former's income. The poor man does not have to buy cigarettes, or he may buy a cheaper brand, but if he does smoke he is voluntarily bearing a disproportionate tax burden, and reducing the degree of horizontal equity in the system. Against this, how-ever, such taxes do spread the burden of taxation and bring into the tax net people who would otherwise remain outside.

(*b*) The discriminatory nature of excise duties imposes a disproportionate burden on those who consume alcohol or tobacco, or who drive cars. In 1971 almost 70 per cent of Customs and Excise duties were raised via drinking, smoking and the tax on petrol. Thus the man who drinks orange juice, takes snuff and rides a bicycle has a lighter fiscal burden than the man who is more self-indulgent!

(*c*) The pattern of expenditure taxes affects the allocation of resources as we saw in Unit Twelve. The greater the elasticity of demand and supply the greater the effect of outlay taxes on the allocation of resources.

(*d*) Expenditure taxes are likely to be inflationary since their effect is always to raise prices (except in the rare case where the producer absorbs the whole of the tax). We have seen that in the face of rising prices workers respond by demanding pay rises to meet the rising cost of living. This effect is likely to be more pronounced than with income taxes.

(*e*) One advantage of expenditure taxes is that their burden is very often concealed. Consumers are not always aware of the amount of tax that they are being asked to pay. Thus the taxes probably do not have the same dis-incentive effect on effort as income taxes.

(*f*) As they are discriminating, indirect taxes can easily be adjusted to meet certain policy objectives of the Government. First of all the Govern-ment can tax undesirable activities (such as smoking or drinking); secondly, subsidies (negative taxes—see Unit Twelve) may be awarded to 'desirable' activities; and thirdly, the Chancellor has the power to vary indirect taxes by up to 10 per cent overnight by use of the *regulator*. This is a distinct advantage over changes in income taxes which, owing to administrative complexity, may take six months to take effect.

(*g*) The most obvious effect of VAT is to broaden the base of expenditure taxes, so that a given amount of revenue can be raised by taxing a wider range of commodities at a lower rate. The severe rates of excise duties are, of course, retained.

(*h*) In some respect the introduction of VAT increases the regressiveness of the tax system, for whereas purchase tax was levied on luxury goods at a rate of 55 per cent of their wholesale value as recently as 1970, they are now taxed at 10 per cent of the retail value, while seemingly more essential goods, such as toothbrushes, are now taxed for the first time.

(*i*) While it has its own anomalies, VAT does achieve a tidier and more logical tax structure. For the first time goods and services are subject to the same rate of tax: if you buy a television set you pay VAT at 10 per cent, and if I go to watch a football match (i.e. buy the entertainment service provided) I pay VAT at 10 per cent. Previously I paid only an arbitrary amount of Selective Employment Tax, while you paid 25 per cent purchase tax. Furthermore if you purchased a deep freeze, purchase tax was levied if its capacity was not more than 12 cubic feet, but not if it was over this size; now they are both taxed at 10 per cent. Many other such inconsistencies have been eliminated by the introduction of VAT.

(*j*) Since VAT is broader based than purchase tax it has less effect on the allocation of resources than previous indirect taxes. This may reduce the control the Chancellor has over the economy unless he is prepared to introduce different rates of VAT, for no industry can be taxed at excessive rates. In fact this problem has already been overcome with the motor manufacturers, who, in addition to VAT, have to pay a special 10 per cent tax on the wholesale value of their products.

(*k*) Exports are zero-rated so exporters are able to reclaim any input tax embodied in them. They were not able to do this in respect of the £60m of purchase tax estimated to have been paid on the constituent parts of exported goods before 1973.

One distinct disadvantage of the tax is the great increase in administrative work it involves. Whereas Purchase Tax was payable at about 70,000 different points, VAT is paid by upwards of a million people and institutions, involving considerable expense for both business and the Customs and Excise department.

29.11 The Selective Employment Tax

This tax was introduced in the budget of 1966. It was levied on employers in respect of each employee, but a system of rebates resulted in only employers in the tertiary (services) sector paying the full tax. The objectives of the tax may be summarized like this.

(a) To redress the balance of indirect taxation between manufactured goods and services, by spreading indirect taxation to services.

(b) To force labour out of service industries and into the manufacturing sector.

(c) To increase efficiency in the service industries and also increase economic growth in the manufacturing sector.

(d) To raise revenue at a time when other taxes were reaching the point of diminishing returns.

(e) By giving preferential treatment to employers in the Development Areas, through the Regional Employment Premium, to reduce the high level of unemployment in those areas.

An inquiry conducted by Professor Reddaway of Cambridge University revealed a noticeable rise in productivity in the retail trade, probably attributable to SET. As the tax has a somewhat arbitrary incidence, and an alleged inflationary effect in some trades, the Conservatives decided in 1971 that it should, like purchase tax, be replaced by a Value Added Tax in 1973.

29.12 The Role of the Budget

The Government exercises the right to raise through taxation the revenue it needs to meet its commitments, and its proposals are normally embodied in the budget statement made in March or April each year. Many adjustments, however, are now made between budgets. The original purpose of the budget was simply to raise sufficient revenue to meet Government requirements, and the Government was expected to set a good example to the community, by living within its income. But it became clear, particularly through the work of Lord Keynes, that the significance of the budget far exceeded its role in the Government's housekeeping. As we observed in Unit Twenty-one the budget is one of the most important economic regulators. The overall fiscal policy—the balance between Government income and expenditure has an important part to play in determining the performance of the economy.

(a) The macro-economic role of the budget

As the level of economic activity depends basically upon the level of Aggregate Monetary Demand the Government may influence the level of activity by varying its own expenditure. In conditions of high unemployment the Government will run a *budget deficit* (i.e. will spend more than it takes in taxation) in an effort to get more people working. The difference between what it spends and what is raised through taxation must be found by borrowing, or, as a very dangerous last resort, by printing banknotes. Deficit financing will also probably involve lower taxation, which will leave consumers with more money to spend than they previously had. From the point of view of employment the taxes that should be reduced are those that will give the greatest stimulus to demand. If the Government wishes to increase consumption then outlay taxes might be cut or income tax adjusted so as to benefit mainly those with a high propensity to consume (i.e. those who only just qualify to pay income tax). On the other hand investment might best be encouraged by reducing company taxation and adjusting investment incentives. The important point is that it is not just a question of reducing taxes and hoping that demand will rise; a careful assessment must be made of the amount by which, and the directions in which, it needs to rise. We must remember that an injection of extra Government expenditure will not have a once and for all effect, but like the injection of investment expenditure discussed in Unit Twenty-one will have extensive multiplier effects. Moreover, an increase in Government expenditure of £500 million is likely to have a more beneficial effect on the level of employment than a reduction in taxation of £500 million since part of the tax reduction will be saved by those who benefit.

When the economy suffers from over- rather than under-activity and inflationary pressures are being generated the Government must budget for a *surplus* and reduce its own expenditure, and that of consumers by raising taxation. The difficulty is that while consumers are likely to be happy to co-operate in expanding the economy by spending their higher after-tax incomes they are not normally so eager to accept lower money and real incomes as taxes are increased. Public reaction to smaller pay packets, when income taxes rise, or higher prices when expenditure taxes rise, is normally to demand (and often secure) wage increases which more than compensate for the higher tax burden and so add another twist to the inflationary spiral. In such circumstances the Government is likely to have to resort to those various policies of direct control already discussed in Unit Twenty-three.

If the Chancellor's only task was to budget either for a surplus or a deficit according to whether inflation or unemployment was his main problem, then his job would be a simple one. The Government must place various macro-economic objectives in order of priority, however, and consider the effects of fiscal policy on each of them. It may be that the main problem is inflation and so the Chancellor decides to run a large budget surplus by increasing taxes. Increases in taxation have a depressive effect on the economy and on business confidence so investment is likely to decline. But investment increases productivity and contributes to the fight against inflation and stimulates economic growth. Furthermore since taxation leads to higher prices fiscal policy may in the circumstances be counter-productive. Enough has been said to indicate the complexity of the Chancellor's task in respect of macro-economic policy, but there are further complications to be considered when he assesses the micro-economic effects of his measures.

(b) **The micro-economic role of the budget**
When framing his budget the Chancellor must consider not only the general level of economic activity but also the effect that his proposals will have on particular sections of the economy. He must assess the effects of tax

increases on incentives, on particular industries, and on the allocation of resources. He must also consider the relative merits of and need for extra expenditure on motorways or infant schools, on hospitals or housing, on pensions or prisons, knowing that he cannot meet the requirements of all sections at the same time, yet being subject to pressure from all sides to assist special causes.

29.13 The National Debt

There are times when the Government cannot raise all the money it requires through taxation. On these occasions it resorts to borrowing, and the total of its accumulated borrowing is the National Debt, which we examined in connexion with the Bank of England in Unit Eighteen.

29.14 Local Taxation

Like the Central Government, Local Authorities provide many services which are financed partly by taxation, partly by grants from the Government and partly by borrowing. Table 29.5 gives a breakdown of the main sources of revenue and the principal areas of expenditure.

Table 29.5 Local authority income and expenditure 1971 (£m)

Current Account			
Receipts		*Expenditure*	
Central Government grants	2858	Current expenditure on goods & services	4095
Rates	2087	Subsidies	126
Gross trading surplus	121	Grants to personal sector	171
Rent, dividends & interest	1185	Interest	1134
		Surplus	725
Total	6251	Total	6251

Source: *National Income and Expenditure*, HMSO

The Authorities receive a variety of grants from the Central Government, some are specifically allocated to particular services (e.g. the police); and some are general grants whose disposal is left basically to the discretion of the Local Authority. There are also *rate deficiency grants*, which are allocated to poor authorities whose rates produce insufficient revenue.

The rating system itself attracts a disproportionate amount of criticism. The rates levied on a particular property depend upon its rateable value which in turn depends upon the annual rental value of the property. This is the amount that a tenant might be expected to pay for its use; thus the rateable value of a Victorian terraced house in the industrial heart of a town might be £50 while that of a five bedroomed detached house set in three acres of land on the outskirts might be £500. The Rating Authority each spring estimates the expenditure to be met from the rates for the following twelve months and divides this by the total rateable value of an area to establish the rate per pound to be levied. Thus if the total rateable value of a town was £1,000,000 and the projected expenditure £500,000 the rate per pound of rateable value would be $\frac{500,000}{1,000,000} = 50p$ in the pound.

The system has been much criticized as a tax on thrift: if a householder improves his property by installing central heating the rateable value rises and so does his tax liability. Countless other examples could be quoted, but while most rating officers would acknowledge the faults of the system, the alternative proposals of a local income tax or local expenditure taxes are fraught with difficulties and would probably cause as much resentment as the present system.

A further weakness of the system is that as it assesses property rather than the occupants it is not progressive. It cannot distinguish between relatively wealthy and relatively poor occupants of similar houses. To compensate for this it has been necessary to introduce a system of rate rebates for those on very low incomes.

29.15 Questions and Exercises

1. Examine critically the main areas of Government expenditure.
2. Consider the desirability of a continuing growth in the proportion of GNP going to the Government.
3. Distinguish between the main forms of taxation in the United Kingdom.
4. To what extent does the United Kingdom fiscal system meet the main principles of taxation?
5. Examine the case for raising an increasing proportion of Government revenue through taxing incomes.
6. Examine the case for raising an increasing proportion of Government revenue through taxing expenditure.
7. Compare the economic effects of different kinds of expenditure taxes.
8. Examine the role of the budget in economic policy.
9. What are the main considerations influencing the Chancellor in the formation of his budget?
10. Consider the economic implications of the following budgetary changes (before answering re-read Section 22.7 on the multiplier).
    (a) An increase in the standard rate of Value Added Tax to $12\frac{1}{2}$ per cent.
    (b) A reduction in the standard rate of income tax to 25 per cent.
    (c) The abolition of preferential treatment for the Development Areas.
    (d) A £10 tax free payment to all Old Age Pensioners.

Planning the Economy

30.1 Introduction

It is quite clear from much of what has been said that the United Kingdom economy is far removed from the kind of unfettered free-enterprise capitalism implied by the model of competition. Even in the great period of free trade from 1846 to 1931 there was an increasing amount of Government interference in various sectors of the economy. The general problems of macro-economic management became a matter of minute Government concern only from the 1930s onwards. There is room for considerable disagreement over the desirable amount of Government interference or planning. On the one hand are those who believe that the Government's role should merely be that of providing the right economic atmosphere, possibly by fine control of the money supply. At the other extreme are those who believe that State influence should predominate in every field, to the extent in the fully planned economies of determining what is produced, the level of investment, the level of consumption and the general allocation of resources. In this country, however, we have a compromise between the two extremes and this concluding Unit considers the extent of Government economic planning and the difficulties associated with it.

30.2 The Government and Micro-economic Policies

The economic history of the nineteenth century is littered with legislation restricting the economic freedom of various groups in the economy. Trade Union laws, Factory Acts, parliamentary legislation, and the beginning of modern Company Law are all to be found by a cursory glance at the period. In every case the legislation was conceived (even if sometimes wrongly) as protecting the interests of the majority against exploitation by the minority. The same applies today. Every year areas of conflict emerge which the Government seeks to resolve by legislation. Thus the monopoly legislation is designed to prevent powerful groups of companies from exploiting consumers or other businesses. This is achieved not by handing the means of production over to the Government but by establishing a framework within which cases can be examined. Similarly the Industrial Relations Act became necessary because of the apparent chaos prevailing in employer–employee relationships. It was possible for the production and supply of all kinds of goods and services to be disrupted by any small group of workers striking over any grievance they might have. The fact that the problems may not be solved by legislation is not at issue here: the point is that the Government of the day felt it necessary in this way to safeguard the interests of the community. Once again there is no question of *planning* in the sense of directing.

Regional policy is pursued in the same way and the problems discussed in Unit Nine are tackled. The Government defines its objectives and then tries to persuade or induce firms to co-operate in achieving those objectives. Sometimes there may be more direct interference such as when the Government instructs the nationalized industries to accelerate their expenditure in areas of high unemployment, or when it directly subsidizes those firms in Development Areas which are in danger of going out of business. In general, micro-economic planning is of an indicative nature: the Government indicates the objectives and strives to create a situation in which it can obtain the co-operation of firms in achieving them. Macro-economic planning is carried out in a similar way.

30.3 Macro-economic Planning

As has been shown earlier, the achievement of a single macro-economic objective need not present great difficulties to the Government. Prices can be stabilized, unemployment reduced, or the Balance of Trade moved into surplus but each of these activities is likely to occur at the expense of the others. Successive Governments have sought to reconcile the conflicts by pursuing a faster rate of economic growth through *indicative planning*. This involves rather more than the mere planning of the level of Aggregate Monetary Demand, and at the heart of British economic planning is the National Economic Development Council (NEDC), which was established in 1961 with the task of laying plans for a faster rate of growth. The NEDC, which has a membership drawn from Government, industry and the Trade Unions, published in 1963 a survey of the likely performance of the economy on the assumption of a 4 per cent growth rate. Like the Department of Economic Affairs' National Plan of 1965 which assumed a slightly lower growth rate, the NEDC plan was really a forecast of the best that could occur if everything went well. To achieve the 3·8 per cent annual growth rate set by the 1965 plan it would be necessary to avoid further Balance of Payments crises, to achieve increases in productivity and the level of employment, and to allocate a greater proportion of resources to investment at the expense of consumption. If the desired rate of growth could be achieved then the other objectives of economic policy would, it was hoped, fall into place. The co-operation of the whole of the private sector was needed to achieve the desired growth rate, and was to be obtained through the network of National Economic Development Committees. Each of these was to be related to a particular industry, and charged with the identification of the particular problems of that industry, and with the most appropriate ways of facilitating the growth of the industry. In addition other attempts were made especially by means of fiscal policy to generate the required level of investment, and to raise the level of employment in the Development Areas.

As it happened each of the 'plans' foundered because of the Balance of Payments difficulties of the 1964–7 period and their consequences. Failure to achieve the targets led to disillusionment with the very idea of planning in many circles. There does, however, seem to be some merit in the processes involved. Perhaps the most important fact is that the formulation of a plan (even if it is more a forecast than a plan) enables the short-term economic tactics of the Government, prompted by immediate problems of unemployment or the Balance of Payments, to be seen against the background of long-term economic strategy, with its emphasis on general expansion rather than particular problems. The task of writing and publishing the plan has the useful function of drawing attention to areas of the economy where problems exist; and the implementation of the plan involves all kinds of Government bodies and agencies with representatives from all sides of industry. Even the contact between them and the mutual discussion of economic problems is helpful in advancing the cause of a coherent economic policy.

Recent experience suggests that the problems which develop if market forces are left to themselves are even greater than those associated with planning. The Prices and Incomes Board was abolished in 1971 and the determination of wages was left essentially to the relative bargaining power of Trade Unions and employers. One of the outcomes was that strategically placed groups of workers were able to force up wages very rapidly. Others followed suit and pressed their employers for further increases in pay themselves. In 1972 wages rose approximately 17 per cent, prices 10 per cent and output $3\frac{1}{2}$ per cent.

In such circumstances the losers are those whose incomes don't rise by 10 per cent, and they are usually the least well organized and least well paid. To protect their interests the Government was forced to resort to a formal prices and incomes policy. After a four-month pay freeze during which no increases were payable, the Pay Board and the Prices Commission were established to vet future pay and price increases. Pay increases for any particular work group were limited to £1 plus 4 per cent per person on average, though there was nothing to prevent the lowest paid obtaining a higher increase provided the better paid were happy to accept less. In any case the £1 flat payment meant a higher percentage increase for the lower paid.

But whatever the details of such policies, the essential point is that we have passed the time when the Government can stand back and leave large monopoly groups to exploit the community. The only body that can supervise the wages jungle is the Government or a Government-established agency. Other incomes too must be supervised and all this can only be done by having a permanent body to make an objective assessment of potential pay and prices and incomes. Only if these important items are under control can the Government formulate any coherent plans for economic growth and development.

However no amount of planning, however optimistic, can conceal the fact that the whole process of economic forecasting is still in its infancy and this necessarily makes the formulation of economic policy a hazardous task.

30.4 Economic Forecasting

Every unit in the economy makes some sort of forecast of important economic trends over the next few months. It is generally true that the smaller the unit the easier the forecast. The wage earner when assessing whether he can afford the rental of a colour television rather than his existing monochrome set, has a very firm idea of the wages he is likely to receive during the next twelve months and of the various claims on those wages—housekeeping allowance, rent, rates, car running expenses and so on. He can accordingly determine very quickly whether or not he is able to meet the higher rental. The proprietor of the small corner shop can forecast his sales fairly easily. The policy of a large Public Corporation or a multi-product Public Company, however, will depend on much more complex forecasts of demand, costs, new inventions and changes in tastes. Even more complicated are the problems facing those who seek to forecast the trend of the whole economy. There are two particular problems facing them.

The first difficulty they face is that any forecast has to be based on existing data, and the existing data are out of date and not necessarily absolutely reliable. The figures themselves may be so far out of date that the forecast must be based on estimates of what has probably happened since the last recorded figures. The retail distribution census for example is taken only every five years, but may take two or three years to process and publish. The Budget, the most important single instrument of economic policy, is formulated in February or March on the basis of figures relating to perhaps October or November of the previous year. As for the reliability of the figures the inclusion of an item for 'residual errors' or 'errors and omissions' at least raises doubts about their value as the basis of a forecast.

The second and more important difficulty facing the economic forecaster is that while it is an interesting exercise to draw up an economic model, and analyse the effects of, for example, an increase in taxation, there are a hundred political or sociological occurrences that can prevent the forecast being translated to the real world. A forecast made in December 1971 of the growth of the Gross National Product of the United Kingdom in the first quarter of 1972 could hardly have taken account of the prolonged miners' strike in that period.

The Government employs a large number of highly talented people to make its economic forecasts. Normally the events deny the forecast. The reader will find it instructive to keep the economic forecasts made by some national Sunday newspapers on a monthly basis, and compare them with what actually happens. The inconsistencies that arise reflect not the incompetence of the forecasters but the infancy of the science. The author hopes that this book will have given its readers an appreciation of some of the problems that are involved and will encourage some of them to participate in their solution.

Suggested Further Reading

General Background Reading

Galbraith, J. K.: *The Affluent Society*. Hamish Hamilton (London, 1958).

Shanks, M.: *The Stagnant Society*. Penguin (Harmondsworth, 1972).

The Economist (weekly) provides commentary and analysis of all current economic developments. It is found in all reference libraries.

Barclays Bank Review.

Lloyds Bank Review.

Midland Bank Review.

National Westminster Bank Review.

Unit One

Hanson, J. L.: *A Textbook of Economics*. Macdonald & Evans (London, 1968).

Harvey, J.: *Intermediate Economics*. Macmillan (London, 1969).

Unit Two

Beckerman, W.: *Introduction to National Income Analysis* (Chapters 1, 2). Weidenfeld & Nicolson (London, 1968).

Hacche, J.: *The Economics of Money and Income* (Chapter 9). Heinemann (London, 1970).

Unit Three

Marshall, B. V.: *Comprehensive Economics* (Chapter 11). Longman (Harlow, 1967).

Turvey, R.: *Demand and Supply* (Chapter 1). George Allen & Unwin (London, 1972).

Unit Four

Lipsey, R. G.: *An Introduction to Positive Economics*. Weidenfeld & Nicolson (London, 1966).

Marshall, B. V.: op. cit. (Chapter 11).

Unit Five

Smith, A.: *The Wealth of Nations* (Chapters 1, 2, 3). Penguin (Harmondsworth, 1970).

Unit Six

Marshall, B. V.: op. cit. (Chapters 1, 12).

Unit Seven

Beacham, A. and Cunningham, H.: *The Economics of Industrial Organization* (Chapter 1). Pitman (London, 1970).

Branton, N.: *Economic Organization of Modern Britain* (Chapter 2). English University Press (London, 1968).

Thornhill, W.: *The Nationalized Industries* (Chapters 1, 5). Nelson (London, 1968).

Unit Eight

Branton, N.: op. cit. (Chapter 12).

George, K. D.: *Industrial Organization* (Chapter 2). George Allen & Unwin (London, 1971).

Pratten, C.: *Economies of Scale in Manufacturing Industry* (Chapters 1, 2). Cambridge University Press (London, 1971).

The Times 1000. Times Newspapeas Limited (London, annually).

Unit Nine

Beacham, A. and Cunningham, N.: op. cit. (Chapter 4).

Beckerman, W.: *The Labour Government's Economic Record* (Chapter 6). Duckworth (London, 1972).

Turvey, R.: op. cit. (Chapter 10).

Unit Ten

Turvey, R.: op. cit. (Chapter 4).

Units Eleven, Twelve, Thirteen and Fourteen

The appropriate chapters of Lipsey: *Introduction to Positive Economics* or Marshall: *Comprehensive Economics*, will be found most useful.

Unit Fifteen

Allen, G. C.: *The Structure of Industry in Britain* (Chapter 4). Longman (Harlow, 1970).

George, K. D.: op. cit. (Chapters 8, 9).

Turvey, R.: op. cit. (Chapter 5).

Unit Sixteen

Hacche, J.: op. cit. (Chapters 1, 2).

Morgan, E. V.: *A History of Money* (Chapters 1, 2). Penguin (Harmondsworth, 1969).

Unit Seventeen

Darby, D. J.: *Financing of Industry and Trade* (Chapter 6). Pitman (London, 1970).

Hacche, J.: op. cit. (Chapter 2).

Unit Eighteen

Hacche, J.: op. cit. (Chapter 4).

Unit Nineteen

Hacche, J.: op. cit. (Chapters 2, 5).

Units Twenty, Twenty-one and Twenty-two

Marshall, B. V.: *Comprehensive Economics* (Chapter 12). Longman (Harlow, 1967).

Unit Twenty-three

Hacche, J.: op. cit. (Chapter 14).

Unit Twenty-four

Evans, D.: *Destiny or Delusion (Britain and the Common Market)* (Chapters 1, 2, 3). Gollancz (London, 1971).

Unit Twenty-five

Credland, I. R.: op. cit. (Chapter 19).

Unit Twenty-six

Evans, D.: op. cit. (Chapter 5).

Harrod, Sir R.: *Reforming the World's Money* (Chapters 1, 2, 3). Macmillan (London, 1965).

Johnson, B.: *The Politics of Money* (Chapters 11, 12). John Murray (London, 1970).

Paish, F. W.: *How the Economy Works* (Chapter 7). Macmillan (London, 1970).

Unit Twenty-seven

Beckerman, W.: *Introduction to National Income Analysis* (Chapters 6, 7). Weidenfeld & Nicolson (London, 1968).

Hacche, J.: op. cit. (Chapter 12).

Unit Twenty-eight

Marshall, B. V.: *Comprehensive Economics* (Chapter 6). Longman (Harlow, 1967).

Newlyn, W. T.: *The Theory of Money* (Chapters 6, 9). Oxford University Press (London, 1962).

Unit Twenty-nine

Credland, I.: op. cit. (Chapter 21).

Hacche, J.: op. cit. (Chapter 8).

Sandford, C. T.: *Realistic Tax Reform* (Chapters 3, 7, 8, 9). Chatto & Windus (London, 1971).

Unit Thirty

Beckerman, W.: *The Labour Government's Economic Record* (Chapter 4). Duckworth (London, 1972).

Shonfield, A.: *Modern Capitalism* (Chapter 6). Oxford University Press (London, 1965).

Answers to Questions

(where appropriate)

Unit Two
   4. (c).
   5. a, b, d.
   6. (a) F; (b) T; (c) F; (d) T; (e) F.
   7. (a) 300; (b) 330.

Unit Three
   6. 1; 3.27; 0.
  11. A: +2; B: −4; C: 0; D: +0·28; E: 180.
  12. When price of B changes from 20–15, cross elasticity is 28·5.
      When price of B changes from 15–12, cross elasticity is 5.
      When price of B changes from 12–10, cross elasticity is 1·5.
  13. (a) 212·3 and £21.23.
      (b) 137·67 and £19.27.
  15. (a) 8p; (b) 12p.

Unit Four
   8. $11\frac{1}{2}$ meat; 5 drink.
   9. He should increase his purchases of records at the expense of books.
  10. (a) 11 bread, 16 milk; (b) bread rises in price till only 12 units can be bought; (c) $5\frac{1}{2}$ bread, 21 milk.

Unit Eleven
   7. (a) Supply moves up.
      (b) Supply moves down.
      (c) Supply moves up.
      (d) No move.
      (e) Supply moves down.
   8. (a) −0·66; (b) 1·33; (c) 1.

Unit Fourteen

8a.

AFC	AVC	MC	AC
100	20	—	120
50	25	30	75
33·3	30	40	63·33
25	28·75	25	54
20	28·80	29	48·80
16·6	29·10	31	45·70
14·3	30·00	35	44·30
12·5	37·50	90	50·00

   (b) 2000.

Unit Twenty-two
9. £15m; £7·5m.
10. £2m.
12. GNP rises by £2,000,000.

Index